The Transformation of Academic Health Centers

The Transformation of Academic Health Centers
Meeting the Challenges of Healthcare's Changing Landscape

Edited by

Steven A. Wartman, MD, PhD, MACP
Association of Academic Health Centers
Washington, DC, USA

AMSTERDAM • BOSTON • HEIDELBERG • LONDON • NEW YORK • OXFORD • PARIS
SAN DIEGO • SAN FRANCISCO • SINGAPORE • SYDNEY • TOKYO

Academic Press is an imprint of Elsevier

Academic Press is an imprint of Elsevier
125 London Wall, London EC2Y 5AS, UK
525 B Street, Suite 1800, San Diego, CA 92101-4495, USA
225 Wyman Street, Waltham, MA 02451, USA
The Boulevard, Langford Lane, Kidlington, Oxford OX5 1GB, UK

Library of Congress Cataloging-in-Publication Data
A catalog record for this book is available from the Library of Congress

British Library Cataloguing-in-Publication Data
A catalogue record for this book is available from the British Library

ISBN: 978-0-12-800762-4

For information on all Academic Press publications
visit our website at http://store.elsevier.com/

 Working together
to grow libraries in
Book Aid developing countries
International

www.elsevier.com • www.bookaid.org

Publisher: Mica Haley
Acquisition Editor: Mica Haley
Editorial Project Manager: Shannon Stanton
Production Project Manager: Julia Haynes
Designer: Mark Rogers

Typeset by TNQ Books and Journals
www.tnq.co.in

Printed and bound in the United States of America

Contents

Section I
The Evolution of the Academic Health Center

1. The Changing Ivory Tower: Balancing Mission and Business

*Paul B. Rothman, Edward D. Miller,
Landon S. King and Ellen F. Gibson*

2. Presidential and Academic Health Center Leadership within the Modern University: Opportunities and Challenges

Christopher C. Colenda and Ricardo Azziz

3. How Academic Health Centers are Transforming in Leadership, Administration, and Management: A Case Study

Jeffrey R. Balser and William W. Stead

4. The Changing Roles and Expectations of Faculty

*Robert N. Golden, Marc K. Drezner, Jeffrey E.
Grossman, Richard L. Moss, Elizabeth M. Petty
and Patrick L. Remington*

Section II
Educating the Future Health Workforce

9. The Growing Integration of Health Professions Education

Frank B. Cerra and Barbara F. Brandt

10. Advancing Collaborative Global Education Programs

Yang Ke, Qiudan Sun, Lei Zhang, Jianlin Hou and Weiwei Du

11. The Future of Graduate Medical Education: Is There a Path Forward?

Philip A. Pizzo, Clarence H. Braddock III and Charles G. Prober

12. Guiding the Future of Medical Education through "Enlightened" Accreditation Policy

Deborah M. DeMarco, Terence R. Flotte, Michael D. Kneeland, Paulette A. Seymour-Route and Michael F. Collins

17. Bridging Science and Practice—A Case Study: The Military Translation of Innovative Responses to Urgent Military Medical Needs into Widespread Clinical Practice

Kent E. Kester and Charles L. Rice

18. How Research Can and Should Inform Public Policy

Claire Pomeroy and Fred Sanfilippo

Section IV
Preparing for Health System Change

19. Population Health and the Patient

Jay A. Perman, C. Daniel Mullins and Robin Newhouse

20. The Changing Delivery of Patient Care

*Katy A. Stevenson, Sarah E. Peyre,
Katia I. Noyes and Bradford C. Berk*

List of Contributors

Ricardo Azziz, MD, MBA, MPH President, Georgia Regents University; CEO, Georgia Regents Health System, Georgia Regents University, Augusta, GA, USA

Jeffrey R. Balser, MD, PhD Vice Chancellor for Health Affairs, Dean, School of Medicine Vanderbilt University, Nashville, TN, USA

Jeremy M. Berg, PhD Pittsburgh Foundation Chair and Director, Institute for Personalized Medicine; Associate Senior Vice Chancellor for Science Strategy and Planning, Health Sciences; Professor of Computational and Systems Biology, University of Pittsburgh, Pittsburgh, PA, USA

Bradford C. Berk, MD, PhD Senior Vice President for Health Sciences, Chief Executive Officer, Medical Center and Strong Health System, University of Rochester, Rochester, NY, USA

Ron Berkman, PhD President, Cleveland State University, Cleveland, OH, USA

Clarence H. Braddock III, MD, MPH, MACP Vice Dean for Education, David Geffen School of Medicine at UCLA, UCLA Medical Center Office of Graduate Medical Education, Los Angeles, CA, USA

Barbara F. Brandt, PhD Director, National Center for Interprofessional Practice and Education; Assistant Vice President for Education, University of Minnesota, Minneapolis, MN, USA

Frank B. Cerra, MD Professor, Department of Surgery; Senior Advisor, National Center for Interprofessional Practice and Education; Former Senior Vice President of Health Sciences, Academic Health Center University of Minnesota, Surgery Department, Minneapolis, MN, USA

Rex L. Chisholm, PhD Associate Vice President for Research, Vice Dean for Scientific Affairs and Graduate Education, Northwestern University Feinberg School of Medicine, Chicago, IL, USA

Carolyn M. Clancy, MD Assistant Deputy Undersecretary for Health for Quality, Safety and Value, Office of Quality, Safety and Value (10A4), U.S. Department of Veterans Affairs, Washington DC, USA

Robert Clarke, PhD, DSc Dean of Research, GUMC Professor, Department of Oncology; Co-Director, Breast Cancer Program, Georgetown University Medical Center, Washington DC, USA

Christopher C. Colenda, MD, MPH President and Chief Executive Officer, West Virginia United Health System, Fairmont, WV, USA; Former Chancellor for Health Sciences, Robert C. Byrd Health Science Center, West Virginia University, Morgantown, WV, USA

Michael F. Collins, MD, FACP Chancellor, University of Massachusetts Medical School, Worcester, MA, USA

C. Donald Combs, PhD Vice President and Dean, School of Health Professions, Eastern Virginia Medical School, Norfolk, VA, USA

Elliott Crooke, PhD Chair and Professor, Department of Biochemistry and Molecular and Cellular Biology; Senior Associate Dean, Faculty and Academic Affairs, Georgetown University Medical Center, Washington DC, USA

Rebecca S. Crowley, MD, MS Professor of Biomedical Informatics, Department of Biomedical Informatics, University of Pittsburgh, Pittsburgh, PA, USA

Deborah M. DeMarco, MD, FACP Associate Dean of Graduate Medical Education, Senior Associate Dean of Clinical Affairs, University of Massachusetts Medical School, Worcester, MA, USA

Marc K. Drezner, MD Senior Associate Dean for Clinical and Translational Research; Director, Institute for Clinical and Translational Research; Professor of Medicine, Section of Endocrinology, Diabetes and Metabolism, University of Wisconsin School of Medicine and Public Health, Madison, WI, USA

Weiwei Du, PhD Program Officer, Office of International Cooperation, Peking University Health Science Center, Beijing, China

Victor J. Dzau, MD President, Institute of Medicine, Washington DC, USA

William F. ElLaissi, MBA, MHA Partnership Development, Duke Institute for Health Innovation, Duke University, Durham, NC

Leslie Fall, MD Professor of Pediatrics, Geisel School of Medicine at Dartmouth, Hanover, NH, USA

Howard J. Federoff, MD, PhD Executive Vice President for Health Sciences, Executive Dean of the School of Medicine, Georgetown University Medical Center, Washington DC, USA

Terence R. Flotte, MD Executive Deputy Chancellor, Provost, and Dean, Celia and Isaac Haidak Professor of Medical Education, School of Medicine, University of Massachusetts Medical School, Worcester, MA, USA

Arthur Garson Jr., MD, MPH Director, Center for Health Policy, University Professor, University of Virginia, Charlottesville, VA, USA

Jay Gershen, DDS, PhD President, Northeast Ohio Medical University, Rootstown, OH, USA

Ellen F. Gibson Office of the Dean and CEO, Johns Hopkins University School of Medicine, Baltimore, MD, USA

Robert N. Golden, MD Dean, University of Wisconsin School of Medicine and Public Health; Vice Chancellor for Medical Affairs, Madison, WI, USA

Jeffrey E. Grossman, MD Senior Associate Dean for Clinical Affairs, Professor of Medicine, University of Wisconsin School of Medicine and Public Health; President/CEO, University of Wisconsin Medical Foundation, Madison, WI, USA

Randolph Hall, PhD Vice President of Research, University of Southern California, Los Angeles, CA, USA

Sonja Haywood, MD College of Medicine, Northeast Ohio Medical University, Rootstown, OH, USA

Jianlin Hou, PhD Assistant Professor, Institute of Medical Education, Peking University Health Science Center, Beijing, China

Craig Johnson, MBA Vice Dean for Finance and Administration, Northwestern University Feinberg School of Medicine, Chicago, IL, USA

Larry R. Kaiser, MD, FACS Senior Executive Vice President for the Health Sciences; President and CEO, Temple University Health System; Dean, School of Medicine, Temple University, Philadelphia, PA, USA

Yang Ke, Professor Executive Vice President, Peking University Health Science Center, Peking University, Beijing, China

Kent E. Kester, MD Associate Vice President and Head, Clinical and Translational Sciences, Sanofi Pasteur, Swiftwater, PA, USA

Michelle L. Kienholz Department of Medicine, University of Pittsburgh, School of Medicine, Pittsburgh, PA, USA

Landon S. King, MD Executive Vice Dean, Johns Hopkins Medicine, Baltimore, MD, USA

Michael D. Kneeland, MD, MPH Associate Dean of Allied Health and Inter-Professional Education, Interim Associate Dean of Student Affairs, University of Massachusetts Medical School, Worcester, MA, USA

Tim Lahey, MD, MMSc Associate Professor of Medicine, Associate Professor of Microbiology and Immunology, Geisel School of Medicine at Dartmouth and Dartmouth-Hitchcock Medical Center, Lebanon, NH, USA

Arthur S. Levine, MD Senior Vice Chancellor for the Health Sciences, John and Gertrude Petersen Dean of the School of Medicine, Professor of Medicine and Molecular Genetics, University of Pittsburgh, Pittsburgh, PA, USA

James Merlino, MD Chief Experience Officer, Cleveland Clinic Health System, Cleveland, OH, USA

Bertalan Meskó, MD, PhD Medical Futurist, Webicina LLC, Budapest, Hungary

Edward D. Miller, MD Dean and CEO Emeritus, Johns Hopkins Medicine; Professor of Anesthesiology and Critical Care Medicine, Johns Hopkins School of Medicine, Baltimore, MD, USA

Richard L. Moss, PhD Senior Associate Dean for Basic Research, Biotechnology and Graduate Studies; Professor of Cell and Regenerative Biology, University of Wisconsin School of Medicine and Public Health, Madison, WI, USA

C. Daniel Mullins, PhD Professor and Chair, Pharmaceutical Health Services Research Department, University of Maryland School of Pharmacy, Baltimore, MD, USA

Eric G. Neilson, MD Lewis Landsberg Dean, Feinberg School of Medicine; Vice President for Medical Affairs, Northwestern University, Chicago, IL, USA

Robin Newhouse, PhD, RN, NEA-BC, FAAN Professor and Chair, Organizational Systems and Adult Health, University of Maryland School of Nursing, Baltimore, MD, USA

Katia I. Noyes, PhD, MPH Professor of Surgery and Public Health Sciences, Scientific Director of Surgical Health Outcomes and Research Enterprise, University of Rochester School of Medicine, University of Rochester Medical Center, School of Medicine and Dentistry, Rochester, NY, USA

Greg Ogrinc, MD, MS Associate Professor of Community and Family Medicine, Associate Professor of Medicine, Associate Professor of The Dartmouth Institute, Lebanon, NH, USA

Jay A. Perman, MD   President, University of Maryland, Baltimore, MD, USA

Elizabeth M. Petty, MD   Senior Associate Dean for Academic Affairs, Professor of Pediatrics, University of Wisconsin School of Medicine and Public Health, Madison, WI, USA

Sarah E. Peyre, EdD   Assistant Dean for Interprofessional Education; Associate Professor of Surgery, Surgery Education/Minimally Invasive, University of Rochester Medical Center, Rochester, NY, USA

Philip A. Pizzo, MD   Former Dean and the Susan and David Heckerman Professor of Pediatrics and of Microbiology and Immunology, Stanford University School of Medicine, Founding Director, The Stanford Distinguished Careers Institute, Stanford University, Stanford, CA, USA

Claire Pomeroy, MD, MBA   President, Albert and Mary Lasker Foundation, New York, NY, USA

Charles G. Prober, MD   Senior Associate Dean for Medical Education, Stanford School of Medicine, Pediatric Infectious Disease, Palo Alto, CA, USA

Daniel W. Rahn, MD   Chancellor, University of Arkansas for Medical Sciences, Little Rock, AR, USA

Susan Reeves, EdD, RN   Associate Academic Dean for Dartmouth-Hitchcock Partnership Program, The Gladys A. Burrows Distinguished Professor of Nursing, Colby Sawyer College, New London, NH

Patrick L. Remington, MD, MPH   Associate Dean for Public Health, Professor of Population Health Sciences, University of Wisconsin School of Medicine and Public Health, Madison, WI, USA

Charles L. Rice, MD   President, Uniformed Services, University of the Health Sciences, Bethesda, MD, USA

Paul B. Rothman, MD   CEO, Johns Hopkins Medicine, Dean of the Medical Faculty, The Johns Hopkins University, Baltimore, MD, USA

Fred Sanfilippo, MD, PhD   Director, Healthcare Innovation Program, Emory Healthcare, Atlanta, GA, USA

Paulette A. Seymour-Route, PhD, RN   Dean and Professor, Graduate School of Nursing, University of Massachusetts Medical School, Worcester, MA, USA

Rich Simons, MD, MACP   Senior Associate Dean for MD Programs at The George Washington University School of Medicine and Health Sciences, Washington DC

Jeanne C. Sinkford, DDS, PhD   Associate Executive Director and Director, American Dental Education Association, Washington DC, USA

Wiley "Chip" Souba, MD, ScD, MBA   Former Vice President for Health Affairs, Dean of the Geisel School of Medicine at Dartmouth, Hanover, NH, USA

William W. Stead, MD   Associate Vice Chancellor for Health Affairs, Chief Strategy Officer, McKesson Foundation Professor of Biomedical Informatics, Vanderbilt University Medical Center, Nashville, TN, USA

Katy A. Stevenson, MPP   Manager, Strategy, Planning and Development Medical Center, University of Rochester Medical Center, Rochester, NY, USA

Louis Sullivan, MD   Office of the President Emeritus, Morehouse School of Medicine, Atlanta, GA, USA

Qiudan Sun, MA, Professor   Director, Office of International Cooperation, Peking University Health Science Center, Beijing, China

Jeffrey L. Susman, MD   Northeast Ohio Medical University, Dean, College of Medicine, Rootstown, OH, USA

Krishna Udayakumar, MD, MBA   Head of Global Innovation for Duke Medicine and Associate Professor of Global Health and Medicine, Duke University, Durham, NC

Sarita Verma, MD, LLB, CCFP   Deputy Dean, Faculty of Medicine, Associate Vice Provost, Health Professions Education, University of Toronto, Faculty of Medicine, Toronto, ON, Canada

Steven A. Wartman, MD, PhD   President/CEO, Association of Academic Health Centers, Washington DC, USA

Catharine Whiteside, MD, PhD, FRCPC   Dean of Medicine, Vice Provost, Relations with Health Care Institutions University of Toronto, Faculty of Medicine, Toronto, ON, Canada

Lei Zhang, MD, PhD   Associate Director, Office of International Cooperation, Peking University Health Science Center, Beijing, China

Foreword

Daniel W. Rahn

Academic health centers exist for the overarching purpose of improving health and reducing the burden of illness in society. They accomplish this purpose by providing patient care; educating and training future health professionals; conducting biomedical, translational, clinical, population-level, and health services research; and translating research discoveries into improved approaches to health and disease. The components of the "three-legged stool" of academic medicine (patient care, research, and education) really are different lines of endeavor that act synergistically to advance a unified purpose—that of a healthier future for all.

The platform for advancement of this mission is the patient care enterprise. It serves as our largest classroom. It is where our clinicians practice their professions, teach the next generation of health professionals, and conduct patient-based research. Patients receive state-of-the-art care and serve as our partners in education and research. Through the integration of patient care, education, and research in the clinical setting, we not only model the best in patient care for the next generation of professionals but we point the way to innovations in how to approach health and disease.

The business model that supports this complex enterprise relies on a high degree of integration between mission areas and the sources and uses of funds. The model has been extraordinarily successful. Over the past 50 years, academic health centers have been great idea factories for new biomedical discoveries and for the development of new medical technologies. Advances emanating from academic health centers have transformed our understanding of human biology and the pathophysiology of disease, refining our approach to most diseases that affect the population at large.

In the process, academic health centers have become centers for the provision of advanced care and the education of an increasingly broad array of health professionals following a model of increasing specialization in patient care and research—specialization that has mirrored our advancing knowledge. We have grown in size and complexity in all mission areas. Growth of the clinical enterprise, however, has dwarfed growth of the educational and research missions ([1], p. 15). The ability to support growth of the academic and research enterprises has been dependent on this continuous clinical enterprise growth and the associated patient care revenue. But now, after decades of robust growth, the operating environment of the academic health center has changed in fundamental ways with downward pressure on all sources of funds on which these centers of advanced knowledge and care rely. Academic health centers now face the requirement to reengineer all aspects of their complex operations to adapt to a rapidly changing operating environment that is constraining the fiscal resources needed to advance the interdependent mission components.

The challenges to the education, research, and patient care enterprises are considerable. Health professions education is uniquely expensive because of the requirement that students, faculty, and patients be present simultaneously. Knowledge transfer by itself is insufficient to educate and train a health professional; students must acquire skills with graduated levels of supervision until they not only have the knowledge but also the technical skill and decision-making capability to practice independently. Clinical education relies heavily on cross-subsidization from the patient care enterprise, because the fiscal resources available to fund the educational enterprise from other sources are increasingly inadequate to fully cover attendant costs. Reductions in state funding for public higher education and concerns about ever-increasing student debt both place an increasing financial burden on the clinical enterprise. These financial challenges are occurring at the same time that we must essentially reinvent health professions education if we are to meet the broad health needs of society in the future. We must educate students differently if they are to be prepared for new patient- and family-centered, team-based care models, including the incorporation of integrated health information systems into routine practice and new payment models emphasizing value over volumes.

Simultaneously, society is increasingly demanding that research investments translate into a demonstrably positive impact on population health. Knowledge in and of itself is no longer accepted as an adequate outcome of investments in biomedical research. In addition, constrained federal research funding requires that institutions provide increased amounts of internal financial support to maintain and advance research programs. A benchmarking study recently conducted by the Association of Academic Health Centers found that for every dollar of total extramural research funding received, an additional 52 cents is needed from internal institutional sources ([2], p. 12). The only source of funds for this institutional cost sharing for most institutions is, once again, patient care revenue.

Historically, academic health centers have benefited from their ability to expand their clinical enterprises and grow revenue in excess of the direct expenses associated with providing patient care to cross-subsidize and cover the unfunded components of education and research. What has been a recalcitrant problem for the United States' health system—uncontrolled cost growth—has provided the financial fuel to grow all components of the AHC. Over the past 6 years, however, since the onset of the great recession, the trend of uncontrolled health care cost growth that has been present in the United States for 50 years has come to an abrupt end. Since 2007, per capita health care expenditures have been brought to a virtual standstill when adjusted for inflation ([3], p. 11). This is very good news for the US economy, but it has placed the entire financial model on which the modern academic health center is based in jeopardy. Suddenly there is decreased ability for clinical earnings to cross-subsidize education and research at the very time that funding for support of education and research is shrinking. The Association of American Medical Colleges has estimated that cuts to Medicare alone, associated with the Affordable Care Act, will cost major teaching hospitals more than $5 billion annually by 2017, assuming normal penalties for budget neutral programs [4]. States, the federal government, and private business are all engaged in explicit strategies to ensure that this bending of the curve in health care cost growth is not a temporary phenomenon but is a long-term durable change.

As we adapt to this changing fiscal environment, there has been increasing recognition that the very health care system we participate in, and for which we educate and train future health professionals, is failing to meet the health needs of society in fundamental ways. Our health care system has been primarily focused on high impact interventions for individuals with serious disease. Through this lens we have experienced dramatic success, but this success has been accompanied by an unsustainable rate of cost growth, unacceptable variation in quality of care, and growing public dissatisfaction with the complexity of our health system. The recognition of these issues has led to the current widespread consensus that our health system is in need of fundamental overhaul to achieve a triple aim of higher quality care, better patient experience, and lower cost. The passage of the Patient Protection and Affordable Care Act has made it possible to envision a health care system that achieves these goals but not without far-reaching system redesign.

One further issue for academic health centers to tackle is the role that social determinants play in individual and population health. Although access to high quality, patient-centered, affordable health care is a major determinant of overall population health, there are many influential factors at play. The United States has been progressively lagging behind other developed nations with regard to basic societal-level health indicators. From infant mortality rates to life expectancy and healthy life years, the US is falling behind [5]. There is a fundamental disconnect between the dramatic advances made in the treatment of serious illness and the overall health of the populace. We must reconcile these disparate measures of impact because there is simply no good future one can envision without a healthier populace. Pursuit of better health in the future requires that academic health centers embrace the complexity of all of the issues that impact health. Biology and genetics play a large role in determining an individual's health risks, but so do economic status, educational level, political forces, geography, social support systems, environmental factors (both urban and rural), cultural norms, nutrition, race, ethnicity, and behavioral choices ([6], p. 4). To lead society toward better health, academic health centers have a responsibility to address all of these challenges. If we are to fulfill our societal purpose, we simply must tackle the complex array of internal operating issues, the rapid changes in our external operating environment, and the evolving complex factors impacting population health. Everything is indeed connected to everything else.

The long and short of it is we have a dire need to reengineer our organizations—to transform them in fundamental ways in order to preserve our ability to accomplish our public purpose as organizations leading society toward a healthier future. This transformation is particularly challenging for academic health centers that have been loosely organized and structured around islands of excellence. As universities, we are in the knowledge business with a long tradition of, and value placed upon, the concept of academic freedom. Simultaneously, however, as academic health centers we are in the highly competitive health care business. We rely on success in the latter to support the academic and research programs conducted in the former. The dynamics and cultures that characterize these two endeavors sometimes seem like parallel universes. Our job is to merge the best of both, align all aspects of the academic health center mission and create an engine for innovation in how our society approaches issues related to health and disease.

The rapid change in our external operating environment is not a temporary phenomenon but is a long-lasting shift. The pace of the change is a particular threat because our organizational structures and culture are not well suited to rapid adaptation and change. But change we must and we must do this in real time. There is little time for reflection and debate. We are at risk of being overtaken by events. The traditional, galvanized, academic culture that is faculty centric and involves a high degree of independence, must shift to a patient- and student-centric model. We need a culture of collaboration with tighter alignment between clinical, academic, and research missions based on an interprofessional model of care and education designed to achieve better outcomes at lower cost, and a research agenda that includes implementation of advances in the community setting and a focus on overall population health.

In this book, the forces at work impacting academic health center operations and mission attainment in this challenging and rapidly changing environment are reviewed in detail. The potential approaches and solutions are discussed by experienced leaders of academic health centers. The book is organized in a logical fashion beginning with chapters exploring the evolution of the modern academic health center, including its relationship to parent universities; demands on governance, leadership, management, and faculty; and future directions of the academic health center enterprise. This is followed by chapters in which issues impacting health professions education, research and discovery, and the patient care enterprise are discussed successively.

Education program redesign to incorporate emerging fields—such as clinical informatics, interprofessional education, population health management, and the social and cultural determinants of health—is discussed. The impact of accreditation and regulation of educational programs, including competency-based advancement, is also reviewed.

The list of research challenges discussed includes the changing spectrum of biomedical, translational, and clinical research; how to accelerate the incorporation of new discoveries to improve care and influence public policy; how to improve the efficiency of the research enterprise; and the impact of big data.

The challenges to the clinical enterprise are perhaps the most daunting and are explored in depth. The impact of market consolidation is requiring the forging of new partnerships. Payment reform is driving us toward rewarding outcomes at the individual patient and population level and moving away from payment for the volume of services provided. How do we improve quality, patient safety, and the patient experience at the organizational level while preserving our commitment to education and research? How

do we become truly patient- and family-centered with the competing demands on faculty? How do we leverage the power of information technology to transform care? All of the major issues impacting the academic health enterprise are explored with the result being a comprehensive view of challenges and opportunities.

This is a time of disruptive, transformational change in every aspect of the academic health center mission. Navigating this terrain is particularly hazardous because of our need to provide critical services to patients and learners and conduct life changing research while incorporating fundamental change in the structure and operation of our organizations. We are redesigning the airplane while it is flying. But that is both the challenge and opportunity for leadership. The ideas shared in this important volume point the way toward successful advancement of our unchanging purpose: to be institutions that lead toward better health for all—in a rapidly changing environment.

REFERENCES

[1] The Blue Ridge Academic Health Group. Report 18. A call to lead: the case for accelerating academic health center transformation. Atlanta: Emory University; 2014.

[2] AAHC. The state of research funding: AAHC briefing and member dialogue. CEO forum. Washington, D.C.: AAHC; 2014. p. 12.

[3] Council of Economic Advisors. Trends in health care cost growth and the role of the affordable care act. Washigton, D.C.: The White House; 2013.

[4] AAMC. AAMC database. Washington, D.C.: AAMC; 2014.

[5] OECD. OECD family database. Paris: OECD; 2014.

[6] Center For Rural Health. The picture of rural health in Arkansas. Little Rock, AR: University of Arkansas for Medical Sciences; 2012.

ABOUT THE AUTHOR

Daniel W. Rahn, MD is the Chancellor of the University of Arkansas for Medical Sciences (UAMS). He served as the President of the Medical College of Georgia and the Senior Vice Chancellor for Health and Medical Programs for the University System of Georgia before assuming his position at UAMS in 2009.

Preface

An academic health center is both a concept and an evolving organizational structure. It is a place where health professions education, biomedical and clinical research, and comprehensive and advanced patient care take place under the name of a single institution. The academic health center may be part of an established university or freestanding, but, regardless of its structure, its mission is to *apply knowledge to improve health and well-being*. Because of its name recognition and economic impact, it also serves another important function: *to build the knowledge economy and apply it in patient care*.

The use of the term "center" to describe the totality of what these institutions do may in fact be more historical than contemporary; in reality, academic health centers might be better described as "systems" or "networks," as they include a growing geographic range of institutions and facilities along with a host of many different kinds of services. In a real sense, an academic health center can be viewed as an organization without boundaries.

The components that constitute an academic health center may vary from institution to institution, but at a minimum consist of at least one medical school, one or more other health professions schools (e.g., nursing, dentistry, public health, pharmacy, allied health, veterinary medicine) and/or biomedical research graduate programs, and an owned or affiliated relationship to a teaching hospital or health system. Under the rubric of an established university or freestanding, these entities are committed to aligning patient care and academics in order to achieve what is called the "virtuous cycle" (Figure 1).

Interestingly, through the Association of Academic Health Center's international branch, we have discovered that regardless of health system, culture, or economics, established and developing academic health centers want to achieve the virtuous cycle by best aligning their education

FIGURE 1 The "Virtuous Cycle." The clinical and academic missions support each other in order to make each better. *From Wartman SA. Towards a virtuous cycle: the changing face of academic health centers. Acad Med 2008;83:797–9.*

and research with patient care. As a result, they share the common goals of:

- restructuring health professions education to meet changing and evolving societal needs;
- linking research to improved health outcomes; and
- transforming patient care based on population needs and priorities.

Neither an academic "ivory tower" nor a pure business, academic health centers are a unique hybrid of both. Blending traditional academic values with the business of patient care requires exceptional leadership. As this book amply demonstrates, academic health centers are in the process of transforming themselves to meet ever-changing societal needs and priorities. While they are responding to numerous public, scientific, and economic forces as they do so, they are also leading the way in educating a new generation of health professionals, making scientific breakthroughs that offer new diagnostic and therapeutic modalities, and providing cutting-edge patient care—all with the goal of improving health and well-being.

Steven A. Wartman, MD, PhD, MACP
President/CEO, Association of Academic Health Centers

Acknowledgment

I am deeply indebted to the fine work of Lynn Bentley, Communications Coordinator at the Association of Academic Health Centers. She has assisted on virtually every aspect of the book, including organization of the chapters, author and publisher communications, and—most notably—chapter editing. Her enthusiasm and excellent work has contributed greatly to the quality of this publication.

The Evolution of the Academic Health Center

Chapter 1

The Changing Ivory Tower: Balancing Mission and Business

Paul B. Rothman, Edward D. Miller, Landon S. King and Ellen F. Gibson

The practice of medicine is an art, not a trade; a calling, not a business.
William Osler, Johns Hopkins Founding Physician [1].

Most health professionals are still drawn to the field of medical care to relieve suffering. However, twenty-first century medical professionals also are confronted daily with the business realities involved in practicing medicine. In hospitals around the country, words like *coinsurance*, *efficiency*, *capitation*, *utilization*, and *deductible* are part of daily conversation.

Many observers still see the topic of money in medicine as taboo—after all, the primary goal for many organizations is to help patients, not make profits. Yet it is important to acknowledge that the cost of health care has been rising at unsustainable rates. There is an oft-repeated saying among nonprofit executives: "No money, no mission." Without appropriate resources, it would be impossible to do the important work that is the academic health center's (AHC's) raison d'etre.

Most readers of this book will have had some familiarity with AHCs—perhaps as an employee at one of the more than 140 AHCs in this country; perhaps as someone whose family member, or who personally, received a life-saving procedure or therapy at a teaching hospital; or, perhaps benefiting from an AHC as a major economic engine in your region. In our home state of Maryland, the Johns Hopkins Medical Institution is the second largest employer [2]. Whatever the personal connection to AHCs, it is a fascinating time to dissect the business of academic medicine and the challenges involved in balancing mission and money in health care.

The key word is *balance*. AHCs need to be fiscally responsible while still acting in the best interests of those we serve. This tension is not unique to medicine—it plays out in every service industry. Google's stated mission is "to organize the world's information and make it universally accessible" [3]; but in order to provide that valuable service, it has to generate revenue with targeted advertising, which some users see as an irritating violation of privacy.

Of course, Google is compelled to make money for shareholders. As nonprofit organizations, AHCs do not have that imperative. However, they must create margins to support clinical, research, and education programs and generate capital to maintain and upgrade sophisticated facilities. That is an exceedingly complex task, made more difficult by ever-mounting competition, legislative reforms, and recent industry trends. And the stakes are high; success is measured in lives saved, not in dividends paid. Johns Hopkins Medicine and its peer institutions around the country play a vital role in society, but are facing financial pressure on three fronts: declines in federal research funding, the nationwide drive to lower health care costs, and the need to curtail higher education debt.

The strategic steps AHCs are taking to chart a mission-driven course while remaining fiscally fit should hold interest for all leaders, not to mention citizens concerned with the prudent use of health care and tax dollars.

DEMYSTIFYING THE ACADEMIC HEALTH CENTER

Before we dig into the complex economics of academic medicine, we will offer a brief overview of the components of an AHC. An elementary understanding of these pieces—and the ways in which they are interdependent—is an essential foundation for this discussion. The typical AHC is a higher learning institution with three components: an academic division that trains learners in medicine and related professions such as public health; a clinical care arm in the form of an affiliated teaching hospital or health system; and a biomedical research enterprise. AHC mission statements vary, of course, but tend to reflect that same three-part structure: educate and train practitioners and scientists across the biomedical enterprise, provide exemplary care to prevent and treat illness, and advance scientific discovery and innovation to improve global health.

In trying to demystify the makeup of the AHC, one runs into a challenge: each institution operates differently. Some AHCs are public and others are private; some are university-based, while others are freestanding. Many systems are highly integrated, while others are decentralized.

The Transformation of Academic Health Centers. http://dx.doi.org/10.1016/B978-0-12-800762-4.00001-3

At Johns Hopkins, we opted in 1997 to unify our academic division and health system under the Johns Hopkins Medicine banner with one chief executive officer. But at institutions such as Columbia University, the medical school and the affiliated teaching hospital—New York-Presbyterian—are separate entities with distinct leadership. Harvard does not own or operate hospitals but places its students at 17 affiliates around the Boston region [4]. Regardless of how integrated the administrative and governance structures are, the components of the AHC are highly interconnected.

For starters, medical education is not confined to a classroom. After 1–2 years of didactic instruction, students at most medical schools begin their formal clinical experiences. Most medical school faculty members hold a dual appointment at an affiliated teaching hospital where they provide instruction "on the ground." Hospital rounds or time spent in outpatient clinics, for instance, are essential facets of medical education.

The benefit flows both ways: In return for the learning opportunities afforded by the clinic, doctors-in-training provide inexpensive labor, making up a critical portion of a hospital's workforce. In fact, residents comprise what is known as the "house staff" at many hospitals, so-called because in the past they lived on the hospital premises.

The research enterprise, too, has a symbiotic relationship with the educational operation. Graduate students staff research labs and perform much of the day-to-day work required. In exchange, they gain valuable hands-on experience and mentoring from the university's research faculty, whose salaries are often supported by grants.

Finally, research and clinical care are closely linked. Academic departments commonly choose to invest their clinical margins to support research. Many clinical faculty are expected to perform research, and they frequently collaborate with basic scientists. When discoveries arise in the lab, AHCs make every effort to translate them into tests and treatments that have real-world benefit for the patients in their care. Similarly, important advances derive from the questions that arise in the management of patients.

All this activity adds up to the fruitful interrelationships of the clinic, classroom, and lab. While these interfaces are a unique strength of AHCs, they can also be a point of vulnerability, as we will discuss in this chapter. For an AHC to execute its mission, it needs to leverage financial and human capital across the organization. If any part of the AHC is overextended, the whole enterprise is strained.

THE EMERGENCE OF COST IN ACADEMIC MEDICINE

While the tension between business and mission in medicine has always existed to some degree, finances were not always so intricate. Early AHCs were hardly the economic behemoths they are today.

At the end of the nineteenth century, medical schools in America were for-profit trade schools with little academic rigor and frighteningly low entrance requirements. The founding of The Johns Hopkins Hospital in 1889, followed 4 years later by the opening of the Johns Hopkins University School of Medicine, marked a new era in medical education: a high threshold for admissions; a curriculum grounded in the scientific method; and, in a new twist, the linking of the school and the hospital through bedside training and joint appointments of the school faculty to the hospital staff [5].

Private Funding

Even in 1893, money influenced academic medicine. The opening of the Johns Hopkins School of Medicine was delayed for years due to financial shortfalls. In light of this, the founders were offered a large sum of money from a wealthy group of Baltimore women led by railroad heiress Mary Elizabeth Garrett. The gift stipulated that the medical program admit women on the same terms as men and set rigorous application requirements, including a bachelor's degree and a background in biology [6]. The medical community deemed these demands unreasonable at the time, but the Hopkins founders had little choice but to comply.

In 1910, the Flexner Report was published, laying the groundwork for modern-day academic medicine. The report castigated U.S. medical schools for their lax standards and called for a restructuring of medical education to resemble the Johns Hopkins model. The Flexner Report had great impact: it eliminated many diploma mills, standardized curricula, and lengthened the training period. As fewer new physicians were minted, the profession gained prestige; doctors were better-trained and more highly paid [7].

Before World War II, the funding of medical education was fairly straightforward. In 1904, the American Medical Association formed the Council on Medical Education, which published a paper recommending that graduate training in hospitals should be a licensure requirement for physicians. In the decades immediately following, hospitals were able to subsidize their own clinical internships by charging patients a little extra.

Government Funding

Many pin the advent of the modern AHC and the big business model of academic medicine to the 1960s, when money started flowing in from federal agencies. The Centers for Medicare and Medicaid Services (CMS) was established in 1965. Prior to that, about half of health care spending was paid by consumers out of pocket. As more people became covered under public and private insurance plans and the responsibility of the medical bill shifted from the individual to the insurer, both health care access and expenditures climbed sharply. Per-person spending on health care exploded from $147 a year in 1960 to $8915 in 2012. Meanwhile, national health expenditures rose from 5% of the country's gross domestic product (GDP) to more than 17% [8].

CMS also enabled the public financing of graduate medical education. Recognizing that society benefits from highly trained clinicians, Medicare began to cover a portion of the cost to train doctors. Initially, the reimbursements were retrospective and the number of residency slots was unlimited. Growth took off during this period as hospitals were free to launch new training programs. To manage costs, Medicare shifted to a prospective system in 1983, whereby direct medical education (DME) payments are made based on the hospital's allotted number of residents and typical per-resident cost, which varies widely by region [9]. Medicare support for DME totals approximately $3 billion a year [10].

AHCs tend to provide care for patients with more health needs than nonteaching hospitals. Recognizing that, Medicare gives teaching hospitals extra funds to compensate for the added costs of handling more severe cases as well as the inefficiencies inherent in educating trainees. These indirect medical education (IME) payments amount to roughly $6.5 billion a year [10].

In the mid-twentieth century, the National Institutes of Health (NIH) also began infusing money into academic medicine. In 1946, the NIH created a Research Grants Office charged with soliciting and reviewing extramural grant applications for studies in the health sciences and with exploring neglected areas of investigation. Between 1959 and 1969, the NIH budget more than tripled from $292 million to $1.1 billion. By 1979, Congressional appropriations to the NIH had ballooned to $3 billion. Currently, the NIH allocates grants to more than 300,000 researchers, amounting to more than $30 billion a year [11].

With the rapid influx of funds from Medicare, Medicaid, private insurance, and the NIH, academic medicine flourished in the second half of the twentieth century. Concurrently, many new hospitals were built in the U.S. with federal subsidies. During this heyday, the union of research, teaching, and clinical services in American medicine led to remarkable advances in the fields of oncology, cardiovascular disease, transplantation, reproductive health, genetic engineering, mental health, medical imaging, and other scientific areas far too numerous to name here.

THE COSTLY BUSINESS OF ACADEMIC MEDICINE

If the latter half of the twentieth century was a boom period in academic medicine, the last two decades have been more challenging economically, and AHCs have had to make tough decisions to ensure long-term viability.

In 1987, a federal law was repealed that had required companies to demonstrate community need before building a new health care facility or purchasing pricy medical equipment [12]. Subsequently, new technologies—pioneered by AHCs—made their way into community hospitals and physician groups, elevating regional competition for patients. Managed care organizations began to exert tremendous influence in health care, forcing providers to discount their rates and shorten hospital stays at the same time Medicare payments to hospitals were being cut. Indirect payments to teaching hospitals were capped in 1997, restricting new residency and fellowship slots. And after peaking in 1998, "real-dollar" NIH funding for health research began to slip. In that unforgiving environment, margins became slim to nonexistent, and at least one large AHC and dozens of smaller hospitals filed for bankruptcy [13,14].

Despite the trend toward managed care, U.S. health care spending continued to rise, outstripping the growth in the GDP and increasing more rapidly than in other industrialized nations. The reasons for this country's exorbitant health care costs are complex and include higher utilization, misaligned incentives, and steep regulatory burdens that drive up administrative costs.

Institutional complexities also produce especially high expenses at AHCs. First, there is technology. Teaching hospitals are often the first to implement and deploy cutting-edge medical technologies, which tend to be pricy at first and then come down in cost. Indeed, teaching hospitals often help *develop* such technologies. Second, medical instruction is costly. Training a resident costs about $100,000 per year, and Medicare picks up just 40% of that tab [15]. Teaching hospitals need a cadre of physicians available to train students in every discipline, including areas where subspecialists command very high salaries. For example, a teaching hospital must pay a neurosurgeon hundreds of thousands of dollars a year to spend as much time supervising residents as performing lucrative procedures.

As safety-net health care providers, AHCs also maintain the capacity to treat anyone who comes in the door. Specialty hospitals can focus on money-making areas—orthopedics, cardiology—but AHCs need to cover the whole spectrum of human illness around the clock. Often, care can be provided less efficiently because AHCs take the time to teach as they treat.

Like education, research is a cost center: the expense associated with doing research is greater than the support provided to pay for the work [16,17]. Overhead costs can be quite high. Money for facilities and administrative (F&A) support—the maintenance of high-tech laboratories and equipment, compliance work, utilities such as heat, and so on—gets tacked on to federal research awards, but grants rarely cover the full cost of projects [17].

While academic research is a windfall for society, it offers little direct financial benefit for universities in the form of licensing fees and royalties. Universities have had the right to patent and commercialize inventions arising from federally funded research since 1980, thanks to the Bayh–Dole Act. While universities spend $63.7 billion total on sponsored research in science and engineering, they take in just $2.6 billion in total licensing income—a 4% return, according to the Association of University Technology Managers [18].

Academics historically have taken an open-source approach to discovery and invention, sharing information broadly via publication. Daniel Nathans and Hamilton Smith, former Hopkins microbiologists, won the 1978 Nobel Prize for their discovery of restriction enzymes, which helped usher in modern molecular biology [19]. They never patented their findings. If they had, the university would have garnered immense royalties, but it may have stifled innovation in the exploding field of genetic engineering. As we will discuss later in this chapter, universities are increasingly tapping tech transfer to compensate for budgetary shortfalls, but in doing so, we must not sacrifice core values.

What AHCs do is expensive—for the reasons listed above and more—but there is a very strong value proposition. Academic institutions conduct more than half of basic research and more than one-third of all research in the United States, according to the National Science Foundation (NSF) [20]. There are not enough pages in this book to list all the spectacular achievements that have sprung from academic medicine, but here are just a few: the polio vaccine, magnetic resonance imaging (MRI), the artificial heart, targeted cancer therapies, and antiretrovirals that allow people with HIV/AIDS to live with the disease. In particular, academic scientists have waged an aggressive battle against cardiovascular disease; consequently, the age-adjusted death rate from heart disease fell 60% from 1950 to 2010 [21]. Another example is life expectancy. Thanks in large part to new discoveries and advances in medicine, Americans' life expectancy has been increasing steadily over the past 50 years; the average American is now expected to live 78.7 years, compared to 70.2 years in 1964, the year before CMS came into being [22].

Major teaching hospitals, which make up only 6% of hospitals, provide 19% of all hospital care and a much larger proportion of critical care [23]. For instance, AHCs provide two-thirds of pediatric intensive care and 61% of Level 1 trauma treatment for the severely injured. If you suffer a bad accident, you want to receive care at a teaching hospital; national studies have shown that treatment at a designated trauma center boosts survival rates by 20–25% [24].

AHCs are also a social safety net, delivering a disproportionate share of treatment for the underserved and indigent. Major teaching hospitals provide $6.7 billion in charity care each year, or more than 40% of all care provided without expectation of payment [25]. We also account for a quarter of Medicaid hospitalizations.

In addition to healing the sickest patients and addressing health care disparities, AHCs generate substantial economic activity at the local and national level. Medical schools and teaching hospitals account for 3.3 million jobs in this country and spend large sums on goods, services, and construction contracts [26].

Indirectly, each dollar spent in our research labs translates into several dollars for the nation's economy in terms of medicines and devices generated, cost savings afforded by medical advances, and jobs created. The U.S. Department of Commerce reported that approximately 36 jobs are generated for every $1 million in academic research spending. As an example, the University of Pittsburgh estimates that its yearly research and development (R&D) outlay supports more than 23,000 local jobs [27].

Finally, and perhaps most obviously, medical education itself is a public good. U.S. medical schools produce roughly 18,000 new graduates each year—highly skilled trainees equipped to prevent disease, alleviate suffering, and provide the highest level of care. Americans pay 1.2 billion visits to physicians' offices, emergency departments, and hospital outpatient centers each year, and that demand is only expected to grow [28].

RESPONDING TO CHANGE: CHALLENGES AND OPPORTUNITIES FOR AHCs

Today's business and regulatory climate presents challenges and opportunities for all sectors of the nation's health care systems but none more so than academic health systems. Millions of previously uninsured Americans have gained coverage as a result of the Patient Protection and Affordable Care Act, which mandates that individuals have health insurance and facilitates coverage through new online exchanges and the expansion of Medicaid.

While the new insurance exchanges have garnered major attention, health reform entails much more than expanding access to care. Around the country, policy leaders are testing ways to bend the cost curve in health care, while at the same time ensuring a greater focus on quality and prevention. In many cases, AHCs are ground zero for these experiments.

Concurrently, the United States is encountering demographic shifts that statisticians are projecting to result in a surging demand for health services over the coming decades. Not only is the population growing, but it is graying. Roughly 10,000 Baby Boomers enter the Medicare rolls each day, a trend that will continue over the next 15 years [29].

Medicaid and Medicare traditionally have reimbursed providers at 70–80% of the private rate, so health systems are nervously anticipating the steps necessary to manage more chronic conditions for less pay. Against this backdrop of profound change, university and hospital leaders have been huddling together, strategizing about how academic medicine can continue to lead with less reliance on clinical margins to cross-subsidize education and research.

Cost-cutting is a given, and it is happening across the board. A number of institutions are already announcing plans for spending reviews and budget cuts: Cleveland Clinic reported that it would cut $360 million from its budget; Vanderbilt plans to trim $250 million; Yale–New Haven intends to cut $125 million each year for the next 4 years.

But the transformation of AHCs goes beyond simply trimming fat from each budget line. Many academic health systems are engaged in a radical rethinking of the business model, implementing sustainable, forward-looking moves that will ensure they do more than merely survive—they will thrive. We will inventory some of those actions here, and they will be explored in greater depth in this book's remaining chapters.

Clinical Innovation: New Payment Models, Big Networks, and Focus on Quality

As stated earlier, misaligned incentives have driven waste in our health care system. Historically, the United States has operated under a fee-for-service model, wherein the more procedures a provider bills, the more that provider is paid. Quality outcomes were rarely factored in, and there was scant motive for physicians to limit health care expenditures.

Legislators, insurers, and other health care players have responded with efforts to realign the incentive system using new models that reward quality and value. The labels vary—bundled or value-based payments, patient-centered medical homes, accountable care organizations (ACOs)—but the principles are the same. Providers are expected to work as a team, reviewing whether services are medically necessary, and pursuing the most proven and cost-effective options along the care continuum.

In anticipating these shifts, AHCs are taking steps to ensure that they are well-positioned. One popular strategy involves forming a tightly integrated network of physicians via acquisitions and affiliations in the university's surrounding communities. For instance, Vanderbilt University recently formed a clinical consortium called Vanderbilt Health Affiliated Network, which includes 32 academic and community hospitals, dozens of independent physician groups, and even two insurers [30].

A large, diversified network of clinical sites ensures that a health system will have sufficient patient volumes, even if inpatient demand drops off. It offers an avenue for what we call *upstream* interventions—reaching patients before diseases and illnesses progress. It also serves as a feeder system for academic subspecialists who work on referrals.

Johns Hopkins, the University of Virginia, and the University of Michigan health systems are just a few AHCs that have recently formed their own ACOs, as created by the Affordable Care Act. In the team-based approach emphasized by the ACO or medical home, a group of providers collaborates to manage illness and receives a bundled-sum payment if it meets certain quality measures. The approach is intended to ensure less duplication, fewer errors, and better outcomes—all at a lower price. The more efficient we can become at keeping populations well and managing chronic conditions, the healthier our margins will be.

In this competitive environment, other AHCs are opting to unify the hospital and medical school leadership the way Johns Hopkins has done. For instance, in 2007, the New York University (NYU) School of Medicine and NYU Hospitals began functioning as an integrated center with one Dean/CEO. A unified governance structure facilitates alignment across the clinical, teaching, and research missions, and leaders have a panoramic view of the whole organization when making decisions.

A highly-integrated system also allows for operating efficiencies and greater purchasing power. As part of our five-year strategic plan, the Johns Hopkins Health System has begun consolidating all six hospitals' contracts for supplies and services, including linens, computers, fuel, food, medical equipment, and waste management. By utilizing economies of scale, we can realize significant cost savings.

As competition for market share intensifies, the diversified academic health network can take several additional steps. It can steer routine cases to community hospitals and perform more procedures on an outpatient basis. It can delve into home care and devote resources to community initiatives that focus on population health. And, clinical affiliations can extend beyond one's zip code: international partnerships represent a growing opportunity for some AHCs.

Another feature of health care's new frontier is the highly discerning patient. As transparency around costs and outcomes grows, a consumerist attitude toward health care is emerging. Widespread access to information via the Internet and insurance plans that shift more costs onto patients are driving heightened price sensitivity and more doctor- and hospital-shopping. Even Medicare has started publicizing its payments to individual physicians.

Catering to this consumer mentality, some AHCs are seizing on a popular marketing theory, which holds that an organization needs to identify and advertise a single core competency—a skill that adds value and sets it apart from its peers. Apple's core competency is design. Zappos excels at customer service. To differentiate his organization from the competition, David Feinberg, president of the UCLA health system, has deemed "people-first" care as the chief priority. Every patient who calls the center is offered a same-day appointment, and faculty who provide excellent clinical care reap rewards. This strategy seems to be working. Independent surveys now place UCLA in the 99th percentile among the nation's hospitals in some categories pertaining to the patient experience [31].

AHCs benefit from excellent brand recognition, but cannot coast on our top rankings in news magazines. If we want to maintain that edge, we need to back up our claims with visible data on outcomes. In April, Johns Hopkins launched a new online dashboard that shows our clinical employees how their division is performing by applying quality indicators such as hospital-acquired infection rates and readmissions; some of this information will be made available to the public.

To improve performance on quality and cost metrics, AHCs are taking strides to establish best-practice protocols and weed out wasteful and potentially harmful procedures. As part of the ABIM Foundation's "Choosing Wisely" program, medical centers nationwide are aiming to devise a list of 250 superfluous tests and treatments [32]. For example, Hopkins doctors have found that the fingertip pulse oximeter, a tool commonly used to measure oxygen saturation, is ineffective in children, leading to pointless follow-up. Academics have flagged many other tests as unwarranted, including CT scans for headaches and stress tests for patients at low risk of heart disease. Students are cautioned to be highly selective about how they intervene as health professionals; even an aspirin can cause harm.

Sharing of electronic health data is essential for synchronizing providers in integrated care coordination systems. Nearly every large health center in the country is currently changing over to a system of electronic medical records. While the transition can feel cumbersome and requires up-front expense, the eventual goal is to streamline record-keeping, saving providers time on tedious administrative tasks. Collating data in a single digital repository offers another advantage: with the right analytics, invaluable insights can be obtained.

Research Innovation: Team Science and Making Up the Funding Shortfall

It is a tremendously exciting—if uniquely challenging—time to be a biomedical scientist. We are at the dawn of a powerful new age of discovery. Technologies that allow us to obtain and analyze complex data sets are enabling huge advances in disciplines such as metabolomics, proteomics, and epigenetics. At the same time, sophisticated brain-imaging tools that can decode neural signals are yielding practical applications such as brain-controlled prostheses.

Once we can overcome some of the associated challenges—shaky research funding, primarily, but also figuring out how to mine vast amounts of data—we are poised to make some tremendous breakthroughs.

Every day, AHCs harvest data through biopsies, lab research, clinical trials, the taking of medical histories, and so on. But our ability to manage and analyze those reams of information is lagging. To keep up, AHCs need to hire more faculty and staff with expertise in data sciences and invest in state-of-the-art computational cores; but new technologies are expensive to acquire, operate, and train people on.

In light of today's massive research opportunities, we believe strongly that it is in the public interest to restore higher federal funding levels. After seeing its share of the federal budget peak in 2003, the NIH has lost roughly one-fifth of its funding in real dollars [33]. As American scientists continue climbing out of the hole caused by sequestration, other countries are allocating more funding for science each year. Universities continue to remind government leaders that the hearty support for science proclaimed by both parties should translate into a matching financial commitment.

In the meantime, AHCs should work to diversify their sources of funding. For instance, Johns Hopkins is accelerating efforts to tap non-NIH federal support in the form of grants from the Department of Defense, the NSF, CMS, and the new Patient-Centered Outcomes Research Institute (PCORI). We are also relying on donations from philanthropists and private foundations more than ever before.

As the NIH pot shrinks, the agency is more prone to fund sure-bet science and established scientists, so AHCs need to find new avenues for underwriting risky, cutting-edge research and product development. At Johns Hopkins, we have started our own internal fund—managed by a faculty research council—to bankroll bold new initiatives and pilot programs. At the University of Virginia, a foundation funded by private interest groups gives starter grants to explore the potential of new technologies such as focused ultrasound. Stanford has launched a new fund to invest in companies that emerge from their StartX Accelerator Program.

One way AHCs are both encouraging funding and maximizing the impact of our research is by promoting team science. Great work happens when AHCs can link up the many islands of excellence that exist on our campuses. Chemists and physicists work side-by-side with nanotechnologists and computer scientists; cardiac surgeons brainstorm with biomechanical engineers. AHCs should continue to break down the silos in academic medicine and replace them with interdisciplinary centers, where experts from many fields can come together to tackle pressing health problems.

The systems approach also helps build a bridge between basic scientists and those at the bedside. More and more, the NIH is choosing to finance research that is closer to the patient. The agency launched the Clinical and Translational Science Award (CTSA) program in 2006 to help accelerate the translation process from discovery to real-world treatments for patients. At Johns Hopkins, as at other AHCs, transformation includes promoting translational research activity and jumpstarting more entrepreneurial momentum.

It is part of the AHC's pact with the public that they will make every effort to translate findings into diagnostics, drugs, and devices that fill unmet needs in medicine. To that end, AHCs are forging partnerships with biotech companies and offering more institutional support for commercialization efforts. As an example, Johns Hopkins kicked off a collaboration with Maryland biotech firm MedImmune, which allows our scientists to share resources in the pursuit of new treatment targets in areas like rheumatoid arthritis and prostate cancer [34]. Last year alone, eight new start-up companies were formed based on Johns Hopkins technologies.

While academic departments can reinvest revenue from tech transfer and commercialization into our research programs, the main goal is not to attract funds but to get inventions out into the marketplace. Industry involvement can give our work that extra push. Even in today's tough funding climate, a steady stream of science that builds on others' work to benefit humanity still overrides the allure of licensing income.

Education Innovation: Tackling Debt and Rethinking Training

Education debt places a tremendous burden on medical school students and graduates. The cost of attending medical school has risen at a faster pace than inflation over the last 20 years, and medical tuition is now at a record high. The median cost of attending a private medical school for 4 years averages $299,000, while 4 years at a public school sets students back roughly $226,000. For the class of 2014, the median education debt amount among those in the red was $180,000, according to the Association of American Medical Colleges (AAMC). While grants and scholarships are available, particularly at private schools, full rides are rare except for combined MD/PhD students; most students (84%) exit medical school with some amount of debt [35].

The high cost of medical education has wide-ranging repercussions. Resident pay has not kept up with medical-school debt, leading many students to choose high-paying subspecialties. This threatens to worsen the primary care shortage at a time when more Americans are getting health insurance coverage and preventive care is becoming a focus for the profession.

One potential solution is to take a scalpel to medical training. The average physician studies for 14 years (4 years of college, 4 years of medical school, and 3–8 years of residencies and fellowships). Bioethicist Ezekiel Emanuel made a stir in 2012 when he published a Journal of the American Medical Association (JAMA) editorial arguing that medical training should be cut by 30%, particularly for those with nonacademic aspirations [36].

While that recommendation remains controversial, it launched a vital conversation about how to create a more expedient, cost-effective path to a career in medicine. Forty schools now offer combined undergraduate-MD programs, which shave the total training time to 6–7 years [37]. Preclinical basic-science education has already been trimmed at several schools, including Duke University and the University of Pennsylvania, with no evidence it impacts board-exam performance. And NYU now offers a three-year

medical school program to some students interested in entering primary-care fields [38].

Other groups argue for an outcomes-based model [39]. Under this paradigm, students would complete their training when they demonstrate competency, not when they finish a prescribed number of years of schooling.

While the specific proposals vary, the Flexner model does need a twenty-first century makeover. With today's system of training, we are looking at a potential shortage of more than 20,000 primary care physicians by 2020 [40]. Over the next 15 years, we need to train 50% more doctors to meet surging demand. Yet, even if medical schools were to accelerate classroom learning and turn out more graduates, frozen DME funding levels have capped the number of residency positions. One alternative is to embed trainees at AHC-affiliated community hospitals and medical practices, and have nonacademics more involved in mentoring the next generation of clinicians.

Academic systems also are lining up reinforcement through other health professionals. In the future, the so-called "mid-level providers"—physician assistants (PAs) and nurse practitioners (NPs)—will help meet demand by delivering a greater proportion of services. (According to the Department of Health and Human Services, the PA and NP workforce is expected to increase much faster than the supply of primary care physicians [40].) To sculpt team players, AHCs are offering more interprofessional training opportunities, bringing together physicians, nurses, and public-health experts in interdisciplinary rounds, simulation centers, and classroom courses.

As we reimagine health care delivery, medical education must be revamped to reflect the new realities. Recognizing that clinical quality and safety go hand-in-hand with patient-centered care, AHCs are instilling in trainees a service-oriented ethos that stresses communication skills. At Johns Hopkins, our relatively new Genes to Society curriculum gets students into the clinic to interact with patients in the first weeks of class. At NYU, the new "Curriculum for the twenty-first Century" stresses the importance of practicing humanistic medicine [41].

We also are using new technological tools at our disposal to enhance and update medical education. Online courses extend our education reach around the globe and present a potential new revenue stream. In a novel pedagogical method known as "flipping the classroom," students watch lectures online, freeing up classroom time for discussion, questions, and hands-on learning. Simulation centers help make our hospitals safer by providing an interactive virtual environment for medical students, residents, nurses, and others to practice procedures.

In trying to modernize medical training, our best intentions are bumping up against economic realities. Since the 2008 recession, states have been making unprecedented cuts to their support for public universities—by as much as half in states such as Arizona [42]. Limits on resident duty hours raise serious questions about the feasibility of shortening residencies. In addition, administrative demands, such as billing compliance, are infringing on the time learners spend on bedside training [43]. Recognizing the need for more solutions, academic health leaders are engaged in the important work of challenging the status quo.

LEADING THE CHANGE IN HEALTH CARE

Prominent statistics have been reported during the debates over health reform: the United States spends more per person on health care than any other developed country, and yet our health outcomes are among the worst. One in three American adults is obese; our life expectancy is in the bottom quartile of Organisation for Economic Co-operation and Development (OECD) countries. The solutions prescribed by politicians are contentious, but most Americans agree that change is needed.

Changes resulting from the Affordable Care Act can reduce treatment disparities, but the law's provisions do relatively little to restrain runaway costs. As we add 30 million newly-insured Americans to the U.S. health care system, we will need to think strategically about how to deliver enhanced preventive and therapeutic services, while at the same time controlling expenditures. The responsibility for reform does not rest solely with policymakers; it lies with health care providers as well.

Who better than this country's prestigious AHCs to find the cures for what ails our health care system? When it comes to solving vexing scientific and medical questions, no one has a better track record in producing outcomes. It is time for our institutions to apply that same level of scientific rigor to producing a better-quality, less-costly health care system.

We must partner with organizations and agencies such as PCORI, the Center for Medicare and Medicaid Innovation, and private foundations to postulate reforms and test their efficacy. We need to be meticulous about eliminating waste and standardizing best practices. Most importantly, we need to apply the spirit of ingenuity that practically defines our campuses to these broad, systemic problems. Innovation is not limited to the laboratory; we have to analyze our processes and continually find new ways to be more efficient and more effective.

Medical schools and teaching hospitals are so steeped in history and tradition that it can be difficult to step back and reevaluate our methods. And, in the past, there was not a pressing need to rethink the way we finance and execute our mission; there was little financial incentive to operate differently. Today's economic exigencies are motivating us to break from tradition and experiment with new ways of delivering health care, conducting research, and educating learners. The transformation is exciting and unnerving at once. In the end, we believe it will result in better health, both physically and fiscally.

Test Case: Maryland's New All-Payer System

Around the country, different groups are piloting innovative new payment models. One such experiment in Maryland—an overhaul of the hospital-payment system—has huge implications for Johns Hopkins Medicine.

Rates at Maryland hospitals are set by a state agency known as the Health Service Cost Review Commission, or HSCRC. Since 1977, Medicare has waived the payment rules it uses in the rest of the country and paid Maryland hospitals according to HSCRC rates. This helped make care more affordable for everyone and funded important public goods such as treatment for the uninsured. The main shortcoming of the old system, however, was its focus on per-admission inpatient costs. The fee-for-service model offered no real incentives to keep people well or to limit health expenditures.

Maryland's new all-payer model, approved in January 2014, moves us closer to a "population health" system. It sets a global budget for hospital spending (both in- and outpatient) in the state. The HSCRC then establishes budgets for individual hospitals and hospital systems based on the number of patients served and the complexity of cases seen. A given hospital is responsible for staying within that budget while charging all patients the same amount for the same service, regardless of their insurance plan.

The new model puts the focus squarely on quality, efficiency, and coordinating care—in and out of the hospital. It creates incentives for reducing duplication, readmissions, and preventable conditions.

Of course, a change this radical poses challenges for a large system like Johns Hopkins. The shift in emphasis from volume to value requires us to think very differently about where and how we treat. Fortunately, we saw these changes coming and have been preparing. The new all-payer test period gives us 5 years to roll out strategic changes, and we expect market forces will compel other AHCs around the country to follow. We hope to provide a model in Maryland.

(Excerpted from Rothman, P.B. (2014). Managing Big Change in Maryland. Hopkins Medicine Magazine. Fall 2014.)

REFERENCES

[1] Hinohara S, Niki H. Osler's "A way of life" and other addresses, with commentary and annotations. Durham, NC: Duke University Press; 2001.

[2] Baltimore Sun website. Top Maryland employers. Available at: http://www.baltimoresun.com/bal-rg-mdbusiness21,0,510201.story [accessed 21.06.14].

[3] Google Company website. About Google. Available at: http://www.google.com/about/company/ [accessed 21.06.14].

[4] Harvard Medical School website. HMS affiliates. Available at: http://hms.harvard.edu/about-hms/hms-affiliates [accessed 21.06.14].

[5] Johns Hopkins Medicine website. Revolution in American Medicine. Available at: http://www.hopkinsmedicine.org/about/history/history3.html [accessed 23.06.14].

[6] Johns Hopkins Medicine website. Women – or the female factor. Available at: http://www.hopkinsmedicine.org/about/history/history6.html [accessed 23.06.14].

[7] Duffy TP. The flexner report—100 years later. Yale J Biol Med September 2011;84(3):269–76.

[8] Centers for Medicare and Medicaid Services website. National Health expenditure fact sheet. Available at: http://www.cms.gov/Research-Statistics-Data-and-Systems/Statistics-Trends-and-Reports/NationalHealthExpendData/NHE-Fact-Sheet.html [accessed 25.06.14].

[9] Association of American Medical Colleges website. Medicare direct graduate medical education (DGME) payments. Available at: https://www.aamc.org/advocacy/gme/71152/gme_gme0001.html [accessed June 25].

[10] Health Resources and Services Administration Council on Graduate Medical Education. Twenty-first report: improving value in graduate medical education. Rockville, MD: HRSA; 2013.

[11] National Institutes of Health website. NIH budget. Available at: http://www.nih.gov/about/budget.htm [accessed 26.06.14].

[12] National Conference of State Legislatures website. Certificate of need: state health laws and programs. Updated July 2014. Available at: http://www.ncsl.org/research/health/con-certificate-of-need-state-laws.aspx [accessed 26.06.14].

[13] Burns LR. Lessons from the Allegheny bankruptcy. LDI Issue Brief 2000;5(5):1–4. [NCBI: Rockville, MD].

[14] Yarbrough A, Landry R. Factors associated with hospital bankruptcies: a political and economic framework. J Healthc Manag 2009;54(4):252–72. [ACHE].

[15] Association of American Medical Colleges. What does medicare have to do with graduate medical education? Available at: https://www.aamc.org/advocacy/campaigns_and_coalitions/gmefunding/factsheets/253372/medicare-gme.html [accessed 27.06.14].

[16] Association of American Universities website. Frequently asked questions (FAQs) about the indirect costs of federally sponsored research. Published October 2013. Available at: https://www.aau.edu/WorkArea/DownloadAsset.aspx?id=14693 [accessed 27.06.14].

[17] Association of Academic Health Centers. AAHC benchmarks & metrics initiative, financial expense and funding source data project. March 2014. Available at: http://www.aahcdc.org/Resources/BenchmarksandMetrics/TheCostofResearch.aspx. [accessed 27.06.14].

[18] Association of University Technology Managers website. AUTM U.S. Licensing activity survey: FY2012 highlights. Available at: http://www.autm.net/AM/Template.cfm?Section=FY2012_Licensing_Activity_Survey&Template=/CM/ContentDisplay.cfm&ContentID=11435 [accessed 27.06.14].

[19] Nobel Prize website. The nobel prize in Physiology or medicine 1978. Available at: http://www.nobelprize.org/nobel_prizes/medicine/laureates/1978/ [accessed 27.06.14].

[20] National Science Foundation. National Center for science and engineering statistics. Arlington, VA: Science and Engineering Indicators; January 2012. Available at: http://nsf.gov/statistics/seind12/c5/c5s1.htm (NSB 12–01).

[21] CDC. Achievements in public health, 1900–1999: decline in deaths from heart disease and stroke – United States, 1900–1999. MMWR/CDC 1999;48(30):649–56. Atlanta, GA.

[22] Hoyert DL, Xu JQ. Deaths: preliminary data for 2011. Natl Vital Stat Rep 2012;61(6). [National Center for Health Statistics: Hyattsville, MD].

[23] Teaching Hospitals Charity Care – Data. AAMC analysis of AHA annual survey database, FY 2007. Available at: https://www.aamc.org/linkableblob/70260-5/data/teachinghospitalscharitycare-data.pdf.

[24] MacKenzie EJ, Rivara FP, Jurkovich GJ, Nathens AB, Frey KP, Egleston BL, et al. A national Evaluation of the effect of trauma Center care on mortality. NEJM 2006;354:366–78. [MMS: Boston, MA].

[25] Teaching Hospitals Charity Care – Data. AAMC analysis of AHA annual survey database, FY 2007. Retrieved June 27 from https://www.aamc.org/linkableblob/70260-5/data/teachinghospitalscharitycare-data.pdf [accessed 27.06.2014].

[26] Umbach T. The economic impact of publicly funded research conducted by AAMC-member medical schools and teaching hospitals. Washington, DC: AAMC; 2011.

[27] University of Pittsburgh website. Economic impact report 2009. Available at: http://www.g20.pitt.edu/docs/econimpact09.pdf [accessed 27.06.14].

[28] CDC/National Center for Health Statistics. Selected patient and provider characteristics for ambulatory care visits to physician offices and hospital outpatient and emergency departments: United States, 2009–2010. Available at: http://www.cdc.gov/nchs/fastats/physician-visits.htm [accessed 27.06.14].

[29] Pew Research Center. Baby Boomers Retire (December 29, 2010). Available at: http://www.pewresearch.org/daily-number/baby-boomers-retire/ [accessed 30.06.14].

[30] Kirk P. Vanderbilt University Medical Center forms Nation's largest clinically integrated network that includes its own health insurance offering. Dark Daily; July 8, 2013. Available at: http://www.darkdaily.com/vanderbilt-university-medical-center-forms-nations-largest-clinically-integrated-network-that-includes-its-own-health-insurance-offering#axzz35lAXJ7E3.

[31] CEO of hospital system has mantra: put patients first [News Release]. Los Angeles, CA: University of California Los Angeles Office of Media Relations; July 22, 2009. http://newsroom.ucla.edu/stories/ceo-of-hospital-system-has-mantra-96848 [accessed 09.07.14].

[32] ABIM Foundation website Choosing Wisely. Available at: http://www.abimfoundation.org/Initiatives/Choosing-Wisely.aspx [accessed 30.06.14].

[33] American Association for Cancer Research. AACR cancer progress report 2012. Clin Cancer Res 2012;18(Suppl. 1):S1–100.

[34] Johns Hopkins and MedImmune announce $6.5M medical research collaboration [News Release]. Baltimore, MD: Johns Hopkins Office of Communications and Public Affairs; December 11, 2013. Available at: http://hub.jhu.edu/2013/12/11/medimmune-research-partnership [accessed 30.06.14].

[35] Medical student education: debt, costs, and loan repayment fact card. Association of American Medical Colleges; October 2014. Available at: https://www.aamc.org/download/152968/data/debtfactcard.pdf [accessed 05.11.14].

[36] Emanuel EJ, Fuchs VR. Shortening medical training by 30%. JAMA 2012;307(11):1143–4.

[37] Association of American Medical Colleges. Medical schools offering combined Undergraduate/M.D. Programs. Available at: https://www.aamc.org/students/aspiring/347100/bsmdprogamlists.html [accessed 30.06.14].

[38] Hartocollis A. N.Y.U. and other medical schools offer shorter course in training, for less tuition. New York Times; December 23, 2012.

[39] Harden RM. Trends and the future of postgraduate medical education. Emerg Med J October 2006;23(10):798–802.

[40] National Center for Health Workforce Analysis. Projecting the Supply and demand for primary care practitioners through 2020. Rockville, Maryland: U.S. Department of Health and Human Services, Health Resources and Services Administration; 2013.

[41] NYU School of Medicine website. Curriculum for the 21st Century. Available at: http://school.med.nyu.edu/studentsfaculty/office-medical-education/curriculum [accessed 30.06.14].

[42] Oliff P, Palacios V, Johnson I, Leachman M. Recent deep state higher education cuts may harm students and the economy for years to come. Center on Budget and Policy Priorities; March 19, 2013. Available at: http://www.cbpp.org/cms/?fa=view&id=3927 [accessed 09.07.14].

[43] Doctors-in-training spend very little time at patient bedside, study finds [News Release]. Baltimore, MD: Johns Hopkins Office of Communications and Public Affairs; April 23, 2013. http://www.hopkinsmedicine.org/news/media/releases/doctors_in_training_spend_very_little_time_at_patient_bedside_study_finds [accessed 09.07.14].

ABOUT THE AUTHORS

Ellen F. Gibson is a speechwriter for the Dean/CEO at the Johns Hopkins School of Medicine. She previously was a business reporter for the Associated Press and Bloomberg News.

Landon S. King, MD is the Executive Vice Dean of the School of Medicine of The Johns Hopkins University. Prior to this position, he was selected to be the Director of the Division of Pulmonary and Critical Care Medicine at Hopkins.

Edward D. Miller, MD is Dean & CEO Emeritus, Johns Hopkins Medicine, and Professor of Anesthesiology & Critical Care Medicine, Johns Hopkins University School of Medicine. He previously served as CEO of Johns Hopkins Medicine, Dean of The Johns Hopkins University School of Medicine, and Vice President for Medicine of The Johns Hopkins University.

Paul B. Rothman, MD is the Frances Watt Baker, M.D., and Lenox D. Baker Jr., M.D., Dean of the Medical Faculty, Vice President for Medicine of The Johns Hopkins University, and CEO of Johns Hopkins Medicine. Rothman came to Hopkins in 2012 after having served as Dean of the Carver College of Medicine at the University of Iowa and leader of its clinical practice plan.

Chapter 2

Presidential and Academic Health Center Leadership within the Modern University: Opportunities and Challenges

Christopher C. Colenda and Ricardo Azziz

*The whole point…is to make sure the right decisions happen –
no matter how difficult or painful – for the long-term greatness
of the institution and the achievement of its mission, independent of consensus or popularity.*

*Greatness flows first and foremost from having the right
people in the key seats, not the other way around.*

James C. Collins
"Good to Great and the Social Sectors" 2007

INTRODUCTION

According to the Association of Academic Health Centers
(AAHC), there are approximately 140 academic health centers (AHCs) in the U.S. of varying structures and degrees
of integration and alignment with the university. Two-thirds
of these institutions are part of a comprehensive university.
In this chapter, we explore the role of the AHC within the
larger university; discuss governance issues, organizational
relationships, and varying institutional cultures; and explore
how leaders can work synergistically to create a unified cultural and corporate identity.

The origin of the modern AHC can be traced to the
bequest of Johns Hopkins, a Baltimore merchant and banker
who left an estate of $7 million in 1873 with instructions
to found two institutions, the Johns Hopkins University
and The Johns Hopkins Hospital. Hopkins will stipulated
that "the hospital shall ultimately form a part of the Medical School of that university," thus, integrating patient care
and academics in medical professions training and setting
the stage for the AHC of today [1]. Subsequent federal initiatives funded biomedical research partnerships between
universities and the federal government, such that today
the National Institutes of Health (NIH) is the primary platform and funding agency for university-based biomedical
research in the U.S. and serves as the modern AHC's core
research support.

AHCs may be private or public institutions, and they
may be associated with universities or freestanding. AHCs
comprise a large majority of medical schools in the U.S.
that are ranked in the top 50, whether ranked by NIH
funding [2], best medical schools for research [3], or best
medical schools for primary care [4]. Of the top 50 AHCs
in each ranking, significantly more are associated with
broader nonmedical universities than are freestanding, as
shown in the following Table 1.

Indeed, nearly 70% of Association of American Medical
Colleges (AAMC) members are part of a comprehensive
university [5]. Clearly, there are benefits to this structural
organization, despite significant differences in culture, mission, and financial structure that may define their partner
entities.

OVERVIEW: THE ROLE OF AHCs WITHIN THE BROADER UNIVERSITY

In this section, we will explore the role of the AHC within
the broader university and discuss some of the challenges
and opportunities these blended organizations must navigate. AHCs provide significant value to the broader university in a number of critical areas. AHCs:

- Are the dedicated site for training medical students as
  well as students pursuing other health professions,
  e.g., nursing, dentistry, public health, pharmacy, allied
  health, etc.;
- Enhance the size and relevance of the university's
  research portfolio and foster the development of
  further collaborative research across the broader
  university;
- Generate income to help offset the high costs of medical/
  health sciences training, research, and faculty;
- Broaden the university's branding and ranking through
  unique research, and clinical opportunities and events;

The Transformation of Academic Health Centers. http://dx.doi.org/10.1016/B978-0-12-800762-4.00002-5

TABLE 1 Top 50 Medical Schools* Shows the Percentage of Schools of Each Type in the Top 50 as Ranked by NIH Funding, and U.S. News and World Report Ranking for Research and for Primary Care; e.g., 63% of Related Proximate Small AHCs are in the Top 50 Ranking for U.S. News Research

	Total	NIH Funding	USNWR Research	USNWR Primary Care
AHC: freestanding	28	25%	18%	25%
AHC: proximate/ small	19	58%	63%	47%
AHC: proximate/ large	35	46%	51%	51%
AHC: distant related	16	38%	31%	25%
Non-AHC: freestanding	9	11%	11%	11%
Non-AHC: proximate/ small	19	21%	26%	21%
Non-AHC: proximate/ large	9	33%	33%	44%
Non-AHC: distant related	4	25%	25%	25%
Federal Government Freestanding	*1*	*0%*	*0%*	*100%*

*According to whether they were part of an AHC or not, and whether they were part of a freestanding health sciences university, a distant health sciences campus of a larger university, or a large or smaller proximate university.

- Stimulate competitive, entrepreneurial, and metric-driven administrative and managerial principles across the institution as profit-generating businesses in addition to educational institutions;
- Drive the recruitment and enrollment of high-caliber undergraduates who may pursue medical and other fields of health study;
- Provide significant economic and health benefits to the community; and
- Are the sites for graduate medical education training.

AHC's Relative Impact on the Overall University

These attributes raise the university's reputation and enhance its brand. Rankings of top universities associated with AHCs track closely with the size of clinical platforms and research programs. Overall, most schools that ranked in the NIH and U.S. News & World Report top 50 also had the most clinical (60%) and the most research (72%) funding [2–4,6]. Thus, greater AHC success can be expected to spur overall university success. Additionally, revenue from AHC clinical operations is generally critical to the fiscal health of the university as a whole, particularly with the university's overall bond rating. AHC budgets typically account for a large part—often a significant majority—of total university resources, and AHC revenues may cross-subsidize the university's nonmedical education mission.

Consequently, as leaders at academic institutions safeguard and maximize their AHC's reputation and income as an institutional priority, real or perceived inequities between the health sciences and non-health sciences faculty, staff, and students can arise. Medical faculty compensation can be significantly greater than that of nonmedical faculty, fostering additional perceptions of inequality. The perception of preferential treatment is also exacerbated by the frequent siloing of AHC operations from the rest of the university in order to reduce potential fiscal risk to the university, allowing the AHC the necessary flexibility to be competitive in the health services market.

Differences between AHC and University Academic Cultures

To get the most out of the AHC–university relationship, institutional leaders need to navigate cultural idiosyncrasies that typically exist within and between the AHC and the general academic environment. If not, these differences can be barriers to smooth operation and administration. On the other hand, if navigated well, they may actually be leveraged to the benefit of the institution as a whole.

Both AHC and non-AHC faculty typically share common traits. For example, higher education faculty are by nature skeptical—a highly desirable trait in scholarly inquiry, but one that must be overcome to create a unified enterprise that can successfully pursue common goals. Faculty are generally more apt to preserve traditional ways of doing things than embrace cultural innovation, despite often being the drivers of innovative thinking outside academe. But, there are also differences between these faculty groups, particularly clinical faculty and nonclinical faculty in nonmedical disciplines (nonclinical faculty in medical schools are typically a cultural hybrid between the two), that need to be considered by the university and AHC leadership. For example, most medical school faculty are not tenured or on the tenure track, unlike most faculty in the non-AHC environment.

Decision-making in academia is typically a shared function, focused on extensive consultation and widespread agreement before action. The highly regulated compliance environment of higher education in academic medical training and clinical service requires decision-makers to be more risk averse, which leads to error avoidance as a high priority for faculty and faculty leaders. Yet at the same time, clinical faculty, who regularly face potentially life-threatening situations in the clinical setting, must be able to make, and thus become accustomed to making, quick, autonomous decisions in the field.

In traditional academia, faculty mobility tends to be low, tenure is highly desired, and earning capacity in the private sector is generally not a particular draw for departure. Thus, non-AHC faculty tend to identify more closely with the broader university and often adopt a "do not rock the boat" attitude. For AHC faculty, the opposite is generally true. AHC faculty tend to operate apart from the broader university, with more limited emotional investment and connection. They may be less likely to participate in elements of university life, such as sports, student activities, and cultural offerings.

Their teaching responsibilities and pedagogy differ significantly as well. Non-AHC faculty are generally required to teach a predetermined load of classes that meet at least weekly and frequently last the entire academic unit or year. AHC faculty are generally on teaching services for shorter defined times, yet are never actually disconnected from the imperative of teaching as they are often accompanied in all they do by their trainees.

Another cultural difference emerges in faculty and staff perceptions of institutional mission. While the educational component of both the AHC and non-AHC segments value scholarly pursuit and learning for learning's sake, the clinical piece of the AHC is primarily a revenue-generating business. Leadership is certainly acutely aware of bottom-line considerations in the operation of the university, but non-medical faculty can perhaps more easily remain in the ivory tower of intellectual pursuit than an institution's leadership or clinical faculty.

Further, within the clinical enterprise of the AHC there is an increased need for swift and high-stakes decision-making among AHC leaders due to rapid changes in the health care delivery environment resulting from: health care reform; consolidation into larger delivery systems; obligated debt financing of health systems; rapid introduction and adoption of technology advances in medicine; and changing scope of practice for health professionals. Thus, the pace of decision-making between the AHC and non-AHC components of the university may be asynchronous and potentially lead to friction and misalignment of priorities, messaging, and strategy.

Finally, the student populations and their educational needs differ significantly for AHC and non-AHC faculty. Faculty in the nonmedical fields, particularly those dealing with undergraduates, must address issues such as retention, progression, and graduation success—imperatives practically unknown to medical faculty, both clinical and nonclinical. Additionally, the broader university faculty face a more diverse population of students in maturity and academic preparedness, while medical faculty have the privilege of teaching more mature students culled from among the best of their colleagues. And consequent to their study load, stage in life, and frequent geographic separation from the parent university, AHC students are less engaged with university culture than are nonmedical students.

Consequences of Cultural Differences

All in all, the differences in the culture, responsibilities, opportunities, and students between AHC and non-AHC faculty may exacerbate divides within the university that university leaders must continually be mindful of in order to build bridges across the real and/or perceived gulfs.

Outcomes of these cultural differences are severalfold. Because external career opportunities are typically more readily available to AHC than to non-AHC faculty, AHC faculty may feel less reliant on, and thus less concerned with, the university's overall reputation and success. In turn, they may be less motivated to initiate or join efforts to improve it, such as focusing on student access, retention, and success. And with a generally higher return on investment for the medical students they are teaching, they may also be less concerned about affordability issues.

Real and perceived differences in priority and resources afforded to the AHC vs. non-AHC segments of the institution can foster sensitivity to inequities in academic or administrative dialog and can complicate the necessary reconciling of institutional egos [7].

A culturally siloed AHC within the university leads to lost opportunities for enhanced and unique educational and research programing. The university is less likely to leverage the rich experiences and administrative innovations inherent in academic medicine, which limits the development of transdisciplinary, interdisciplinary, and interprofessional programs in education and research. Such siloing also creates significant risk for the AHC as the need for greater alignment, shared administrative services, and the leveraging of economies of scale will continue to increase in higher education as external and state support decline. Additionally, siloing ensures that few medically trained leaders are, or are viewed as, sufficiently skilled for management of higher education enterprises.

These challenges can be overcome if leaders on both sides are committed and willing to take specific action. For example, proactive tactics may be employed to encourage and facilitate AHC faculty and student engagement in broader university activities, e.g., promoting athletic and cultural events and providing transportation to them. AHC and non-AHC leaders can be educated about each other, an idea explored further in the next section. AHC faculty can be assigned to serve in university-wide committees and vice versa. And, critically, successful navigation of natural institutional disparities requires the right mix of university and AHC leadership who themselves view the enterprise as a single unit with shared mission, values, and goals.

Priorities for University and AHC Leaders

Shared Priorities

Building the right leadership team is a key requirement for any administration and trust is critical when forming a high-functioning team. When leadership positions are filled by separate individuals, the university president and the AHC leader must trust one another. Trust built upon shared vision and values for the university and adaptive interpersonal engagement leads to leadership effectiveness and personal chemistry. Trust within the leadership team can be thought of as the sum of three key elements: loyalty, shared ethics, and reliable competency. Trust must be consistent and reciprocal. Both leaders must be seen by internal and external constituents as individuals with integrity, character, and shared purpose. Both must work toward bettering the university's reputation and its core missions of education, scholarship, research, and service.

Complementary approaches to decision-making, high political acumen (not only with government and external leaders and forces, but also within the organizational culture of the academy), and healthy self-esteem and optimism that can tolerate debate and setbacks are shared attributes for the successful partnership. Placing others and the institution above self and allowing for and encouraging healthy debate are also important attributes. However, at the conclusion of any debate, the AHC leader must recognize that the president is the ultimate authority in the university, which includes all the academic units and, at times, even the clinical components of the AHC; and the AHC leader must support the president's ultimate authority [8].

There are also needs and expectations specific to each role. AHC leaders require presidents to provide direct access to key administrators; ensure clear and unambiguous communication of strategic plans, operations, and finances; recognize the AHC's value to the institutional reputation; and provide the necessary flexibility when developing policies and procedures in order to accommodate specific AHC requirements.

Presidents need AHC leaders to: (1) clearly and proactively communicate AHC performance and potential issues, avoiding unnecessary reputational or financial surprises; (2) understand the importance of academic and research excellence in establishing institutional reputation and brand, and consistently deliver excellence in both; (3) prioritize cross-campus collaboration on research, scholarship, and program development to the benefit of the wider university; and (4) be responsive to cultural differences and sensitivities between medical and non-medical faculty and staff.

To achieve synergy across the AHC and non-AHC components of the university, a common and shared strategic plan is vitally important in the era of interdisciplinary, transdisciplinary, and collaborative educational and programmatic research. A university-wide leadership team should have regular, strategic, and operational engagement to establish priorities and resource allocation. Universities and AHCs with aligned goals and resource investment will have the highest likelihood of success and sustainability during times when extramural funding is constrained.

Establishing a Strong Personal Relationship

The personal relationship between the president and AHC leader is a very important and nonquantifiable factor that predicts institutional success or failure. The university and AHC community recognizes a solid personal chemistry. The same can be said when the personal chemistry is unstable.

Key characteristics that predict a positive working relationship between the university president and the AHC leader include sharing of core personal values toward higher education, scholarship, and the strategic direction for the university. Complementing these personal and mission-driven values is the ability for the two leaders to have mutual trust and respect for each other that is built upon effective communication, strategic, and tactical transparency, and lack of guile and intrainstitutional competitiveness. Both leaders must have competencies in institutional advancement; shared understanding of the accreditation and regulatory environment within each sector of the university; and shared expectations of an institution's strategic, government, and community priorities. From the perspective of a university's strategic plan, the AHC's strategic plan must complement that of the university, and the university must acknowledge that the AHC's strategic plan may have different emphases than that of the broader university, especially as it relates to the AHC's clinical mission.

AHC Leadership Priorities

First, AHC leaders must have clear and direct access to the university president as well as the key administrators in the president's office; among the most important is the chief financial officer. Leaders must find ways to: effect open communication with set institutional priorities; have mechanisms that vet strategies and options to achieve those priorities; establish clear policies over who has authority to make decisions; identify clear accountability measures; and develop protocols to revisit decisions and strategies at set time intervals.

Second, the operational pace of AHCs requires timely decisions, even when not all data are available to fully inform the process. This requires well-honed coordination between university and AHC leaders and tolerance for uncertainty and risk. There must be regular and effective face-to-face meetings; coordinated financial operations and trust among administrators; and clear and unambiguous strategic plans, operations, and financials. Alignment of the university with its AHC is enhanced when the same clear message is relayed to each governance body.

Third, AHC leaders need recognition of the significant prestige that AHC programs can contribute to the university's reputation to ensure sufficient priority and resources are forthcoming for optimal performance of clinical and research functions.

Fourth, AHC leaders desire flexibility with developing, interpreting, and implementing elements of university policies/practices. For example, faculty physicians frequently receive most, if not all, of their compensation through physician practice plans, and faculty compensation may include market benefits that are not available to general university faculty or staff. These benefits are necessary to recruit and retain high-performing physician faculty.

In a similar fashion, university policies and procedures—such as workweek duties, vacation policies, promotion and tenure guidelines, and teaching assignment accountability—may not map well to professionalism standards for health professions faculty. Resulting differences in personnel practices can create strife and are best mitigated through positive working relationships among senior administrators, including general counsel offices, and through AHC faculty participation in the university's faculty governance structure.

Last, AHC leaders have unique needs, e.g., for information technology, health professions curricular resources, facilities and scheduling, and requirements to meet accreditation standards for a particular program. These can challenge the patience, partnership, and credulity of the best leaders in a university. When not managed well, miscommunication and disruptive relationships can result.

University Leadership Priorities

In return, university presidents have requirements for AHC leaders. For AHC medical schools alone, average budgets were $987.9 million for private schools and $605.8 million for public schools in 2012 [9], often dwarfing budgets of the non-AHC side while serving a much smaller number of students. Thus, anomalies and inconsistencies in AHC flow of funds will have an outsized impact on institutional finances. At a minimum, AHC leaders should ensure that medical schools are self-sufficient and, in most cases, provide revenue to cross subsidize the wider university. It is up to the AHC leader to update the president on any potential or impending fiscal concerns on a proactive and timely basis.

Second, the president needs the AHC leader to understand the potential negative impact on the greater university if clinical and research functions are not consistently delivered at the highest levels of excellence.

Third, in order to maximize the benefits of transdisciplinary, interdisciplinary, or interprofessional organizational, educational, and research opportunities, presidents look to AHC leaders to prioritize cross-campus collaboration on research, scholarship, and program development to the benefit of the wider university.

Finally, presidents expect AHC leaders to understand the sensitivities between the two sets of faculty and students and work to encourage medical faculty and student engagement, fostering a unified identity that will best serve the interests of the university as a whole.

Leadership Engagement in Clinical Enterprise Governance

Governance structures of AHCs are rapidly evolving in response to dramatically changing operating environments in both higher education and health care. Budget pressures in both arenas; demographic shifts that result in a new pool of future students; technological advancements that require institutions to keep up and figure out how to pay for them; dramatic health care system reforms that are just now having an impact; and new delivery models coming in both education and health care—these are but a few of the challenges with which today's AHC leaders are grappling.

Devising a governance structure that manages clinical, research, and educational goals of multidimensional AHCs is difficult in the best of circumstances and more so in times of external transformation. When AHCs are part of a broader university, aligning mission, strategies, and goals across the enterprise is further complicated by differences in culture and priorities between the medical and nonmedical sides of the institution.

While there are variations in governance structures, they can all be viewed according to where they fall on a continuum of alignment and integration—the degree to which academic and clinical functions operate under a single administrative structure [10]. In this section, governance considerations for AHCs that exist as part of a broader university are discussed, with emphasis on the main leadership roles of key university officials including the president, the AHC leader, the CEO of the health system, and provost.

The critical requirement and biggest challenge is to create a structure that achieves strategic alignment and facilitates identifying, concurring on, and achieving university-wide goals. Dr Steven Wartman, president and CEO of the AAHC, describes current AHC leadership as working to harness "the power of a virtuous cycle, whereby clinical revenue and academic performance support each other by being strategically and tactically aligned. The 'virtue' is that each makes the other better." Leaders are "committed to the philosophy 'What is good for one is good for both'" [11].

There is no single "right" governance structure for institutions with AHCs as part of a broader university; there are successful institutions with varying setups. Two examples are provided to illustrate the differing degree of organizational complexity: Georgia Regents University (GRU) in Augusta, Georgia and West Virginia University (WVU) in Morgantown, West Virginia.

GRU Governance Structure

GRU is part of the University System of Georgia (USG) and thus reports to the USG chancellor and Board of Regents. The president of GRU is also CEO and chair of the board of the health system, which includes the faculty practice group, and chair of the board of the medical center. The university provost, executive vice president for clinical affairs, and the chief business officer (who has responsibility over both university and health system finances) serve on the president's Executive Cabinet, reporting directly to him.

WVU Governance Structure

In contrast, the president of WVU answers to the Board of Governors and, by legislative statute, serves as chair of the Board of Directors for WVU Hospital (WVUH) and chair of the Board of Directors for the larger West Virginia United Health System (WVUHS).

The chancellor for health sciences: reports to the president of WVU, is the chief administrative officer for WVU's Robert C. Byrd Health Science Center, is a nonvoting ex officio member of the medical and dental practice plans—which are separate corporate structures, and is a member of the Executive Committees of the WVUH and WVUHS governance bodies. Of note, the CEO of the WVUH reports dually to the chancellor and to the CEO of the WVUHS.

Thus, as it is often said, "If you see one AHC, you have seen one AHC." No matter the structure, the AHC leader must be prominently positioned in the president's immediate cabinet due to the size of the AHC portfolio and its potential impact on the university's fiscal health. And while the president is usually an ex officio member of the university's governance body, the AHC leader may or may not be a member of this body. When she/he is not, there is often

a subcommittee of the board with responsibility for AHC activities. In these cases, the AHC leader should be an ex officio member of that subcommittee and should make routine updates to the board.

It is incumbent upon the AHC leader to provide insight, mentoring, and advice to the president on health care matters, especially in the transforming environment of health care system reform—particularly when the president does not have a strong background in health professions or health care delivery.

Another important relationship is that between the AHC leader and the university provost. Often the AHC leader has responsibilities as the chief academic officer for all health professional schools and, as such, the health professional school deans and programs report to that individual. However, in many cases some or all of the health professional schools or programs report instead to the provost with some aspects of clinical program infrastructure support remaining under the management of the AHC leader. In these instances, close collaboration between the two leaders and their management teams is necessary to avoid competition and discord and preserve necessary academic support.

In summary, all leaders need to be strong proponents and champions for the academic mission of the clinical enterprise, especially when operating budgets and margins are strained due to the changing economics of higher education and health care. If not, the academic mission may suffer as clinical margin improvement plans need to be activated. Finding common ground and synergy should be the expectation of all three top university officials—the president, the AHC leader, and provost.

PATHWAYS TO LEADERSHIP, CORE COMPETENCIES, AND LEADERSHIP DEVELOPMENT OPPORTUNITIES

Pathway to AHC Leadership

A systemic examination of the current leadership of AHCs in the U.S. finds that these individuals have had progressive leadership responsibilities in schools of medicine, often including departmental chair positions, center directorships, and deans of medicine. Given the complexity of the clinical enterprises that support AHCs, this pathway may be most appropriate. Additionally, many AHC leaders have substantial scientific *bona fides* that provide intimate understanding of funding agency policies and priorities, as well as the regulatory environment of the basic, clinical, translational, and population health sciences. They have often served on granting agency study sections, advisory councils, and national advocacy groups. However, becoming a successful AHC leader requires a broader set of experiences than that provided by medical education, clinical practice, or scientific inquiry.

Pathway to University Leadership

The American Council on Education (ACE) provides detailed data on university presidents: educational and professional backgrounds, sex, ethnicity, and much more. It found that between 1986 and 2011, very little changed in the profile of presidents of our higher education institutions. Today, university presidents are likely to be white men in their 60s with doctorates in education. Most have risen from within the ranks of higher education, with about one in three coming directly from the chief academic officer position. Most have served their entire career in academia and have been full-time faculty. There are very few differences from the first college president study conducted in 1986 [12].

Demographic Trends in Leadership

Thus, current pathways to leadership in higher and medical education consistently produce a strikingly homogenous group of presidents and other high-level administrators. A clear example of this lack of diversity rests in the ethnic-racial makeup of university presidents. Our society and our students have become more diverse, and demographic trends are clear that the future pool of higher education students in the U.S. will come increasingly from minority populations [13]. Similarly, in medical education, nearly half of U.S. medical school applicants in 2011 were nonwhite, with Asians being the second largest group [14].

In contrast, white faculty comprise about 60% of U.S. medical school faculty, while only 13% are Asian, 4% are Hispanic, and less than 3% are African-American. This lack of diversity is greatest among the highest ranking faculty [14]. In a worsening trend, the 2012 ACE survey of American college presidents found that the racial diversity of college presidents actually decreased between 2006 and 2011 [12]. Overall, university and medical school leadership increasingly does not reflect the diversity of the student body being led.

Lack of Physician Presidents

As for racial/ethnic diversity, medical professionals, specifically MDs, are significantly underrepresented in the population of U.S. research university leaders. According to the ACE survey, nearly 90% of presidents of doctorate-granting institutions have PhDs or EdDs, compared to 2% who have an MD [12]. Yet professional medical degrees account for about 23% of all graduate degrees attained [15–18]. Clearly, presidents with MDs are disproportionately underrepresented among university presidents.

We can postulate a number of reasons why this situation exists. AHC faculty disengagement from the broader university and other cultural differences between medical and nonmedical faculty may make medical faculty less desirous of university leadership positions. Clinical work requires continuous practice, making it difficult to pick up again after a term in leadership. Physicians also likely have more income opportunities and more leadership opportunities to pursue. Finally, clinician faculty tend to have few mentors in higher education leadership positions to look to and learn from.

However, MD and AHC leaders could add valuable experience and perspective to medical education governance; developing programs designed to increase their numbers is an option that could strengthen the AHC/university governance structure.

Core Management Competencies for AHC and University Leadership

Along with understanding the broader challenges of university regional accreditation, a successful AHC leader must have significant familiarity with the accreditation agencies for health professional programs within their scope of responsibilities.

Managing a comprehensive AHC requires sophisticated financial judgment and acumen and the ability to recruit and retain talented financial managers. A recent trend has aspiring AHC leaders seeking additional formal education in finance and management. Mentoring and coaching relationships is also an important part of the pathway to leadership. Professional leadership development programs are found across the disciplines of the health professions and in higher education as a whole. Those who aspire to senior management should be encouraged to seek these career development opportunities; and, importantly, institutions should financially support aspiring faculty.

The modern university president and AHC leader must have experience and capabilities in philanthropy and donor cultivation and stewardship, including the capacity to inspire and connect with alumni, donors, and corporate partners to generate private support for important AHC initiatives. While some skills may be innate for leaders, institutional advancement mentoring and skills development are essential for the twenty-first century AHC leader.

In addition, research on effective leaders in general is largely applicable to higher education leadership. There is substantial evidence that effective leadership is dependent upon personality factors such as openness, conscientiousness, extraversion, agreeableness, and neuroticism—the so-called big five personality traits [19]. These core traits are behaviorally manifested by an individual via evidence of a high degree of openness, transparency, tolerance for uncertainty, conscientiousness, and a healthy sense of "self-esteem," that can withstand controversy without retaliation [20].

Personal leadership qualities for organizations similar in complexity to universities and AHCs also focus on "leadership character," and were recently summarized as composed of six characteristics: courage, integrity, selflessness, empathy, collaboration, and self-reflection [21]. Of these characteristics, collaboration, empathy, and self-reflection may carry special significance in the highly matrixed relationship between the president and AHC leader.

Leadership Development

Industry has learned through experience that the pathway to effective leadership requires education, mentoring, and diverse experiences in order to broaden the pool of leaders and to hone their skills to meet the diverse challenges they will face.

Higher education, including academic medicine, has begun to recognize this expectation as well. Universities and their AHCs require a diverse and highly-qualified leadership workforce. Many institutions have initiated career development pathways for their faculty and management teams, encouraged by university governance and fostered by enlightened university leaders.

National education and professional organizations have also built leadership development tracts and opportunities for promising mid-career faculty. Unfortunately, these programs are somewhat siloed, and there are precious few opportunities for cross-career development. Membership associations that focus on leadership development in academic health include the AAMC [22] and the AAHC [23]. Membership associations that focus on the broader world of higher education include the Council on Education (ACE) [24], the American Association of State Colleges & Schools [25], and the Association of American Colleges and Universities [26], among others. Together they offer a wide range of programs designed to train and develop effective higher and medical education leaders.

What is missing? As discussed in the previous section, medical professionals are significantly underrepresented among American university presidents compared to other doctoral-level degree earners, and current leadership development efforts are not producing a more diverse group of higher education leaders.

In addition, while these membership organizations are productive and goal-oriented in helping their members develop the next generation of leaders, there are few opportunities for university presidents and AHC leaders to engage in meaningful dialog and career development. Engagement across the medical and nonmedical sides of these organizations becomes more critical as both health care and education become increasingly more competitive and the need to have an aligned and integrated organization becomes even more important. An integral part of the process includes training physician faculty to understand the broader landscape of higher education and university life, both to function more cohesively with the university and as a step in encouraging MDs to pursue university leadership positions.

CONCLUSION

While the matrix of organizational structures and reporting relationships in the modern university is richly varied, the dynamic relationship between the university president and the AHC is one of the most, if not the most, important relationship for the academic success and reputation of the modern university.

Key characteristics necessary for leadership success include a high level of management and financial acumen, strong personnel management skills, and an understanding of and proficiency with philanthropy and development. Presidents and AHC leaders must share expectations of accessibility and aligned vision. Critical to the successful relationship is mutual trust and respect for each other's values and principles. Leadership pathways differ for each leader, reflecting the different cultures from which their professional careers emerge, and there are few opportunities for shared professional developmental experiences that can offer better understanding for each other's day-to-day challenges. This deficiency may contribute to troubled relationships and misalignment of priorities for some university leaders.

There is clear evidence that physicians are underrepresented in the ranks of university presidents, and once presidents or AHC leaders are appointed there are few opportunities for both leaders to simultaneously network among their peers. For universities to reach their highest potential, presidents and AHC leaders must be synergistically engaged in order to lead the university through environmental, academic, cultural, and financial challenges of the future.

REFERENCES

[1] Johns Hopkins University. Who was Johns Hopkins? [brochure]. (n.d.) Baltimore (MD): Johns Hopkins Medicine Marketing Communications Office. http://www.hopkinsmedicine.org/about/history/_docs/who_was_johns_hopkins.pdf [Last accessed 21.08.14].

[2] Blue Ridge Institute for Medical Research. Table 2: Total NIH awards to each medical school in 2013. http://www.brimr.org/NIH_Awards/2013/NIH_Awards_2013.htm [Last accessed 07.04.14].

[3] U.S. News and World Report. Best medical schools: research. http://grad-schools.usnews.rankingsandreviews.com/best-graduate-schools/top-medical-schools/research-rankings. 2014 [Last accessed 24.08.14].

[4] U.S. News and World Report. Best medical schools: primary care, 2014. http://grad-schools.usnews.rankingsandreviews.com/best-graduate-schools/top-medical-schools/primary-care-rankings [Last accessed 24.08.14].

[5] Association of Academic Health Centers. Academic health centers: creating the knowledge economy. Washington, DC: Association of Academic Health Centers; 2009.

[6] Association of American Medical Colleges. AAMC medical school profile system, 2012. Washington, DC: AAMC; 2012.

[7] Their SO, Kelley WN, Pardes H, Knight AW, Wietecha M. Success factors in merging teaching hospitals. Acad Med 2014;89:219–23.

[8] Trachtenberg SJ, Kauvar GB, Bogue EG. Presidencies derailed. Why university leaders fail and how to prevent it. Baltimore (MD): Johns Hopkins University Press; 2013. p. 2,3.

[9] Association of American Medical Colleges. FY2012 medical school financing highlights. https://www.aamc.org/download/344870/data/fy2012_trends.pdf [accessed 06.04.14].

[10] Barrett DJ. The evolving organizational structure of academic health centers: the case of the university of Florida. Acad Med 2008;83(9):804–8. [AAMC: Washington, DC].

[11] Wartman SA. Toward a virtuous cycle: the changing face of academic health centers. Acad Med 2008;83(9):797–9. [AAMC: Washington, DC].

[12] American Council on Education. The American college president 2012. Washington, DC: ACE; 2012.

[13] Hussar WJ, Bailey TM. Projections of education statistics to 2021. Washington, DC: National Center for Education Statistics; 2013.

[14] Castillo-Page L. Diversity in medical education: facts & figures 2012. The Association of American Medical Colleges; Fall 2012.

[15] Association of American Medical Colleges. Total graduates by U.S. medical school and sex, 2012. https://www.aamc.org/download/321532/data/2013factstable27-2.pdf [Last accessed 24.08.14].

[16] American Association of Colleges of Osteopathic Medicine. Graduates by osteopathic medical college and gender, 2000–2012. [Last accessed 24.08.14].

[17] Gonzales LM, Allum JR, Sowell RS. Graduate enrollment and degrees: 2002 to 2012. Washington, DC: Council of Graduates Schools/Graduate Record Examination Board; 2013.

[18] American Dental Association. 2010–2011 survey of dental education. Academic programs, enrollment and graduates, vol. 1. Washington, DC: ADA; 2012.

[19] McCrae RR, Costa PT. Validation of the five-factor model of personality across instruments and observers. J Pers Soc Psychol 1987;52(1):81–90.

[20] Kernberg OF. Internal world and external reality: object relations theory applied. p. 261–7. New York: Jason Aronson, Inc; 1985.

[21] Kail E. Leadership character: a six-part series by West Point's Col. Eric Kail. The Washington Post; 2014. http://www.washingtonpost.com/blogs/guest-insights/post/leadership-character-a-six-part-series-by-west-points-col-eric-kail/2011/04/04/AGSg1DPH_blog.html. [Last accessed 24.08.14].

[22] Association of American Medical Colleges. https://www.aamc.org/initiatives/ [Last accessed 23.03.14].

[23] Association of Academic Health Centers. http://www.aahcdc.org/About/ExecutiveLeadershipGroups.aspx [Last accessed 23.03.14].

[24] Emerging Leaders Group. American Council on education. http://www.acenet.edu/leadership/Pages/default.aspx [Last accessed 23.03.14].

[25] AASCU Leadership Development. Principles across programs. http://www.aascu.org/LeadershipDevelopment/ [Last accessed 06.04.14].

[26] Association of American Colleges and Universities. https://www.aacu.org/index.cfm [Last accessed 06.04.14].

ABOUT THE AUTHORS

Ricardo Azziz, MD, MPH, MBA is founding President of the Georgia Regents University (GRU) and CEO of the Georgia Regents Health System. He previously served as president of the Georgia Health Sciences University (GHSU), and before that as Assistant Dean, Clinical & Translational Sciences (CTS) and Deputy Director for the CTS Institute, UCLA, and Director of the Center for Androgen Related Disorders, Cedars Sinai Medical Center in Los Angeles.

Christopher C. Colenda, MD, MPH has served as the Chancellor for Health Sciences at West Virginia University since November 2009. As of January 2014, he also serves as President and CEO of West Virginia United Health System. Before coming to West Virginia University, he served as the Jean and Thomas McMullin Dean of the College of Medicine of Texas A&M Health Science Center and Vice President for Clinical Affairs.

Chapter 3

How Academic Health Centers are Transforming in Leadership, Administration, and Management: A Case Study

Jeffrey R. Balser and William W. Stead

INTRODUCTION

Historically, academic health centers (AHCs) have been built on a foundation of individual faculty, creativity, and entrepreneurship. As vision and areas of excellence emerge—one faculty member, one department, or one center at a time—senior leaders have guided the evolution of the AHC in making strategic investments to support these programs to further growth and excellence. This leadership model worked well when resources were plentiful and choices centered on growth.

AHCs across the nation face a mounting crisis as all funding sources, but particularly clinical revenue, come under downward pressure. Despite rapid aging of the population and growing demand for health care, the growth in revenue from health care in the nation's not-for-profit sector has reached a 50-year low (Figure 1).

At the same time, the national conversation reflects broad dissatisfaction from the public, government, and industry as US health outcomes and health care access lag well behind the indexed performance of most of the world's leading economies. Awareness is growing that the costs of AHC education and research programs—historically heavily subsidized through large transfers from clinical care margins—are not sustainable. Choices now center on how to increase the effectiveness of the collective AHC enterprise, and how to cut costs to free resources to support that work. These choices require unprecedented levels of coordination and alignment throughout the AHC enterprise while preserving an environment that nurtures individual creativity—a cornerstone of scientific progress and innovation.

Leading the AHC through disruptive change is a process, not an event. It includes development of shared vision; development of organizational and governance models that reduce barriers to coordinated action; and a capability to harness the energy and creativity of a large, diverse enterprise to nimbly address unfamiliar organizational challenges. A dedicated focus on the substance and form of

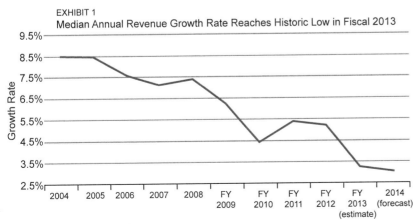

EXHIBIT 1
Median Annual Revenue Growth Rate Reaches Historic Low in Fiscal 2013

FIGURE 1 Decline of the median annual revenue growth rate for not-for-profit hospitals. (*Moody's Investors Services. Industry Outlook: 2014 Outlook—The U.S. Not-for-Profit Hospitals, November 25, 2013.*) The authors received permission to use this figure from Moody's Investors Service.

The Transformation of Academic Health Centers. http://dx.doi.org/10.1016/B978-0-12-800762-4.00003-7

what leadership really means, and how leaders will be held accountable for organization-wide goals, must become a far greater focus for the AHC than in the past.

In this chapter, we describe some of the key leadership tactics used over the past five years to propel the Vanderbilt University Medical Center (VUMC) along this journey—a journey that is ever evolving—requiring from all leaders a commitment to resilience, optimism, and willingness to embrace change.

DEVELOPMENT OF SHARED VISION

Articulation of a shared vision that resonates across the three key missions of research, education, and clinical care allows leadership to far better align the administration and management activities of the AHC to achieve improvements in efficiency and effectiveness. Although discussions of vision and strategy can occur at any time in an organization's history, a change in leadership can be a natural time to initiate strategic dialog. In 2008–2009, VUMC experienced key changes in leadership, structure, and personnel. Among other changes, the Vanderbilt chancellor appointed a single individual to serve as both dean of the medical school and as vice chancellor to oversee the entire health system (dean/VC). This appointment stimulated a system-wide discussion of shared vision that meaningfully included all components of the medical center. Led by the new dean/VC, discussion groups were formed that were highly diverse, including faculty leaders in research, clinical care, and education—as well as administrative leaders responsible for broad operational areas, including the hospitals and clinics, general administration, finance, marketing, development, human resources, and public affairs. Over 100 faculty and staff leaders from every area of the medical center were brought together over a 6 to 8-week period for this purpose.

In articulating vision, AHCs have both the strength and challenge of a highly diversified portfolio of activities across the tripartite missions. As such, classic corporate visioning exercises that build around a single product or class of products are difficult to translate to the AHC setting. The new leadership team took an approach that was tailored to the highly diverse and complex nature of a large AHC. *Strategy themes* are a broader concept than strategic plans, and are best conceptualized as broad ideas, or touchstones that can then be used to anchor the creation of explicit strategic plans throughout many areas of an institution.

Strategy themes should be inclusive, as they must both touch and excite each of the thousands of people that shape or form the health care, biomedical science, and education activities of the medical center; and they should answer the question "What do we want to be?" in a manner everyone can communicate in their own microenvironment. The most valuable themes should also leverage the distinctive strengths of an organization in a manner that allows it to further distinguish itself from peers, giving the institution and the people serving it a sense of pride and shared purpose. At the same time, the themes should respect and support the individual aspirations and diverse, independent scholarship and passions of the entire AHC community.

The Vanderbilt Strategy Theme

Through this process, our community coalesced around a small number of strategy themes to pursue, ranging from "the developing human" with a renewed focus on child health, to "therapeutic discovery" with a focus on the institution's historic strength in basic and clinical pharmacology. In retrospect, the theme that seems to have achieved the greatest resonance across the enterprise and leveraged Vanderbilt's distinctive strengths was "personalized medicine."

This theme leveraged Vanderbilt's well-recognized national leadership in biomedical informatics, and also attracted the many faculty and staff who placed a high value on the personable and collegial culture long appreciated in the organization. The theme was easily translated to innovation in all mission areas—clinical care, research, and education; and, faculty across broad disciplines took advantage of research opportunities tied to "personalization" through resources made available for pilot grants and through shared scientific core resources.

Among other accomplishments guided by this theme, over the period from 2008–2014, the medical center greatly expanded the size and accessibility of its DNA repository linked to health information, which now has over 191,000 adult and pediatric samples and millions of de-identified health records. Now the largest and most comprehensive resource of its kind, BioVU supports unbiased searches of the electronic health record to discover the significance of DNA variations in therapy and disease [1]. The medical center also became the first institution to implement drug selection and dosing at the point of care through its advanced health informatics system using a "preemptive genomics" approach Pharmacogenomics Resource for Enhanced Decisions in Care and Treatment (PREDICT) [2]. The medical center created a Web-based information system, My Cancer Genome (www.mycancergenome.org), which is hosted at VUMC and utilized worldwide to guide oncologists, academic researchers, health care-related companies, and cancer patients to information about targeted therapies matched to patients' individual tumor mutations [3].

FORMING THE LEADERSHIP TEAM AND SHAPING ITS ACCOUNTABILITY

Leadership teams form and evolve over many years, and in AHCs like all large organizations, these teams can often suffer from lack of cohesion. Academic centers have a tendency to assume that leaders who are brilliant and accomplished as individuals will naturally perform well under conditions of organizational stress; in fact, key predictors of team performance across all industries, including sports and music, transcend the talent individuals may possess and relate more closely to the training the team regularly undergoes and trust established over time within the team [4].

As part of the launch of our new leadership team at Vanderbilt in 2009, we initiated monthly off-site half-day leadership retreats with a view to "exercising" our leadership teams to better enable cohesive work, in good times and through challenges. These retreats bring together 15–20 senior leaders responsible for diverse clinical and academic enterprise arenas (chairs, center directors, associate deans, hospital CEOs, finance leaders, etc.)—each group cycling roughly once per quarter—working to develop and grow both individual and group leadership competency, while surfacing major enterprise-wide opportunities or challenges. An expert organizational psychologist facilitates the retreats, which often begin with a segment of a classic movie to set up group discussion of a common leadership challenge, such as *"managing conflict"* or *"managing the commons"*—particularly useful topics given the growing challenges leaders face in managing the growing scarcity of resources at AHCs. The dean/VC attends all retreats and seeks to participate as a self-critical team member, encouraging honest feedback and open dialog on recent or upcoming decisions relating to the topic.

Trust can be established within leadership teams through shared experiences and frequent group interaction. However, the trust dynamic of a leadership team, particularly in times of stress, can be facilitated by making the expectations for team-member behavior explicit. While it is true that "good people will try to do the right thing," the challenges of leadership are complex and anxiety provoking; the entire team can collectively articulate, endorse, and account to each other in administering a more solid foundation for trust.

Like many organizations, VUMC has a longstanding "credo" that reflects its commitment to key behaviors expected of all faculty, staff, and students. The elements of the credo speak to professional conduct, sense of ownership, commitment to colleagues, effective communication, serving others, and respecting privacy. For purposes of providing leaders a set of guiding principles consistent with the credo, but tailored to the real-world challenges of leaders, we adapted the credo elements into a leadership development tool that articulates specific desired leadership practices and behaviors. For example, under "professionalism" we enunciate *"demonstrates integrity in interactions with others, including accepting appropriate ownership for adverse outcomes, rather than blaming others."* These behaviors are discussed and even "play acted" in facilitated discussions in monthly retreats and at large leadership

assemblies (quarterly gatherings of ~1000 managers and leaders). The process is also now used in annual performance review discussions with nearly 250 leaders throughout the organization. Our initial experience suggests that the process stimulates rich, candid conversations that members of the leadership team had never experienced and would not have engaged in without prompting through a structured format.

Having a hard-wired retreat process has also established a reliable and safe environment that molds and reinforces the normative teamwork behavior desired by our leaders across the enterprise. We find the monthly retreats provide a natural setting to engage difficult problems with the key stakeholders on the management team in a dynamic and timely way, without using "called meetings" that can look and feel more like crisis management. Leadership retreats also provide a natural setting to reinforce key organizational values, such as *"academic distinctions are necessary but not sufficient for medical school chairs and other leaders—personality must be first and foremost."* Having a setting where leadership teams can reinforce and continuously recommit to fundamental standards shapes an organization for years to come.

ENTERPRISE ORGANIZATION AND GOVERNANCE

Vanderbilt is among the few remaining private U.S. research universities that owns and operates not only medical and nursing schools, but also a large diverse health system with $3.5 billion in annual revenue, all under a single financial structure (most medical schools in private universities affiliate with hospital systems that are independently managed). As Vanderbilt was simultaneously managing the financial challenges of the great recession, a major thrust was to establish a new management structure that could bring greater alignment between the academic mission areas in the medical school and throughout the university and the growing health system.

While VUMC's hospitals, clinics, and medical school were legally integrated as one entity since its founding, in 2008–2009 the organization moved to a management structure that allowed greater integration in the activities of these entities. Before then, the dean of the School of Medicine and the vice chancellor for health affairs had been separate positions, although the dean reported to the vice chancellor. Like all AHCs, the research and educational missions were dependent upon subsidies from the clinical care activities of both the physician practice and the hospitals to flourish. Foreseeing increasing pressure on the finances of the research, educational, and clinical missions, and with an understanding that the reputation of the clinical enterprise was in many ways dependent upon the health of the research and educational missions, the positions of dean and vice chancellor were merged to assure continued balance in resource allocation and oversight activities.

To ensure that integrated accountability for research, education, and patient care permeated the entire organization, dual titles and accountability were also assigned for key senior reports to reflect their combined responsibilities to the organization. For example, the VUMC clinical services are overseen by an individual with dual responsibility and titling, as deputy vice chancellor for health affairs and senior associate dean for clinical affairs. This signals accountability for both clinical services and the training activities that are undertaken within those services. As such, in a situation where a decision-maker assesses resources related to a clinical program that impact residency positions, this individual must consider the codependent welfare of the resident education and clinical care programs. Likewise, the chief administrative officer for VUMC manages operations and business support functions for the medical center and is also senior associate dean in the medical school, assuring dual accountability for the business and infrastructure needs of both the clinical and academic programs. For example, in the not-infrequent situation where there is a competing need for space, perhaps by research and clinical programs, this individual has a broad understanding of the strategic priorities in both mission areas and is therefore in a stronger position to identify and broker "win-win" solutions.

Beyond aligning academic and clinical objectives, aligning the clinical goals and priorities of faculty practice— housed in school of medicine clinical departments—with hospital and clinic inpatient and outpatient programs has historic challenges [5]. A vast array of strategies are used across the country to align the activities of faculty practices with the highest priorities of the hospitals, from patient care centers that functionally and structurally align physicians, nurses, and staff into disease-themed organizational units to the creation of "specialty" institutes and even specialty hospitals.

VUMC, while employing the above strategies to varying degrees, has created further alignment by designing an explicit set of annual goals for each clinical department chair. These goals focus on the quality, safety, and access priorities of the health system in the clinical domain they manage as service chief. These goals, and progress toward achieving them, are reviewed in a public setting with the health system executive leadership at a quarterly half-day event called "Goal Fest." Examples of goals are hospital length-of-stay, where the chair of internal medicine would have the goal of reducing inpatient length-of-stay in the medical inpatient units. In many cases, multiple department chairs will "share a goal" to increase cross-department teamwork and focus on improving performance on a particular measure. For example, inpatient length-of-stay in internal medicine units is partly dependent upon the degree to which timely consultation services from neurology or psychiatry can be provided; all three department chairs might share the same internal medicine length-of-stay goal.

Our experience has been that, in most cases, failure to make progress on a key health system performance measure is rarely due to the failure of a single individual, but rather is due to a system failure where goals are misaligned or communication and understanding are simply inadequate among multiple stakeholders. Hence, a lynchpin in the success of integrated governance, particularly between the health system and academic clinical departments, is ever-improving relationships between and among the leadership teams. "Goal Fest" has been a case study in leadership development by alignment across silos. By providing the department chairs and health system executives a regular and structured forum to articulate and resolve misalignments and misunderstandings—where civility and open dialog are the primary ground rules—problems can be depersonalized, and trust among the individuals grows.

ADDRESSING UNFAMILIAR ORGANIZATIONAL CHALLENGES

While VUMC experienced steady revenue growth over many decades, forecasts indicate that the impact of state and federal payment changes—as well as downward pressure on commercial insurance payments—will result in revenue reductions of over $200 million (8% of net revenue) over two years by the end of fiscal year 2015. Because the VUMC market position remains very strong, revenue reductions would not result from reduced clinical care volume, but rather from a variety of payment factors—ranging from an increasingly older patient population shifting to Medicare from commercial insurance products, to failure of home state Tennessee to opt-in to the Medicaid expansion, to the shift toward new higher-deductible products on state health care exchanges.

In preparation for what were long-predicted changes in health care reimbursement, the medical center had already greatly reduced its dependence on debt to manage capital expenses, with a view to retiring as much debt as possible over the 2008–2014 period through managing nearly all capital expenditures using cash flow. At the same time, VUMC enacted more conservative expense protocols in categories ranging from discretionary expenses to supply procurement in order to rebuild cash liquidity following the recession of 2008. Nonetheless, it was clear that these measures, even with robust growth in patient care services, would not be sufficient to offset forecasted reductions in revenue per unit of clinical service.

Given over 60% of the medical center's cost structure was in manpower, and recognizing the indexed workforce density at VUMC per unit of patient care was above national medians when corrected for acuity, major workforce reductions were judged as essential. Hence, in May 2013, VUMC launched "Evolve to Excel," a program with the goal of identifying approximately $100 million in cost savings through a portfolio of workforce reduction efforts. Workforce reduction at a large scale was a new experience for VUMC. Studies indicate that strategic workforce reduction efforts involving reengineering of workflow, in contrast to sudden and across the board "layoffs," have a more sustainable impact with reduced risk to operational effectiveness [6]. The financial imperative to enact substantial cost savings inside six months made it clear that tremendous focus would be required to accomplish strategic workforce reduction at the scale required. Hence, to frame and execute this effort, a program management office (PMO) was created to guide the process. The PMO was delegated the responsibility to assure timelines were met, legal and human resources guidelines were followed, system-wide communications were well-timed and coordinated with actions, and the functional stability of the research, clinical, and educational missions were maintained. The organization of our PMO is shown in Figure 2.

To staff the PMO with 25 capable leaders across all mission areas, it was necessary to delay dozens of other projects to free leadership time and effort. It was also essential to establish and communicate decision rights for the PMO. AHCs typically work within complex decision matrices that require endorsement of decisions from multiple committees and/or operating units. While advantageous to assuring long-term stability, these practices may not be sufficiently nimble to support prompt and large-scale change management efforts. As such, leadership team meetings involving an array of department chairs, hospital executives, and senior managers were held to inform the organization on the nature of the decision authority granted to the PMO. Feedback from these leadership team meetings was critical to crafting plans for communication flow and checks and balances surrounding the PMO as decisions were effected throughout the organization.

Our approach to workforce reduction was to focus on administrative and support areas, avoiding direct patient care providers (nurses, physicians, etc.) as much as possible. We included units supporting clinical, education, and research, while emphasizing in regular communications that all mission areas were highly valued and would be treated fairly. We supplied managers with templates showing the number of positions in their unit, by job category and average compensation (not employee name), and then tasked them to identify a workforce model that would provide cost reductions ranging from 10% to 20%. These models were vetted through health care quality specialists in the PMO, as well as health system physician and nursing leaders (chief nursing officers, chiefs of staff, chief medical officers) in full-day sessions to assure changes in selected administrative and management support areas could be accomplished without impacting patient care quality, service, or access. In some cases, staffing reductions were delayed or discarded. Finally, managers force-ranked their employees into the agreed-to workforce models, based on their judgments about employee performance, with consideration to seniority. All decisions were vetted through compliance, legal, and human resources specialists populating the PMO.

It was the highest organizational priority to minimize impact on existing employees. By implementing a hiring freeze, an early retirement program, and 3–4 months of intensive strategic workforce reengineering to eliminate open and unfilled positions, the medical center succeeded in eliminating nearly 1000 positions from its staffing templates. Nonetheless, by fall of 2013 it was clear that VUMC would need to initiate a reduction in force impacting approximately 600 employees. Despite the publicity and anxiety created by messaging these actions prospectively, management literature strongly supported transparency before and during the entire strategic downsizing process in order to build and maintain the trust of the organization in the months and years following the downsizing effort [7,8]. The dean/VC published a series of pieces in the *Vanderbilt Reporter*, gave interviews to local newspapers, and along with other members of the leadership team held town hall meetings throughout VUMC during the summer of 2013 making the case for change and outlining the difficult steps to come in the fall.

By December of 2013, the "Evolve to Excel" effort was declared "ended" in a public forum, the PMO was

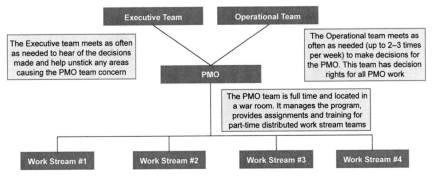

FIGURE 2 Vanderbilt's Program Management Office (PMO) structure depicts the relationships between PMO, its executive and operational teams, and the distributive teams assigned to key workstreams.

disbanded, and over $90 million in sustainable cost reductions had been achieved. VUMC continues to reduce costs to reach its two-year goal through a wide variety of approaches complementary to workforce reduction. However, we do find that achieving an institutional understanding that reductions in workforce density are accomplished while maintaining safe, responsible, and cost-effective levels yield greater legitimacy and "buy-in" for other cost-reduction initiatives. Moreover, the institutional competency gained through the creation and management of a PMO is reusable, and we have begun to utilize new PMOs to accomplish other short-timeline, high-complexity projects, such as deployment of new health informatics programs to achieve "meaningful use" targets.

CONCLUSION

While the tripartite missions of AHCs are timeless, the demands placed on AHC leadership are escalating as the business complexity and cost-sensitive nature of the U.S. health care escalates. As such, the individual and collective management skills required for AHC leadership teams to succeed have become comparable to those of Fortune 500 companies. While the historic, independent nature of faculty and academic departments in AHCs are an abiding strength that stimulates creativity and in many ways stability, these features also require distinctive approaches to management and administration as external forces increasingly require AHCs to act in a concerted, integrated way due to rapidly changing economic forces. AHC leadership teams must be nurtured, mentored, and encouraged to adapt behavioral attributes that allow them to work enjoyably and productively in a dynamic environment that, while differing from the past, offers the high-functioning team a breathtaking array of challenges and opportunities.

REFERENCES

[1] Denny JC, Bastarache L, Ritchie MD, Carroll RJ, Zink R, Mosley JD, et al. Systematic comparison of phenome-wide association study of electronic medical record data and genome-wide association study data. Nat Biotechnol (Online) 2013;31(12):1102–13. http://dx.doi.org/10.1038/nbt.2749.

[2] Van Driest SL, Shi Y, Bowton EA, Schildcrout JS, Peterson JF, Pulley J, et al. Clinically actionable genotypes among 10,000 patients with preemptive pharmacogenomic testing. Clin Pharmacol Ther 2014;95(4): 423–31.

[3] Levy MA, Lovly CM, Pao W. Translating genomic information into clinical medicine: lung cancer as a paradigm. Genome Res 2012; 22(11):2101–8.

[4] Katzenbach JR. Teams at the top: unleasing the potential of both teams and individual leaders. Boston: Harvard Business School Press; 1998.

[5] Heyssel RM, Gaintner JR, Kues IW, Jones AA, Lipstein SH. Decentralized management in a teaching hospital. N Engl J Med (Special Rep) 1984;310(22):1477–80.

[6] Freeman SJ. The Gestalt of organizational downsizing: downsizing strategies as packages of change. Hum Relat 1999;52(12):1505–41.

[7] Cameron KS. Strategies for successful organizational downsizing. Hum Resour Manag 1994;33(2):189–211.

[8] Cascio WF. Employment downsizing and its alternatives. SHRM foundation's effective practice guidelines series (Online). 2004. www.shrm.org/foundation.

ABOUT THE AUTHORS

Jeffrey R. Balser, MD, PhD is Vice Chancellor for Health Affairs and Dean, Vanderbilt University School of Medicine, overseeing one of the nation's fully integrated academic health centers. His prior positions include Chair of Anesthesiology and Associate Vice Chancellor for Research at Vanderbilt.

William W. Stead, MD is Associate Vice Chancellor for Health Affairs and Chief Strategy Officer at Vanderbilt University Medical Center (VUMC). He came to VUMC in 1991 and holds appointments as the McKesson Foundation Professor of Biomedical Informatics and Professor of Medicine.

Chapter 4

The Changing Roles and Expectations of Faculty

Robert N. Golden, Marc K. Drezner, Jeffrey E. Grossman, Richard L. Moss, Elizabeth M. Petty and Patrick L. Remington

INTRODUCTION

Faculty form the core of the academic health center (AHC). Each of the tripartite missions of clinical service, research, and education are largely shaped and put into operation by the faculty. Thus, the successful evolution and transformation of AHCs require a parallel evolution and transformation in the functions and behaviors of their faculty, as well as in the reward systems that reinforce their behavior.

The days of the Oslerian ideal of the "triple-threat" faculty member, who effortlessly moved from their productive research laboratory to the patient's bedside, where a gaggle of trainees quietly and passively observed their professor's thoughtful and methodical history taking and physical examination, are long gone. Each of the interrelated academic mission activities have become highly competitive, and now require such intensity of effort (not to mention intensity of documentation and paperwork) that it is increasingly rare to find individuals who can keep up with two, let alone all three, areas of focus. Another bedrock of traditional academic life, tenure, has morphed away from its original intention (to protect the freedom of academic pursuits from politically motivated restrictions) to an unsustainable guarantee of financial support that is often at odds with the need to connect compensation with productivity.

Despite the accelerating changes in AHCs, faculty roles and expectations have remained largely frozen in a seeming time warp. We believe this is because new faculty learn from their mentors not only skills, but also expectations and roles, that were derived from earlier, radically different academic realities. We will summarize below our thoughts about what needs to change among the major categories of faculty in evolving AHCs, as well as potential strategies for promoting such change. We will then focus on institution-level changes that are required in order to support and effect such changes.

An important caveat should be emphasized. There is an ongoing evolution of new genres of medical schools, born in different epochs with different mandates, priorities, and visions [1]. "If you've seen one medical school, you've seen one..." may only slightly overstate the degree of differentiation. There clearly are schools that have at their core a major focus on research and a tight coupling to a comprehensive research intensive university. Others, including many that are at their earliest stages of development, have as their major mission the creation of a new pipeline of primary care physicians for underserved populations. This newest generation of medical school has primary relationships with community-based health entities rather than a university, and little if any emphasis on traditional biomedical research. There is also considerable variability in mission focus within and across the other categories of key health professions schools that comprise AHCs (e.g., schools of nursing, pharmacy, dentistry, public health, and other health professions). Thus, faculty roles and expectations will vary considerably across the broad spectrum of AHCs. As with clothing, a "one size fits all" approach to faculty development will have very visible limitations, and we hope that our recommendations will be tailored to fit the specific needs and requirements of a given situation.

THE EDUCATOR

Education remains a core defining mission of the AHC. It is a growing and rapidly evolving mission with multiple demands, requirements, challenges, and opportunities. Faculty at AHCs educate a broad array of learners at vastly different levels of training across a wide variety of both university and community-based educational programs, and thus face increasingly robust requirements as mandated by external bodies involved with accreditation, professional licensing, and federal funding. We anticipate that, in the future, learning within narrow disciplines and limited to a single profession will become increasingly supplanted by interdisciplinary and interprofessional education. Educational programs in AHCs will be guided by the growing evidence base that is being created by education experts engaged in scholarly approaches to higher education, and will incorporate theories on the optimization of adult learning.

The Transformation of Academic Health Centers. http://dx.doi.org/10.1016/B978-0-12-800762-4.00004-9

Future Trends

Faculty educators must embrace the future trends affecting both the content and process of health education. As the world becomes smaller, the emphasis on global health issues will continue to grow. Technology has afforded unprecedented opportunities to engage learners in new ways, from asynchronous online distant education to applications of avatar gaming technology in "flipped classrooms." At the same time, the traditional sources of financial support for education are currently experiencing constraints, and this trend is unlikely to reverse itself in the foreseeable future. Taken together, the faculty educators of the future will face increasing opportunities with decreasing resources, which will require greater efficiencies in their approach to their mission.

The days when nationally recognized expertise in a specific clinical field or research accomplishments served as a sole proxy for excellence in teaching are over. Gone too are the days of "the sage on the stage," i.e., the well-published expert pontificating at the podium within their narrow area of scientific or clinical expertise, without regard to clearly-defined learning objectives and assessment outcomes that map to broader specified curricular goals. Instead, clear curricular threads with the goal of achieving competency for the learners are valued and often required by accreditation bodies. In the future, it will no longer be sufficient to learn clinical or technical skills by simply watching and then doing, as was common in the "see one, do one, teach one" model. Goal-oriented, hands-on practice that utilizes paid "patients" or simulated virtual or robotic devices, objective-driven clinical learning metrics, and formal, observed assessment of clinical skills development will be the new health education standards. The teacher-centered approach to education has given way to a more learner-centered approach, and for clinical learners, a more patient-centered focus.

Learning Environment

The learning environment is undergoing dramatic change as well. The emphasis on a tough, competitive "survival of the fittest" learning environment is transforming into an active learning community that focuses on collaboration, cultural sensitivity, and compassion. Curricular flexibility is growing with increased elective time and offerings and consideration of extra time to completion for students with special requests and needs. Accommodations for the increasingly diverse mix of students will grow. Based on the growing body of educational literature, these changes should have a positive impact on learners and create graduates who are well equipped to address future issues impacting health. The confluence of many changes in longstanding educational paradigms and practices, however, will be daunting and unsettling for more traditionally trained health science educators who are accustomed to an earlier generation's approach to education. The successful AHC should offer accessible training programs for seasoned educators who are committed to embracing the new and evolving aspects of teaching and learning.

Basic Science

Traditionally, basic science educators have been expected to teach a broad spectrum of learners in various settings—including seminars, journal clubs, laboratory courses, thesis committees, small group activities, and especially large lecture settings—while maintaining extramural research funding, publishing in high-impact journals, and developing national reputations through their presentations at scientific meetings. Not surprisingly, their comfort zone for teaching often closely paralleled their research expertise and interests. With the emergence of a hypercompetitive biomedical research ecosystem [2], basic scientists will find it increasingly difficult to participate in significant and substantial core teaching roles if their primary focus is on growing and maintaining a well-funded laboratory.

AHCs will need to support a cadre of basic science educators who receive institutional support for teaching and embrace education as their primary—perhaps their sole—area of focus, rather than a "side bar" to their research activities. These core basic science educators need to embrace evolving educational practices, lead, and, at times, collaborate with central curricular or training goals; and they should be recognized and rewarded for their teaching accomplishments. Creative ways to incentivize basic science teaching faculty, whose efforts do not receive the media and public attention bestowed on their research-oriented colleagues, will be important, as will investments in the technology and professional development activities that are crucial to their success. There will still be a role for the occasional input from the active scientists, but we predict that the creation of a vibrant cohort of "master teachers" will be a key element in the evolution of basic science educators at AHCs.

Clinical

The role of the clinical educator is also evolving rapidly. Over the past few decades, clinicians have become increasingly involved in the "preclinical" curriculum in the first few years of medical student education. They offer clinical relevance and context to the basic science materials, and provide early introductions to clinical skill development. We predict that the involvement of clinical educators (and clinical content) in the earliest stages of medical student education will continue to grow, and we hope that the counterpart—i.e., the reintroduction of basic science into the latter stages of the medical student curriculum—will gain traction.

Clinical educators, like their basic science colleagues, will need to embrace new approaches to education. Large lecture formats are rapidly fading and being replaced by "hands-on" active learning, small group case-based scenario discussions, problem-based learning exercises, team-based learning classes, simulations, and various blended learning activities. Clinical learning will continue its transition from a hospital-based apprenticeship model to active training in a wide variety of ambulatory and community settings, which may be far removed from the AHC both geographically and organizationally. Clinical education must expand into these different settings with the clinical educator still physically present to assess learners' skills. With an increasing emphasis on professionalism, clinical educators will be seen as the pivotal role models in this regard. The integration of basic science, public health, ethics, and the humanities will be included in the clinical educators' portfolio, and their students will include learners from all of the health science disciplines.

All of these new demands and opportunities are developing in the context of growing pressures for increased clinical productivity and an emphasis on quality and patient satisfaction. We fear there may be a growing tension at many AHCs which will pit the clinical service mission against the clinical education mission. Wise leaders will develop strategies for minimizing a "Sophie's Choice" situation, including the development of a cadre of "master clinical teachers" who receive institutional support, perhaps in the form of lower clinical volume demands in exchange for greater clinical teaching opportunities. Technology will also play a role in increasing the efficiency and the quality of clinical training through the use of simulation technology, for example.

THE CLINICIAN SCIENTIST

The defining era for the clinician scientist in the United States was the early twentieth century, when biomedical researchers were primarily physicians [3]. During this period, clinician scientists advanced the concept of "bench to bedside" research with profound results, including the discovery of insulin in 1923 by Frederick Banting and the description by Fuller Albright of "menopausal" osteoporosis in 1940. Seminal discoveries were encouraged by the tight linkage between basic science and the pace of clinical research, which were accelerated in the mid-century following the creation of large new streams of support from the National Institutes of Health (NIH) [4].

In the second half of the twentieth century, as biomedical research became a recognized field of medicine, clinical and basic research began to diverge and the number of clinician scientists decreased [4], while the number of PhD biomedical researchers grew substantially. As a consequence, the relative number of NIH research grants awarded to clinician scientists compared to basic scientists declined.

Between 1985 and 2003, the number of clinician scientists in the United States decreased by 36% [5], and the number of clinician scientists in academia declined by 25% [6]. It can be argued that the relative absence of the clinician scientist on the research frontier during this period stagnated laboratory discoveries in the gap between bench research and clinical application, a period often termed the "valley of death" [7].

The decline in the number of clinician scientists continues for a variety of reasons. In addition to the tightening of NIH funding during the past decade, other factors may discourage young physicians from pursuing careers in clinical research. These include the following: (1) the rising costs and debt burdens confronting medical students; (2) the extended postdoctoral training required to become an independent investigator; (3) pressure on academic physicians to increase their clinical productivity and revenue production; (4) the paucity of clinical research training programs; (5) traditional academic views which value individual research achievements over "team science"; and (6) less frequent and robust interactions between clinical and basic scientists in large complex AHCs.

The preservation of the clinician scientist and the resurrection of translational science require radical changes in policy in AHCs and at the national level. AHCs must find ways to support the clinician scientist in order to fulfill its mandate for biomedical discovery. We must find successful treatments for the growing epidemic of "paralyzed academic investigator's disease syndrome" [6], in which the growing and all-consuming pressures of patient care, teaching, and research are etiologic factors. Left untreated, this syndrome often leads to an inability to function as a dedicated clinician and a redirection of research efforts to basic science, rather than an incorporation of translational research into a rewarding clinical practice of medicine [7,8].

AHCs should consider innovations in medical student, resident, and postdoctoral fellowship programs that will encourage trainees to pursue careers in clinical research. Some recent efforts in this area have been limited by several challenges, including heightened clinical responsibilities for faculty, shrinking operating margins for the clinical enterprise, and limited federal funding for the clinical research enterprise. With growing recognition of this issue, the NIH established awards and programs to encourage physicians to enter or remain in clinical research [5]. The NIH funds 43 medical scientist training programs for future MD-PhDs, with a total of 932 trainees; and 75 other medical schools privately fund similar MD-PhD programs [9,10]. In addition, the NIH created a series of loan repayment programs for scientists struggling with student loan debt. Not surprisingly, however, these relatively limited attempts at addressing the shortage in clinical scientists have had limited success, and the clinician scientist population continues to shrink and age.

In the past decade, the creation of the NIH Clinical and Translational Science Awards (CTSA) program at 62 AHCs has been a welcomed step forward in support for translational research. These programs provide infrastructure for translational research, but apart from small pilot award funding, support for clinical scientists requires the development of funding sources that target their needs.

Other federal agencies and programs have begun to recognize the importance of clinical and translational research and offer support for clinical investigators. The Agency for Healthcare Research and Quality (AHRQ)—with a charter to produce evidence that makes health care safer, of higher quality, more accessible, equitable, and affordable—provides a vital setting in which clinician scientists can function and gain support for their efforts. The funding priorities of AHRQ extend across a wide variety of interests that can support the role of the clinician scientist in contemporary team science, include the following: (1) implementation of research and interventions that aim to reduce health disparities in underserved populations; (2) research to improve diagnostic performance in ambulatory care; (3) health information technology; and (4) methodologies and research in translation, implementation, and diffusion of research into practice and policy. More recently, the creation of the Patient Centered Outcomes Research Institute (PCORI) offers another source of support for clinical investigation. The clinical scientist can now receive funding for the assessment of the benefits and risks of preventive, diagnostic, therapeutic, palliative, or health delivery system interventions, which in turn will provide an essential contribution to translational research and team science. Such funding allows clinician scientists to expand their ability to participate in pragmatic clinical trials and community-engaged research, thereby enhancing the value of practice-based research networks to provide "real-world conditions" for research.

In retrospect, the role of the clinical scientist has been in decline for some time now, and the reversal of this dangerous trend will require ongoing innovations in programs. New strategies, such as the NIH CTSA program, and the establishment of AHRQ and PCORI grants, should be expanded. At the same time, AHCs must rededicate themselves, and their resources, to the rejuvenation of clinical investigation, even as financial pressures persist. Otherwise, we may lose a pivotal component in the "academic" mission of the AHC.

THE BASIC SCIENTIST

Traditionally, basic science faculty have spent the majority of their time focused on their individual research with some additional effort dedicated to teaching and perhaps administrative and leadership activities. This model is a vestige of past times in which basic scientists were largely able to fund their individual research programs with a single federal grant, such as an R01, and possibly an additional smaller grant, which did not cover all or even most of the costs of their salaries. Teaching responsibilities were often limited to a few annual formal lectures—including a stack of slides—to large groups of students, as well as direct supervision of graduate students and fellows in their laboratories. Over the past decades, with the flattening of the previously growing NIH budget, it has become increasingly difficult to maintain extramural federal funding and, at the same time, other streams of support have also become incredibly constricted. Like Alice, basic researchers find themselves in a wonderland where they must run faster and faster just to stay in the same place.

With increasing competition for extramural research funding, and the evolution of teaching expectations, there is a clear trend toward greater specialization of basic scientists. Increasingly, the tripartite missions will be addressed in the aggregate at AHCs, and basic scientists will focus more narrowly and intensely on their science and the training of young scientists in their laboratory.

Several interrelated factors have contributed to the increased competition for federal funding for basic research, even as the costs of research have escalated. Since 2003, there has been an erosion in funding available for basic research due to across-the-board cuts in budgets for the NIH, NSF (national science foundation), and other government agencies with substantial extramural grant portfolios. The situation has been aggravated by internal reallocations of budgets for the initiation of new programs of research, along with a reduction in purchasing power due to increased costs of research. At the same time, NIH initial review groups (IRGs) have tended to cap direct costs at the modular maximum of $250,000 annually, presumably in an attempt to sustain the number of funded grants. As a predictable consequence of inadequate funding, investigators now spend increasing amounts of time and effort writing grant applications, often at the expense of hands-on research and participation in teaching and institutional administrative activities. Adding to their time pressures, scientists must devote increasing effort in fulfilling reporting requirements and obtaining institutional approvals in relation to safety, animal care and use protocols, and other regulatory activities.

Other factors have affected the life of the basic scientist. The NIH has placed increased emphasis on research with near-term, practical applications in the prevention or treatment of disease. This has led to a broadening of the research plans proposed in basic science grant applications, often requiring the participation of collaborators with expertise not found in the principal investigator's laboratory and further stretching limited budgets. In addition, IRGs have continued to narrow their definition of "innovation," an important scoring criterion in the review of a grant, placing greatest emphasis on the use of new technologies rather than novel applications of existing technologies or conceptual innovation. While new technologies can be used to gain novel insights into research problems, it seems as likely that novel ideas that challenge existing paradigms will do the same even if tried-and-true (and presumably less expensive) technologies are utilized to test these ideas.

Research programs rely in large measure on trainees for the hands-on work needed to achieve the specific aims of research projects. However, it is now evident that fewer of these trainees will ultimately be employed as independent, principal investigators. In 2012, the NIH Biomedical Workforce Working Group issued a report, titled "Biomedical Research Workforce Working Group Report"[1], summarizing an increasingly bleak outlook for employment of doctoral-level scientists as tenured or tenure-track university faculty, while also noting there is currently little or no growth in suitable positions in industry or government. Greater competition for available jobs has led to laudable proposals that trainees receive more systematic professional development and career counseling. Trainees could then receive additional instruction, either in the classroom or in the workplace, to better prepare them for alternative, nonacademic career paths. Importantly, the Working Group recommended that a greater percentage of PhD graduates be employed permanently as staff scientists working within research groups led by an established principal investigator. This staffing model would increase the number of trained professionals formally engaged in biomedical research and would presumably reduce the reliance of the research enterprise on trainees, thus decreasing the current oversupply of new scientists trained in many areas.

There is increasing emphasis on "team science" in an effort to address the great breadth and depth of biological issues relevant to human health and disease. Multi-PI grants provide opportunities for expansion of the scope of a project—i.e., inclusion of both mechanistic studies and identification of therapeutic targets—to improve the competitive position of a grant during review while maintaining the ability to propose a budget that is sufficient to achieve the specific aims. The most common of these grants are multi-PI R01s (annual direct costs of up to $500,000, and more with NIH staff approval) and program project grants, or P01s (annual direct costs of $600,000–$1.5 million, depending on the institute).

Multiinvestigator teams offer other advantages over the individual basic scientist. Some multiinvestigator awards, particularly NIH P- and U-series awards, provide support for scientific infrastructure and core facilities. These can accelerate the pace of research for the project investigators as well as for other NIH-funded investigators at the same institution, in the same geographic region, or even nationally. Furthermore, the potential for success is amplified by the complementary expertise and experimental approaches of team members; and the interdependence and interactions of project directors provide motivation for higher levels of individual performance and greater likelihood of group success. In the future, basic scientists can anticipate the continued growth of collaborative "team science."

THE POPULATION HEALTH ACADEMICIAN

The demands of the U.S. health care system have continued to evolve over the past few decades, from simply the delivery of high-quality health care to improving the health of entire populations. This new paradigm is clearly described in the Institute of Healthcare Improvement's "Triple Aim" of improving care, reducing cost, and improving the health of populations [11]. Accomplishing the aim of improving the health of populations will require "disruptive changes" by AHCs and their faculty [12–14]. To meet this demand, the faculty at AHCs will need to embrace a "public health" approach across the research, education, and service missions, as they extend their reach from patients to populations and address growing health and health care disparities.

Improving the health of populations requires new research paradigms, focused on learning effective ways to translate the findings of basic and clinical research into demonstrable improvements in the health of populations. Much has been written about this translational research continuum, and various terms have been used to describe such research, including Type II translational, applied, dissemination, and implementation research [15]. Evolving AHCs will need to develop a cadre of faculty who are skilled in population health in order to address these opportunities and challenges.

Leading AHCs should have a cohort of faculty who conduct research that supports the implementation of entirely new and innovative models for care, such as accountable care organizations and partnerships with community-based organizations. In addition, researchers will need to go beyond improving the quality and efficiency of health care, to conducting public health research that seeks to prevent disease and promote health. The evidence base for prevention science is growing, as reflected by the work of the U.S. Community Preventive Services Task Force, which provides evidence-based findings and recommendations related to community preventive services, programs, and policies to improve health. More prevention research is needed, not only to expand the breadth of topics covered, but also to increase our understanding of effective dissemination and implementation strategies.

In order to accomplish these aims, researchers must develop new skills for building partnerships and understand the perspectives of community practitioners. Public health and health systems research invariably occurs in a community or clinical practice setting. Thus, traditional models of

---

[1] http://acd.od.nih.gov/Biomedical_research_wgreport.pdf.

academic research—single principal investigators within vertical organizational relationships—may be ineffective. More faculty will need to develop "community-academic partnerships" and adopt community-based participatory research models. This research must also address the inevitable health disparities that are created as clinical and public health advances occur [16].

AHCs are also adapting education programs in response to the increasing demand to improve the health of populations [17]. Contemporary public health problems—such as the epidemic of obesity and related chronic diseases—require a public health approach [18]. To address these changing needs, public health academicians should lead the integration of public health perspectives into existing basic science and clinical courses, with topics such as systems design and engineering, quality assurance, and the principles and practice of effective leadership. Students and trainees must not only understand the organization and financing of the U.S. health care system, but also will need the skills to lead needed changes in the future—rather than simply being a player on the field.

One effective approach for bridging the divide between medical and public health perspectives that has been developed and applied is the use of "integrating cases" that focus on contemporary public health problems while incorporating clinical and basic science concepts. Early evaluations of this approach confirm that these experiences offer a more expansive view of medicine and public health and make connections across basic science, medicine, and health [19]. Faculty need to also support the development of a subset of health professionals to be "fully trained" in all the public health competencies, such as through dual-degree programs [20].

Population health academicians must also address the longstanding calls from policy makers and the public to meet the needs of patients and society [13,21,22]. This "service" function extends beyond the delivery of traditional clinical services to participation with community partners on projects designed to make measurable improvement in population-based health outcomes [23]. Faculty can bring important skills to these partnerships by supporting program planning and evaluation efforts and by using health information systems to track changes in health outcomes.

THE CLINICIAN

AHCs are differentiated from other health systems by their teaching, research, and safety net missions. Since all of these missions increasingly represent "cost centers," the AHCs clinical delivery system has taken on an increasing burden of support. This is a daunting proposition for an enterprise that at times appears to be ponderous, inefficient, and underresponsive to its patients. The very characteristics of academia that have been so fundamental to its success—individual achievement and internal focus—might be counter-productive to the organization of modern health care delivery. The AHC clinical enterprise is also plagued by practical inefficiencies, including those imposed by graduate medical and medical student education.

Challenges and opportunities for the AHC clinician involve embracing and exploiting the resources of academia in ways that will enhance clinical marketplace excellence. This will require significant changes in institutional thinking and culture. In *Built to Last* [24], Collins describes the "Tyranny of Or" and the "Genius of And," pointing out the unfortunate consequences of an organization's focus on a single mission to the exclusion of other critical endeavors. Translated to the AHC, we must pursue the "genius" of great academics and the delivery of great health care. At the same time, we must avoid what Kotter has described as the "interdependency" of organizational missions [25]. Simply put, a busy research or teaching agenda is not a valid reason for showing up late for clinic; or, while there, paying inadequate attention to patient concerns. Being "in the moment" and giving adequate "share of mind" to the work at hand are important attributes for the academic clinician.

It is essential to embrace, and for leadership to reinforce, the concept that health care delivery is not a "lesser" mission for the AHC. As an intellectual undertaking, the calculus of great health care delivery is every bit as challenging as the most complex laboratory science. As former President of Dartmouth University, Jim Yong Kim, MD, PhD, said, "The real rocket science now in health care is cost and quality" [26]. AHCs should embrace the science of delivery from "bench to community," with the same enthusiasm that has been afforded traditional basic science.

While those who have spent their careers in the AHC environment may have a special affection for this setting, the marketplace shares no such fondness. The current health care environment offers no dispensation for AHCs, and incremental sources of AHC support—such as GME and disproportionate share funding—are decreasing. We cannot rely on our cachet with consumers; recent surveys show that AHC name recognition is a low priority when selecting their health care system. AHCs must compete head-to-head with delivery systems that have less complex missions, lower-cost structures, and more homogeneous physician groups.

The AHC is increasingly populated by diverse groups of physicians who differ not only in their clinical focus, but also in their increasingly more specialized mission focus. AHCs continue to have traditional physician scientists and clinical educators who will need to become more fully committed to their primary mission. At the same time, AHCs depend on full-time clinicians, both in the core academic clinical sites as well as in local and regional community-based settings. This results in a diversity and complexity of physicians with different skills, interests, and expectations within the same organization.

One distinct aspect of this diversity in many AHCs is the recognition that a base of primary care physicians is necessary to meet the goals of population health management. A commitment to embrace primary care requires a major exercise in carefully planned and executed cultural assimilation. The plan must include suitable formulations for titling and career advancement of physicians largely devoted to the practice of clinical medicine.

For some faculty members, the perceived value of patient–care activity may be subordinated to academic pursuits. At the University of Wisconsin, the "culture" of an organization—the attitudes and behaviors that members are expected to manifest—is critical. Clinical excellence, not self-proclaimed, but reflected in standardized metrics, must be given the same emphasis, value, and comparable rewards as academic excellence. We know of no single formulation of best practice in this area, but some combination of performance transparency (down to the level of the individual physician) and financial incentives, is one path toward the creation of a culture of clinical excellence. We publish, internally, patient experience data of individual physicians, and celebrate programs and individuals with exceptional performance. We share part of our operating margin with clinical departments based on the achievement of institutional clinical quality goals, and build quality and satisfaction metrics into our primary care compensation plan, soon to be incorporated into subspecialty compensation plans as well.

For physicians whose principal responsibility is clinical care, there must be demonstrable value in being associated with an AHC. This begins with an affirmation that they are not "second-class citizens," made credible by active roles in organizational and departmental leadership and the palpable culture of the organization. Out of respect for the work to be done, we embrace our responsibility to provide a microenvironment for patient care that, as Berwick has described, makes it "inescapably easy to do the right thing" [27]. In addition to investments in the optimization of the electronic health record, this includes large-scale "reengineering" of primary care that combines organizational modeling of care with the provision of "toolkits" for clinical innovation and quality to local teams.

Case Example: University of Wisconsin-Madison

At the University of Wisconsin-Madison, we have also promoted the development of a "dyad" system for clinical leadership that pairs a physician leader with an administrative partner, and we have created a Center for Clinical Knowledge Management to inform our practice. "Compacts" between primary care and subspecialty clinicians to streamline care protocols are in development, and we are focusing on the "bundling" of amenable episodes of care. Industrial engineers at our university are vital partners in many of our clinical change initiatives. Our clinicians are given the opportunity to participate in health services research through a Health Innovations Program, which provides mentoring, core services, and infrastructure support for those physicians who wish to study and improve the ways in which we deliver care. A nationally recognized Maintenance of Certification Program allows all clinicians to meet their board recertification requirements while doing meaningful quality improvement projects that improve patient care. The concept of documented quality improvement as part of a faculty member's promotion portfolio has taken hold in several AHCs; and the study of such applied areas as medical informatics, quality, safety, or population health can create a substantive body of scholarly work.

A successful approach to the new world of health care delivery in the AHC demands an unprecedented degree of physician leadership at all levels of the organization. Physicians with both the commitment and the requisite skills for effective leadership are needed. Such skills can be acquired in MBA, MMM, or MHA degree programs, and the University of Wisconsin has provided institutional support for a few "rising stars" to earn such degrees. In addition, our eighteen-month, in-house Physician Leadership Development Program has greatly extended our capacity to provide skills to cohorts of about 25 physicians. In addition to providing highly relevant knowledge and skill sets, the experience forges cross-disciplinary relationships that penetrate traditional interdepartmental barriers. A condensed version of this curriculum will soon provide all of our faculty members a basic toolkit for working in the modern health care world.

Interdisciplinary Collaborations

The "jurisdictional" barriers that limit interdepartmental collaboration in clinical care delivery have become more increasingly evident and less acceptable. Just as interdepartmental research collaborations are crucial for the advancement of science, so are collaborations across departmental lines in pursuit of health care excellence. Clinical department chairs can no longer operate as CEOs of their own enterprise, and deans should not recruit new chairs who hold such expectations. Modern chairs must be willing to manage up and across the AHC. For all of its venerated contributions to the scholarship of health and disease, the clinical department structure offers little to the efficient delivery of high-quality modern health care delivery. The great challenge for all AHCs lies in the creation of an effective new model which integrates the traditional intellectual and academic leaderships, that have been the province of department chairs, with the crosscutting clinical leadership that is required for effective modern health care delivery.

INSTITUTIONAL CONSIDERATIONS

AHCs are embarked on a rapidly accelerating course of transformation, and changes in the roles and expectations of their faculty are, of necessity, following in its wake. Unfortunately, the "academic" components of AHCs have not always kept pace with innovations. Faculty policies and practices may increasingly impede the progress of evolution unless attention is paid to the updated adage: "Medical school: health thyself." Deans and other AHC leaders must promote modifications in university and school policies, so that institutional processes support, as much as possible, the evolutionary progress in the medical school and the AHC.

Clinical care has always been the bedrock mission of AHCs, but in the past (and to a degree at present), clinicians and their work have received less praise and promotion than their research counterparts. This must change.

As research becomes more focused on "team science" and on applied areas that are relevant to health care, the metrics used to gauge the performance of researchers must evolve. An overemphasis on senior authorship and individual R01 grants impedes the progress of modern multidisciplinary research. Indeed, tenure policies at many institutions need to be reevaluated in response to the clear trends that we have outlined above.

The vital teaching mission will require new technologies and approaches if it is to remain effective and efficient in the modern world of AHCs. As with all of the tripartite missions, it will rely increasingly on "specialists" who focus on a primary, perhaps single mission, rather than all three.

Within the context of rapid transformation, it is easy to find a few faculty who are frustrated, frightened, and at times even angry with the substantial changes that are underway. It is our belief, however, that there has never been a more exciting time to be a faculty member at an AHC, and the opportunities for making meaningful differences in the lives of our patients and communities have never been greater. It is incumbent on the leadership of the AHC to instill a sense of excitement and inspiration among the faculty, as we embrace the evolving changes in all of our roles and expectations.

REFERENCES

[1] Bondurant S. Lessons from an Epic. In: McLendon WW, Blythe WB, Denny Jr FW, editors. Norma Berryhill lectures 1985-1999. Chapel Hill, NC: The Medical Foundation of North Carolina, Inc.; 2000. p. 181–200.

[2] Alberts B, Kirschner MW, Tilghman S, Varmus H. Rescuing US biomedical research from its systemic flaws. Proc Natl Acad Sci 2014;111:5773–7.

[3] Waldman M. Medical research: them and us no longer. Nature 2006;439:779–80.

[4] Butler D. Translational research: crossing the valley of death. Nature 2008;453:840–2.

[5] Ley TJ, Rosenberg LE. The physician-scientist career pipeline in 2005: build it and they will come. JAMA 2005;294:1343–51.

[6] Hauser SI, McArthur JC. Saving the clinician-scientist: report of the ANA long range planning committee. Ann Neurol 2006;60:278–85.

[7] Roberts SF, Martin JD, Frischhoff A, Sakowski SA, Feldman EL. Transforming science into medicine: how clinician-scientists can build bridges across research's "valley of death". Acad Med 2012;87:266–70.

[8] Goldstein JL, Brown MS. The clinical investigator: bewitched, bothered and bewildered – but still beloved. J Clin Invest 1997;99:2803–12.

[9] Rosenberg LE. MD/PhD programs–a call for an accounting. JAMA 2008;300:1208–9.

[10] National Institutes of Health, National Institute of General Medical Sciences. Medical scientist training program. http://www.nigms.nih.gov/Training/InstPredoc/Pages/PredocOverview-MSTP.aspx [accessed 07.04.14].

[11] Stiefel M, Nolan K. A guide to measuring the triple aim: population health, experience of care, and per capita cost. IHI Innovation Series white paper. Cambridge, MA: Institute for Healthcare Improvement; 2012.

[12] Wartman SA. Toward a virtuous cycle: the changing face of academic health centers. Acad Med 2008;83(9):797–9.

[13] Newton WP, DuBard CA. Shaping the future of academic health centers: the potential contributions of departments of family medicine. Ann Fam Med September 2006;4(Suppl. 1):S2–11. Available at http://www.ncbi.nlm.nih.gov/pmc/articles/PMC1578669/.

[14] Roper WL, Newton WP. The role of academic health centers in improving health. Ann Fam Med September 2006;4(Suppl. 1):S55–7. Available at http://www.ncbi.nlm.nih.gov/pmc/articles/PMC1578666/.

[15] Woolf SH. The meaning of translational research and why it matters. JAMA 2008;299(2):211–3.

[16] U.S. Department of Health and Human Services. HHS action plan to reduce racial and ethnic disparities: a nation free of disparities in health and health care. Washington, D.C: HHS; 2011. Available at http://www.minorityhealth.hhs.gov/npa/templates/content.aspx?lvl=1&lvlid=33&ID=285.

[17] National Research Council. Training physicians for public health careers. Washington, DC: The National Academies Press; 2007.

[18] Maeshiro R, Johnson I, Koo D, Parboosingh J, Carney JK, Gesundheit N, et al. Medical education for a healthier population: reflections on the Flexner report from a public health perspective. Acad Med 2010;85(2):211–9 (AAMC: Washington, DC).

[19] Schapiro R, Stickford-Becker AE, Foertsch JA, Remington PL, Seibert CS. Integrative cases for preclinical medical students: connecting clinical, basic science, and public health approaches. Am J Prev Med 2011;41(4 Suppl. 3):S187–92.

[20] Institute of Medicine. Who will keep the public healthy: Educating public health professionals for the 21st Century. Washington, DC: National Academy of Sciences; 2002.

[21] Lashof JC. Building partnerships for healthy communities–the role of the academic health center. Am J Public Health 1994;84(7):1070–1.

[22] Awasthi S, Beardmore J, Clark J, Hadridge P, Madani H, Marusic A, et al. The future of academic medicine: five scenarios to 2025. New York: Milbank Memorial Fund; 2005.

[23] Boex JR, Keck CW, Piatt E, Nunthirapikorn TN, Blacklow RS. Academic health centers and public health departments: partnership matters. Am J Prev Med 2006;30(1):89–93.

[24] Collins JC, Porras JI. Built to last: successful habits of visionary companies. New York: Harper Business; 1994.

[25] Kotter JP. Leading change. Boston: Harvard Review Press; 2012.

[26] Kim JY. Available at: http://www.pbs.org/moyers/journal/09112009/transcript2.html [last accessed 17.10.14].

[27] Berwick D. John Eisenberg lecture: health sciences research as a citizen of improvement. Health Serv Res 2005;40(2):317–36 (HRET: Chicago).

ABOUT THE AUTHORS

Robert N. Golden, MD is the Dean of the School of Medicine and Public Health and the Vice Chancellor for Medical Affairs at the University of Wisconsin-Madison. He also serves as the Chair of the Board of the University of Wisconsin Medical Foundation and as a member of the Executive Committee of the Authority Board of University of Wisconsin Hospital and Clinics.

Marc K. Drezner, MD is the Senior Associate Dean for Clinical and Translational Research and Professor of Medicine at the University of Wisconsin School of Medicine and Public Health. He also serves as the Executive Director of the Institute for Clinical and Translation Research and the Principle Investigator of a National Institutes of Health Clinical and Translation Science Award.

Jeffrey E. Grossman, MD is the Senior Associate Dean for Clinical Affairs at the University of Wisconsin-Madison School of Medicine and Public Health and is the President and CEO of the University of Wisconsin Medical Foundation. He has served in other administrative positions for the University of Wisconsin Health, including Vice President for Medical Affairs at University of Wisconsin Hospital and Clinics, Physician-in-Chief, Chair of the Department of Medicine, and Medical Director of the Trauma and Life Support Center.

Richard L. Moss, PhD is the Senior Associate Dean for Basic Research, Biotechnology and Graduate Studies at the University of Wisconsin-Madison School of Medicine and Public Health. Dr. Moss previously served as the Chair of Department of Physiology from 1988 to 2009 at the University of Wisconsin.

Elizabeth M. Petty, MD is the Senior Associate Dean for Academic Affairs and a Professor of Pediatrics at the University of Wisconsin-Madison School of Medicine and Public Health. Prior to joining the faculty at the University of Wisconsin in 2011, Dr. Petty was a Professor of Internal Medicine and Human Genetics at the University of Michigan.

Patrick L. Remington, MD, MPH is the Associate Dean for Public Health and Professor of Population Health Sciences at the University of Wisconsin-Madison School of Medicine and Public Health. Prior to joining the faculty at the University of Wisconsin in 1997, Dr. Remington worked at the Centers for Disease Control and Prevention and the Wisconsin Division of Public Health.

Chapter 5

Universal Lessons for Academic Health Science Centers—Recognizing the Value of Integration

Catharine Whiteside and Sarita Verma

ACADEMIC HEALTH SCIENCE CENTERS: INSTITUTIONAL SETTINGS AND CHALLENGES

The Academic Health Science Center (AHSC) has emerged in different parts of the world in a variety of institutional configurations, but universal elements are evident among all of them. To illustrate lessons learned in applying the AHSC model, we take a situational narrative approach in which each section is presented in three parts: the context giving the orientation background; example(s) of the academic leadership or management experience; and, the universal lesson learned. Since our direct experience is in the Canadian and, more specifically, the Toronto context, we have focused on the federated model of AHSCs, where integration among institutions is a particular challenge.

The Canadian Context

In 2010, the National Task Force on the Future of Canada's Academic Health Science Centers, commissioned by Health Canada, published a report entitled *Three Missions—One Future … Optimizing Performance of Canada's Health Sciences Centers*. This report states that the integration of patient care, education, and research uniquely defines the AHSC and differentiates it from other organizations that focus mainly on the provision of health care services. The overall objective of AHSCs is to provide access to world class patient care, well-trained health care professionals, and state-of-the-art research. This is accomplished through the synergies arising when health care educators, students, highly skilled practitioners, and researchers work closely in concert. According to this report, AHSCs have the potential to discover and apply innovations that could improve patient and population health outcomes; enhance the benefits of research, innovation, and commercialization; and make the system as a whole more cost-effective [1].

In Canada, AHSCs have emerged as multi-institutional academic partnerships between research-intensive universities with Faculties of Medicine and regional quaternary and tertiary academic hospitals. Universities and their affiliated hospitals are government funded within a universal health care system managed by each province. Academic practice plans are frequently supplemented by the provincial Health Ministries to support the remuneration of university-appointed physicians. External funding for research and infrastructure from philanthropy and the private sector has created a public–private environment for the AHSCs across Canada. A number of the examples in this chapter are taken from the largest AHSC in Canada, the Toronto Academic Health Science Network (TAHSN). This academic partnership encompasses the University of Toronto Health Science Faculties, nine general and specialty academic hospitals, plus four academic community hospitals. In addition, the University affiliates another 18 community health care institutions in the Greater Toronto Area. Since the University of Toronto has the only Faculty of Medicine in an urban population area of over six million, TAHSN is one of the largest and most complex AHSCs in North America.

In the Canadian publicly funded system, the university health professions education programs are dependent on investment from affiliated health care institutions for clinical training. More recently, the integration of teaching and learning in community clinical care settings has necessitated a broadened scope of clinical placements beyond the traditional AHSC institutions, challenging the elite position of the academic hospitals. New community-oriented accreditation standards and increasing reliance on ambulatory interprofessional care models are deconstructing the matrix upon which AHSCs were built.

The sustainability of health and biomedical research in both the university and academic hospital settings, where infrastructure and funding may be institution specific, is perhaps one of the more difficult challenges. Depending

The Transformation of Academic Health Centers. http://dx.doi.org/10.1016/B978-0-12-800762-4.00005-0

on the AHSC, research conducted on the university campus may dwarf hospital-based health and biomedical research. However, the collective research enterprise of the affiliated hospitals, strongly supported by independent Hospital Foundations, may often provide significantly more research infrastructure and funding than the university does for the nontenured faculty employed by these institutions.

Managing these complexities in rapidly evolving health care systems can be profoundly challenging for AHSC leaders. Visionary and creative leadership can be undermined quickly without effective joint governance between university and affiliated hospitals and respectful relationships. Clarity on how to achieve value that is experienced by both internal and external stakeholders of the AHSC is the focus of this chapter.

DEFINING THE VALUE OF THE ACADEMIC HEALTH SCIENCE CENTER

All AHSCs define a mission and vision that articulate engagement in education, research, and advances in clinical care that will contribute to improving health. A deep understanding of the value of the AHSC requires the inclusion of perspectives of all stakeholders who make up, support, and profit from the entity. How is collective value created by the AHSC and for what purpose? How is value enhanced for all stakeholders? The mindset of competitiveness shifts from the individual institution to the synergy and catalytic effect that multiple institutions can achieve together to improve their community's health and economic gain.

Strategic planning for any organization, public or private, begins by clearly defining the value it wishes to achieve or the "why" of the organization. An AHSC has many stakeholders, but in the end it relies on the contributions of the academic health professionals, scientists, and leaders to deliver on education programs, scientific discovery, and translation of new knowledge into clinical practice. The collective achievements of these individuals produce the measurable outcomes and the impact of the AHSC. Hence, the "why" for an AHSC revolves around valuing the academic performance of every faculty member who should be expected to contribute to the stated mission and vision, recognizing that academic freedom remains the cornerstone of our universities. The strategic plan for each institution can be implemented only when the leadership understands how to optimize the contribution of each of its internal stakeholders. Added together, the career outcomes of faculty members become "how" the value of an AHSC is achieved.

Many ASHCs, certainly in North America, rely on external stakeholders—including government, donors, and the private sector—to support their academic mission. The balance between external stakeholder agendas and the vision, mission, and strategic goals of the AHSC is an important challenge for academic and hospital leaders. Governments press

for health care transformation that emphasizes community-engagement and reduced cost. Philanthropists seek impact on curing specific diseases that have affected them or their families. The private sector seeks precompetitive discovery that will ultimately benefit their commercial interests. The management of external stakeholder expectations—while successfully marketing the value of institutional education, research, and clinical care programs—is a daily routine for leaders within the AHSC.

The AHSC creates the environment for bidirectional knowledge flow between the domains of "clinical care and research," "education and research," and "education and clinical care." The UK's Advanced Institute for Management Research presents this concept in a brief entitled; *Going with the Flow—A Systems Approach to How an Academic Health Science Centre Creates Value through Collaboration.* The Institute's research identifies that linkages among care, teaching, and research are crucial for the successful creation of value. The authors claim that in these interactive flows, collaboration maximizes value, achieving outcomes over and above what is possible when the three elements are operating alone. They propose that the role of the AHSC management organization is to focus on and make sure these flows are optimal, and that outcome measures should address systems optimization function (knowledge flow, consequent innovation, and impact) as opposed to independent evaluation of the separate missions of research, education, and care. The interventions recommended to achieve optimized integration are not surprising and include incentives of increased monetary resources and infrastructure, personal rewards, and systematically addressing barriers to collaboration [2].

Example: Valuing Academic Performance— Analysis within a Complex AHSC

In 2009, the dean of medicine of the University of Toronto indicated to the TAHSN Council of Hospital CEOs that the current measures of performance across its academic units and institutions were heterogeneous, rendering collective analysis difficult. Standardized measures of academic input, output, and outcomes relevant to key performance indicators were not uniformly established broadly across the TAHSN members. Although each academic unit and hospital/research institute aspired to achieve its own articulated mission and vision, common to all of their strategic goals were enhanced health care, health services, and knowledge translation aimed at improved health quality.

To analyze how to optimize academic performance, the TAHSN CEOs supported creating a Task Force with the following statement of purpose: "To establish a framework for measuring and recognizing academic performance of University faculty members both on- and off-campus, the Task Force will analyze best practices among the TAHSN partners as well as external academic health science centers of similar stature. Indicators and metrics that measure both research and education performance will be considered."

The Task Force was launched, chaired by the dean of medicine with engagement of leaders across every sector of TAHSN—research, education, and clinical care. The Task Force members experienced a unique opportunity to analyze the contribution of academic physicians to the joint mission and vision of the university and affiliated hospitals. Within 6 months the Task Force reported recommendations that set in motion the emergence of a new framework for the academic partnership. These included, among others, the joint commitment to:

- create opportunities to strategically align common institutional goals among the TAHSN partners for improved performance and measurable outcomes in health services, quality health care, health and biomedical research, and knowledge translation (including commercialization of intellectual property);
- build the TAHSN collective brand, synonymous with leading edge, globally competitive innovation and excellence reflected by these outcomes; and
- implement methods of promoting improved performance of TAHSN through valuing individual faculty member academic achievements aligned with common institutional goals [3].

These recommendations were adopted by the working committees of TAHSN and are reflected in the Academic Strategic Plan of the Faculty of Medicine for 2011–2016 [4]. The Faculty of Medicine has now established many metrics for measuring academic performance, recently reported in the Faculty of Medicine Dean's Report 2013–2014 *Fulfilling our Potential* [5].

> *Universal Lesson #1: Define and acknowledge the value that each stakeholder contributes to the AHSC and optimize the environment to achieve best collective performance.*

ALIGNMENT OF VALUES IN A COMPLEX ACADEMIC HEALTH SCIENCE CENTER

The promise of the AHSC is that by aligning strategic goals among the academic affiliated partners the whole will be more successful than the sum of its parts. Achieving this collective vision is not easy and increases in difficulty with the complexity of the organization. Measures that provide evidence of comparative performance create an incentive for the alignment of goals to achieve the innovation and outcomes desired by all stakeholders. A fundamental question for AHSC leaders is, "What standards of comparison should be used for our AHSC?"

National and international rankings and brand recognition are generally accepted standards and translate into external recognition and reputation building. Nevertheless, many external rankings are fraught with methodological and interpretive challenges. Methods that use publicly available data—such as research funding, publications, and citations—are reasonably accurate as long as institutional attribution appears consistently in the information. If a faculty member who works in an affiliated hospital research institute fails to note their university department on a manuscript, this publication will not be picked up in a university ranking analysis.

Measures of education excellence and performance often rely on qualitative reputational surveys that lack outcome-based evidence. AHSCs are responsible for establishing quantitative and qualitative outcome metrics that enable them to measure performance against strategic goals. If a goal is to move up in international rankings, then establishing a clear understanding and implementation of the right measures is the correct tactic.

Internal to the AHSC, measuring performance is useful for reward and recognition. One of the most difficult challenges is measuring the contribution of creative professional activity (CPA) of faculty members. These activities do not necessarily fit into the standard measures of teaching—publications/citations, and peer-reviewed research funding. These activities may include transforming evidence into practice guidelines, developing new simulation teaching and learning tools, or patenting new molecules and diagnostic devices. Since CPA is highly relevant to the value of AHSCs, these contributions should be recognized and rewarded appropriately.

Measuring performance is necessary but not sufficient to align values in a complex AHSC. Outcomes must be communicated effectively to both internal and external stakeholders. It is common to have world class research being conducted in one institution of a large AHSC with little knowledge of this activity transmitted among the institutional partners. Collaboration opportunities are created when communication is facilitated across disciplines, departments, and institutions. The investment in highly professional and effective communications expertise is invaluable. Website links among AHSC units, newsletters, academic achievement days, and joint institutional media releases all build a culture of recognizing local collective achievement that has global impact. In this era of information and knowledge flow, an AHSC can step forward through effective communication to demonstrate how the cooperative interdependence of its partners is a critical success factor.

University Affiliation Recognition—Necessary for International Recognition

At the University of Toronto, the vast majority of our faculty members—including full time clinical faculty (academic physicians) and PhD scientists—are located off campus in affiliated hospitals and research institutes. For many years the deans of medicine and department chairs have struggled with faculty members failing to recognize their university affiliation on publications.

By 2012, the Faculty of Medicine implemented new methods of tracking every publication in each department and estimated that ~20% of publications were missing the "University of Toronto" recognition. Therefore, the international university rankings were underestimating the university's health and biomedical publications and citations. Each department chair received information about who among their faculty members were not recognizing the University of Toronto on their publications and asked to follow up with these faculty members. Further, assistance was offered from the university to work directly with faculty members who were having difficulty with editors who insist on a limited number of institutional titles.

The result has been a continual increase in university recognition in publications by faculty members located off-campus, with the goal of achieving well over 90% of manuscripts with University of Toronto as a major institutional home.

Academic Promotion—Assessing CPA

Academic promotion through university ranks generally requires a series of internal and external reviews of scholarly achievements based on evidence. The criteria for demonstrating scholarly achievement through CPA are less standard compared to research and teaching activities, but increasingly include scholarship in knowledge transfer and quality improvement.

The Faculty of Medicine at the University of Toronto recognizes CPA contributions for promotion to associate and full professor. In 2012, the *Promotions Manual* was revised to include new guidelines for faculty members to provide evidence of scholarship and impact of their CPA activities. A CPA dossier must be provided to the promotions committees with evidence of scholarship—such as publications—describing novel CPA contributions and their evaluation relevant to improving clinical practice. Patents and commercialization of intellectual property are recognized as important CPA. The *Promotions Manual* indicates what is not considered relevant activity for promotion, such as being a clinically skilled physician [6]. Mentorship is provided from senior academics who were promoted based on their CPA and who are familiar with the criteria for demonstrating scholarship in this field. Currently, ~20% of faculty members in the Faculty of Medicine at the University of Toronto are engaged successfully in CPA as a major contribution to their academic activities.

Communication Integration in an AHSC

The 2011–2016 strategic academic plan of the Faculty of Medicine at the University of Toronto specifically identified improved communications as a target. Therefore, the Faculty invested in two new offices that are achieving much improved communications among all faculty members and affiliated institutions.

The Office of Strategic Communications and External Relations includes an executive director, along with experts in media, communications, writing, design, and strategic planning coordination [7]. With its own in-house expertise, the Faculty has revamped all communications—including Website, social media, UT Medicine magazine, and biweekly communication and announcements from the dean's office. Now, each academic unit has assistance in coordinating its academic strategic planning. The metrics and key performance indicators for the faculty are currently managed and improved on a continuing basis. Importantly, this office is charged with facilitating communication among all the affiliated hospitals, particularly media releases that involve faculty members. The executive director and staff meet regularly with their counterparts in the fully-affiliated hospitals of TAHSN.

Another identified gap was communication with, and academic management of, education operations related to part-time and adjunct academic physicians in community-affiliated sites. In conjunction with the opening of a major community satellite of the undergraduate medical education in 2011, the Faculty of Medicine launched the Office of Integrated Medical Education. This office supported an integration project led by a Steering Group to establish the Toronto Model of Integrated Medical Education. The purpose of this Steering Group was to engage clinical departments, undergraduate medical education, postgraduate medical education, continuing professional development, and the University Center for (Clinical) Faculty Development to implement a new model of integrated medical education with community partners. The group focused on establishing performance excellence of learners and teachers in the short term and improved quality of care in community sites in the longer term. The office is responsible for the administration of physician remuneration, analyzing teaching capacity, tracking performance related to accreditation standard requirements, and enabling communication among all stakeholders [8]. In 2013, the Office of Integrated Medical Education published *Key Performance Indicators for Integrated Medical Education*, which documents how the Faculty of Medicine is measuring the success of the Toronto Model for Integrated Medical Education [9].

> *Universal Lesson #2: Measure performance strategically to identify where alignment of education, research, and clinical care goals—and their implementation—will benefit all stakeholders. Establish effective communication about these achievements throughout the AHSC to facilitate recognition and new opportunities for collaboration.*

VALUE THAT DRIVES IMPACT WITHIN A COMPLEX HEALTH SYSTEM

A critical role of the AHSC is its function as a "hub" for positive change in the health system locally, nationally, and often around the globe. The return on the investment of AHSCs is now focused on driving quality improvement and innovation that reduce the cost of health care and promote healthy lifestyles and disease prevention. As such, the value of the AHSC must be aligned with the goals of public health and primary care to address the most costly management of chronic disease, mental health disorders, and end-of-life care. A newer concept is the expectation that AHSCs will lead networks within their clinical care catchment area with nonacademic health care sites and agencies that service populations and individuals in communities.

The National Health System (NHS) project in the UK, entitled *Innovation Health and Wealth, Accelerating Adoption and Diffusion in the NHS*, is addressing the need to identify barriers to innovation in health care. This project is directed at all leaders of health services, including AHSCs, and recognizes that ill health impairs economic productivity. It has the goal of pursuing innovations in the health care system that add value but not cost. The NHS indicates that the alignment of organizational, financial, and personal incentives and investment to reward and encourage innovation is necessary to attain the vision of achieving invention→adoption→diffusion [10].

The concept of community engagement now permeates academic goals in health professions curriculum development, health services and personalized medicine research, and new models of integrated care. In this setting, value is created when health outcomes in a region are demonstrably improved based on data-driven evaluation. Equally important is evaluative analysis that points to inefficiencies and ineffective systems which lead to less than acceptable health outcomes and lack of return on investment.

Within a community-engagement network, the AHSC is placed in a position of authority and leadership when its academic experts provide solutions to complex health system problems based on evidence. The expansion of scope and scale of this engagement requires alignment of the mission and vision of the AHSC and its stakeholders with communities beyond in-hospital patient populations and acute or specialized institutional care. The potential value is clear, but the incentives necessary to fully engage the faculty and staff in the AHSC in this broader mandate of community engagement are evolving.

Canadian AHSC Trial "at Home/Chez Soi"— Solution to the Health and Economic Cost of Homelessness

In 1999, the Canadian Federal government initiated the $1 billion National Homelessness Initiative after documenting a significant rise in homelessness across all provinces over the previous decade. This program funded community projects and beds in shelters, but failed to reduce the number of homeless people across Canada. To address the difficult challenge of reducing homelessness, a multicenter trial was initiated by clinician-investigators in five AHSCs in different provinces funded by Health Canada under the auspices of the Mental Health Commission of Canada.

Using a randomized control trial design, the goal was to analyze the effect of provision of housing for previously homeless individuals before they were deemed "ready" to re-enter society after institutional treatment. Each randomized participant for intervention (n~1000) was given a choice of apartments to live in, a rent subsidy, and an assigned case worker for support. The treatment as usual group (n~1000) had access to existing housing and support services. The $110 million trial ran from 2009 to 2013. At the completion, about 85% of the intervention participants who were housed were still in their first or second apartment with profoundly positive results. Many were volunteering, enrolling in school, and accepting sustained professional help for their mental illness. For every two dollars spent on this intervention, called *Housing First*, the publicly funded system saved one dollar in costs of police detentions, hospital services, and shelters. For individuals at highest risk who would have used these usual services the most, three dollars were saved for every two dollars spent.

The return on investment, calculated as improved health and economy, was absolutely clear. This groundbreaking research has the potential to immediately improve Canada's challenge of homelessness through improved social policy, strategic investment in housing, and specialized support for the most marginalized in our society [11].

Universal Lesson #3: The strategic networking of AHSCs with community health services and agencies to address complex health system barriers through innovation increases the probability for positive return on investment, as measured by improved health and improved economy across the region.

INCENTIVE FOR INTEGRATION—THE SUCCESSFUL BLENDED MISSION

One of the major challenges in the federated AHSC model is the confusion over brand recognition. This is particularly problematic when individual institutions launch fundraising campaigns or advocate directly to government or the private sector when a joint approach among AHSC members could be more successful. Generally, personal incentives for executive leaders are aligned with their own institutional goals, with little or no incentives to achieve strategic integration with AHSC partners unless shared goals are recognized at the outset. The establishment of innovative, leading-edge programs among institutions creates the opportunity for effective cobranding when the value of collaboration and joint investment is evident.

Therefore, successful integration within complex AHSCs requires leaders who recognize that collaboration creates competitive advantage. When strategic partnering enables better performance by academic physicians, scientists, graduate students, health care workers, and fundraisers, a new culture of networking emerges across the AHSC. Often the initial incentives are external, e.g., opportunities for research funds, but once successfully executed, the tactic of networking becomes more accepted. Internal incentives are necessary and must align with clear evidence that sharing access to infrastructure and expertise benefits all the institutions involved.

Successful networking among segments of an AHSC requires learning how to blend strategic goals without compromising any single brand. The transparent display of multiple institutional identities on signage and media materials can become part of the culture of communication if made easy and feasible for the users. Clarity on acceptable use of more than one visual identity requires careful management and prospective agreement. If cobranding is expected and viewed as advantageous because it reflects the synergistic value of the relationship, it will be adopted by internal stakeholders and recognized externally as evidence of a positive partnership.

Wilson Center—Interdisciplinary Unit Focused on Research in Education

In the 1990s, interest and expertise in research in health professions education led to the launch of a new interdisciplinary Center at the University of Toronto. This was cofunded by the Faculty of Medicine and one of the university's fully-affiliated academic hospital complexes—the University Health Network (UHN). The center was located off-campus at the UHN where space and infrastructure were provided. Strong support for research in education by a former chair of the University Department of Surgery led to the naming of the Wilson Center in his honor. Over the first decade, under the leadership of the inaugural director—who also held the position of vice president education at the UHN—the Wilson Center attracted international scholars and fellows to become a highly successful academic unit.

The Wilson Center has always been recognized as a University of Toronto Center at UHN and a major collaborative resource for all members in TAHSN. Today, drawing on scholars from across many departments and strongly supported financially by the UHN and university, this interdisciplinary center is known worldwide for its innovation, excellence in scholarship, and international reach attracting postdoctoral fellows, graduate students, and faculty from around the world [12].

> *Universal Lesson #4: A blended mission is achieved when both internal and external stakeholders of the AHSC experience the synergistic value of multi-institution investment and effective collaboration.*

SHARED AUTHORITY AND RESPONSIBILITY IN THE AHSC

The AHSC relies on leadership and governance that is committed to fulfilling the tripartite mission within an effective academic partnership between a university and its affiliated health care related divisions. In the consolidated AHSC model, where the university owns the academic hospitals or the academic hospital operates the medical school, single leadership and corporate ownership of the tripartite mission simplifies lines of authority and responsibility. The federated AHSC model, where university and affiliated clinical care institutions are separate corporate entities, requires robust agreement among institutions to set the framework for shared academic authority and fiduciary oversight.

In a multi-institutional, federated AHSC, harmonization of policies and practices, where feasible, creates a framework for operational efficiencies and reduced risk and a stage set for shared responsibility. Another advantage, along with the necessity to meet accreditation standards for health professions education and quality of clinical care, is the standardization of performance expectation of faculty. For instance, if the position description and terms for appointment of a clinician scientist is set by a university department, all the clinician scientists within this department recruited across multiple institutions in the AHSC should sign on to the same terms. Further, the academic practice plans serving this department should subscribe to these terms, e.g., protected time for clinician scientists to engage in research.

Academic promotion, merit-based remuneration, and recognition for academic achievement require uniform opportunities for faculty members across the AHSC based on their position description and local environment. Policies and procedures should enable the leaders of the organization to do the right thing the right way with as much ease as possible.

University–hospital affiliation agreements can be used to create a framework for shared oversight of joint academic activities that enable both faculty members and institutional leaders to experience a harmonized environment. In the federated AHSC model, three types of policies and procedures must be recognized: (1) activities subject to university authority; (2) activities subject to the hospital's authority; and (3) those that are harmonized and agreed upon to be used both by the university and the hospital. Then, faculty members, students/trainees, and administrators can be very clear about who has authority and jurisdiction over any matter arising. A comprehensive affiliation agreement should take into consideration all relevant academic and clinical activities and include terms for conflict resolution in the eventuality that an issue cannot be resolved following policy and procedures.

In the dynamic university–hospital environment, affiliation agreements should be considered living documents and revisited periodically (e.g., 5-year intervals).

This opportunity can be used to evolve effective management of new concepts and understanding of the joint academic mission and relationships among the university and affiliated hospitals. For instance, the concept of joint fundraising among separate institutions in an AHSC may require agreement about standard operating procedures that set out conditions for donor solicitation and stewardship programs that are collaborative and not competitive. Hospital mergers and acquisitions, international partnerships, and relationships with government health authorities require continual attention to the terms in university–hospital affiliation agreements.

University of Toronto—Academic Hospital Affiliation Agreements

Over the past two decades, the University of Toronto has evolved robust university-hospital affiliation agreements. The Preamble of these agreements articulates the joint academic mission shared by the two institutions. The terms in the agreement define the requirements for both parties regarding policies and procedures that facilitate academic partnership development and reduce risk. These include the University Policy for Clinical Faculty, in which full-time university appointment of a physician requires membership in an academic practice plan approved by the dean of medicine and credentialing by and appointment to a university-affiliated hospital. The recruitment of hospital clinical heads of departments requires a search that includes university representation. The appointment of university department chairs and deans requires search committees with affiliated hospital representation.

The commitment of the affiliated hospitals to provide infrastructure for undergraduate and postgraduate medical education that adheres to accreditation standards is clearly articulated. Harmonized policies have been established for joint management of research misconduct, sexual and other forms of harassment, and oversight of intellectual property, to name a few. Specific affiliation agreements have been designed for university fully affiliated hospitals, community-affiliated health care sites, public health units, and independent health facilities—all of which share learners and faculty members with the University of Toronto.

These agreements are negotiated and renewed by the vice provost relations with health care institutions, who is also the dean of medicine, every 5 years with the affiliated hospital CEOs and Board. They are considered living documents and require amendments when policies and procedures evolve.

Universal Lesson #5: Effective shared governance within an AHSC requires institutional leaders who build relationships that facilitate successful integration of education, research and clinical care. A robust university-hospital affiliation agreement creates the necessary operational framework to implement this shared mission.

CONCLUSION

The Canadian experience, and the Toronto one in particular, exemplify the federated AHSC model—perhaps the most complex and challenging of academic health center models. Despite differences in organizational structure, governance, health systems, and culture, common elements that drive success and universal lessons emerge. Notably, the effective leadership of AHSCs requires clarity of vision about the value of integration and the successful implementation of shared strategic goals.

REFERENCES

[1] National Task Force on the Future of Canada's Academic Health Sciences Centres. Three missions, one future…optimizing the performance of Canada's academic health science centres. In: A report from the national task force on the future of Canada's academic health sciences centres. http://www.ahsc-ntf.org/?reports; 2010.

[2] Ng I, Parry G, Davies SM, Lim WM. Going with the flow – a systems approach to how an Academic Health Science Center creates value through collaboration. In: Report from the advanced institute of management research, engineering and physical sciences research council, UK. http://www.aimresearch.org/uploads/file/Publications/Executive%20Briefings%202/AIM_Going_with_the_flow_FINAL.pdf; 2012.

[3] Toronto Academic Health Science Network. Task force report on valuing academic performance. http://www.facmed.utoronto.ca/about/dean/task.htm; 2010.

[4] University of Toronto, Faculty of Medicine Strategic Academic Plan 2011–2016. http://www.facmed.utoronto.ca/about/dean/strategic.htm.

[5] University of Toronto, Faculty of Medicine. Fulfilling our potential. Dean's report. http://www.facmed.utoronto.ca/staff/appointment.htm; 2014.

[6] University of Toronto, Faculty of Medicine. Academic promotions manual. http://www.facmed.utoronto.ca/staff/appointment.htm; 2013.

[7] University of Toronto, Faculty of Medicine. Office of Strategic Communications and External Relations. http://medicine.utoronto.ca/page/communications/oscer.

[8] University of Toronto, Faculty of Medicine. Office of Integrated Medical Education. http://www.oime.utoronto.ca/Page4.aspx.

[9] University of Toronto, Faculty of Medicine. Office of Integrated Medical Education report on key performance indicators for integrated medical education. http://oime.utoronto.ca/Assets/Root+Digital+Assets/KPI/KPI+documents/KPIs+Report.pdf; 2013.

[10] Department of Health, NHS Improvement & Efficiency Directorate, Innovation and Service Improvement. Innovation health and wealth, accelerating adoption and diffusion in the NHS. http://www.institute.nhs.uk/images/documents/Innovation/Innovation%20Health%20and%20Wealth%20-%20accelerating%20adoption%20and%20diffusion%20in%20the%20NHS.pdf; 2011.

[11] Goering P, Veldhuizen S, Watson A, Adair C, Kopp B, Latimer E, et al. National at Home/Chez Soi final report. Calgary, AB: Mental Health Commission of Canada. http://www.mentalhealthcommission.ca/English/node/24376; 2014.

[12] University of Toronto, Wilson Center. http://cre.med.utoronto.ca/.

ABOUT THE AUTHORS

Sarita Verma, MD, LLB, CCFP is a Professor in the Department of Family and Community Medicine, Deputy Dean of the Faculty of Medicine, and Associate Vice Provost Health Professions Education at the University of Toronto. She originally trained as a lawyer at the University of Ottawa (1981) and later completed her medical degree at McMaster University (1991).

Catharine Whiteside, MD, PhD, FRCPC served as Dean of the University of Toronto Faculty of Medicine and Vice Provost, Relations with Health Care Institutions since 2005, completing her final term at the end of 2014. Previously, she was the Graduate Coordinator of the University of Toronto's Institute of Medical Science before she assumed the role of Associate Dean Graduate and Inter-Faculty Affairs in the Faculty of Medicine.

Chapter 6

Future Directions

Victor J. Dzau, William F. ElLaissi and Krishna Udayakumar

Despite their significant societal value, academic health centers (AHCs) face an uncertain future. The health care landscape is changing rapidly. There is consensus that health care costs are too high and that this financial burden is unsustainable if the cost curve is not bent [1]. In the United States, with the passage of the Patient Protection and Affordable Care Act, there will be significant cuts in the Medicare budget, shifts in payments from volume to value, and potential reduction in Indirect Medical Education support. Medicaid budgets are being cut substantially due to difficult state finances. This is compounded by the fact that commercial insurers are taking significant measures to reduce reimbursement and demand lower provider costs. Meanwhile, the National Institutes of Health (NIH) budget has been flat for the past seven years with increasing threat of reductions in the near future. These changes are occurring against the backdrop of a global economic downturn that has resulted in significant loss in endowment value as well as decrease in philanthropic support. With the convergence of these events, US AHCs are in great economic peril. Although specific details may differ, similar issues are confronting AHCs elsewhere in the world.

In order to continue meeting their missions, AHCs must make bold transformative changes. They must extensively reform their systems for care delivery and financing, improve the productivity of research, and reduce the cost of medical education. And they must foster innovation that yields "disruptive" technologies and approaches to reduce costs and/or increase revenues.

UNIQUE CHALLENGES FACING ACADEMIC HEALTH CENTERS

In addition to the changing health care environment, AHCs are facing challenges not experienced by their nonacademic counterparts. Demand for care and services are rising even as already tight government budgets are tightening further. And although resources for AHCs continue to dwindle, there is still an expectation that AHCs will place service to the community over margin [2]. AHCs must care for the uninsured and indigent, but the much-needed Disproportionate Share Hospital dollars are dissipating [3].

Furthermore, AHCs have recently been experiencing major challenges when it comes to their research and education missions. For instance, research is outgrowing the "soft money" available to support it; yet, the NIH budget is likely to be severely constrained in the foreseeable future [4]. And the cost of medical education is continuously rising at a time when Medicare Indirect Medical Education dollars are at risk of being cut [5]. In addition, Graduate Medical Education (GME) is increasingly regulated with restrictions placed on residents' work hours. Moreover, AHCs are dependent on endowments and philanthropy, both struggling in the context of a fledgling global economy.

Finally, AHCs are complex organizations facing internal structural challenges related to hospitals, clinics, faculty practices, schools, traditional departments, historic disciplines, and contemporary disease centers and thematic institutes. Furthermore, the traditional culture of faculty individuality and autonomy are up against the more current framework of teamwork and increasing regulations and compliance [6].

EXAMINING THE FINANCIAL SITUATION

The end result of the above challenges is that the margin from clinical activities used to support the unfunded portions of the research and teaching missions of AHCs is being seriously threatened. How might this change in the "academic subsidy" play out in practice? For illustrative purposes, let us examine the financial picture of a "typical" AHC in the United States (a composite compiled from Association of American Medical Colleges data). A "typical" US AHC (exhibiting a degree of clinical integration across multispecialty ambulatory care and one or more hospitals) in 2010 had $2 billion in total clinical revenues, with a margin from clinical services of approximately 3–4% (extrapolated from individual AHC-affiliated teaching hospital reports). AHCs in general also have a diverse payer mix: 1/3 commercial, 1/3 Medicare, 1/5 Medicaid, and 1/10 other. After 2014, one might expect the proportion of care paid for by Medicaid to increase relative to these other categories [7].

Research support in a "typical" US AHC amounts to about $100 million in NIH funding as well as a similar amount in non-NIH funding. Despite this, in our AHC example, there would still be a shortfall in research support of $20–30 million because the costs of research are not covered by the funding received. Recently, a report on research-intensive medical schools has calculated that for every NIH research dollar, the institution has to provide an additional 30–40 cents, or higher, of support from other fungible sources. This subsidy, or mandatory internal cost-sharing, is net of the recovery of indirect costs as negotiated by each institution with the federal government [8a,b]. A shortfall also exists for education: $15–20 million for Undergraduate Medical Education and a similar amount for GME, which is covered by the hospital(s) and the clinical departments of schools of medicine. These deficits add up annually to an average academic need of $35–50 million, even with endowment payouts and philanthropy ranging between $13–40 million and state government and parent university support totaling $7–50 million—amounts which are also declining [7].

Given the state of AHC finances, what will the "academic subvention" look like in the future? If one were to look ahead to the year 2017, the current gap (i.e., total academic shortfall net of subsidy) will surely widen. These combine to increase the risk of what a recent Price Waterhouse Cooper report cautioned as a "margin meltdown" [9] for AHCs, one that would clearly jeopardize their ability to carry out their academic missions.

In order to continue fulfilling their missions in the wake of these challenges, AHCs must make transformative changes. They must extensively reform their systems for care delivery, prepare for changes in financing, improve the yield on research, and address the costs of health professional training. The future will also require AHCs to foster and embrace innovation that includes "disruptive" technologies and other approaches that can reduce costs and/or increase revenues, such as the development of productive industry and building global relationships.

ACADEMIC HEALTH CENTERS MUST TRANSFORM

How should AHCs respond to these difficult challenges? AHCs must transform themselves into efficient clinical delivery systems, effective translational research engines, and education innovators. This entails evolving into academic health science systems (AHSSs) by developing a "bench to bedside to population" model, which would include a vertically integrated care-delivery system, a research model that achieves a seamless continuum from discovery science to translational research to population health, and education programs that develop the workforce of the future. Doing so will necessitate the commitment of all components of the AHC to an enterprise-wide effort that defines institutional priorities and makes the necessary changes for adaptation to, and success in, a rapidly challenging environment.

Currently in many AHCs, the clinical, research, and educational components often function to varying degrees like separate organizations, as do the centers, institutes, departments, and divisions that comprise the mission-based components. As a result, each entity tends to develop its own administrative, operating, financing, and management structures, which in turn add up to an inefficient organization that lacks transparency and clarity in governance and priorities. AHCs must first understand the true cost of the each mission through a transparent accounting system and then determine what the current and future finances can support. This means that AHCs need to take the difficult but important step of addressing the balance, priorities, and tradeoffs of the missions in an open and participatory manner.

Finally, AHCs must leverage their innovative capacities to develop new programs and activities that are of value to the missions and also generate new sources of revenues. These include clinical, research, and educational initiatives nationally—as well as globally.

VERTICAL INTEGRATION OF CARE DELIVERY, CARE REDESIGN, AND PROGRAMS OF DISTINCTION

Given the reality that current academic subsidies depend largely on clinical margins, AHCs must become higher-performing health systems if the same level of support for their academic missions is to continue. Highly functional, bidirectional, and proactive relationships between hospitals, clinics, and providers are necessary to achieve the improved quality, efficiency, and continuity needed as patients move from care setting to care setting, including their home. To make such relationships a matter of course, AHCs should move towards becoming vertically-integrated care delivery systems with tightly aligned provider organization(s).

Such a system would ideally cover the entire spectrum of care—including home- and community-based care—from primary care to superspecialized quaternary hospital care. There should be an integrated information system that links electronic health records to decision-support systems, work flow, and revenue cycle management, even for unaffiliated providers involved in a given patient's care. Importantly, patients in such a system will be able to get care at the site most appropriate—as well as cost-effective and convenient—for their condition [10].

Indeed, within a network of hospitals, there would be different levels of hospital care for differences in acuity and need for specialized services. The most complex cases (e.g., transplant, brain tumor, premature infants) would be handled by (or referred to) a tertiary/quaternary flagship hospital. Community hospitals with lower cost structures could provide inpatient care for more routine "bread and butter" cases (e.g., chronic obstructive pulmonary disease (COPD) exacerbations, joint replacement, uncomplicated deliveries). This will allow for better capacity management while optimizing the cost of care within the same system. Follow-up postacute care would then be provided at a rehabilitation hospital or a skilled nursing facility. Ambulatory care may be organized along a similar spectrum, from multispecialty clinics to networks of primary care, patient-centered medical homes, as well as outpatient services such as physical therapy intended to restore and preserve patients' functional status. As AHCs become increasingly responsible for population health, they should consider developing clinically integrated networks consisting of hospitals and nonemployed providers that can pursue joint contracts and share dollars directly under arrangements intended to improve collaboration around the care of patients.

To achieve the higher performance required, AHCs must also reexamine and redesign their care delivery approaches. Clinical pathways represent one tool for reducing variability in clinical practice and improving outcomes by managing certain groups of patients with a predictable clinical trajectory that emphasizes preparation and coordination. Such pathways not only emphasize quality and safety within the hospital and ambulatory settings, but also span the entire care continuum by focusing on transitions and home- and community-based care for both acute episodes and ongoing management of chronic conditions [11].

AHCs should seize this opportunity to be innovative and explore new models of care delivery as well. Clay Christensen has advocated that health care providers adopt segmented approaches to care delivery such as "focused factories" and "solution shops" [12]. Focused factories, usually located outside general hospitals (e.g., free-standing, same-day surgery centers), seek to highly standardize more routine medical procedures with the goal of increasing quality and decreasing per procedure or per episode costs. Some of the best models of this approach are found in developing countries, where resources are limited [13]. Solution shops (which could be attached to quaternary hospitals), on the other hand, would focus on diagnostic dilemmas as well as complex chronic illness care. The rationale for this type of segmentation will intensify as competition based on price for more commoditized specialty services (e.g., elective joint replacement, cataract surgery, nonemergent cardiac catheterization) increases.

It is also absolutely imperative that AHCs do not abandon the tertiary and quaternary care programs that have been critical to their past success, but rather identify, invest in, and build up specific "programs of distinction" as world class, cutting edge, specialty care delivery programs that are driven by innovation and that may lead to game-changing medical care. While even specialty health care is becoming increasingly commoditized, the areas of medicine practiced by programs of distinction feature a high level of expertise that are unique to select AHCs—and can often be the best examples of integration of research and teaching missions

with clinical care. Such programs can also have a significant impact on the brand recognition that distinguishes AHCs from nonacademic local and regional peer competitors and attracts patients from regional, national, and international markets.

Leveraging Medical Research and Innovation

Just as AHCs need to transform clinical care, they also need to work to increase the yields from research—speeding translation and having an even greater impact on medicine and health. More specifically, they need to foster a culture of innovation and entrepreneurship by teaching it, creating space for it, and then implementing it. Many of the medications, therapies, and medical devices that we consider conventional today came about because of innovative research. AHCs also need to provide infrastructure support in order to facilitate translation and provide essential services through cores that provide natural economies of scale by concentrating necessary technical expertise, rather than diffusing and isolating it in separate laboratories.

We have proposed the concept of the "Discovery to Care Continuum," i.e., the translation of scientific discoveries to human application and improved health outcomes through the horizontal integration of research and implementation activities across the institution [14]. To achieve this, one approach is to create integrating structures that catalyze interdisciplinary collaborations and facilitate a seamless continuum between discovery science to translational research to implementation science and global health.

For example, Duke Medicine established several such entities that interconnect with one another along the discovery–care continuum, including the Duke Translational Research Institute (bench to bedside translation), the Duke Clinical Research Institute (clinical research), the Duke Center for Community Research (translation into care delivery and population health), the Duke Global Health Institute (global health), and the Duke Institute for Health Innovation. We believe that the horizontal integration of research can speed translation, improve competitiveness, increase research productivity and output, and increase societal impact, including commercialization.

Given the ongoing shortfall in research funding, AHCs should seek to optimally size their basic and clinical research enterprise. This could include emphasizing specific areas of research excellence, reducing and formalizing the support for unfunded research, measuring productivity by developing evaluation metrics, creating dedicated grant application resources, providing more structured mentoring, and undertaking an enterprise-wide approach to prioritizing research areas that complement clinical strengths. The overall efficiency of research management at the institutional level needs to be reexamined as well, including seeking ways to bend the research cost curve.

AHCs, now more than ever, must also capitalize upon their intrinsic assets. For example, they need to leverage their access to biological samples and clinical data and develop research collaborations with diverse partners, including industry. AHCs need to build an infrastructure for "rapid-learning health systems" by making clinical data "research grade," by lowering the costs of data acquisition and new knowledge generation, and making use of the intellectual capital in computer and information sciences from their affiliated universities.

Discovery-translational research needs to become more productive. One way is through novel industry partnerships. Public–private collaborations between AHCs and companies in industries—such as pharmaceuticals, diagnostics, imaging, information technology, and telecommunications—can help address downstream barriers along the discovery–care continuum and diversify research funding sources. Other types of relationships can also be developed, such as broad "first-look" agreements with pharmaceutical companies over multiple therapeutic areas that can bring much larger amounts of resources and funding for research as well as increased potential for future royalties and equity. For example, University of California, San Francisco's (UCSF) Program for Breakthrough Biomedical Research and Sanofi-Aventis have formed two broad research and development partnerships: one focused on oncology and the other on stimulating groundbreaking research in pharmacological science as well as various therapeutic areas [15].

Another critical issue to address is the need for greater efficiency and effectiveness in the technology transfer process. It is well recognized that overly bureaucratic university licensing and tech transfer offices can be impediments to translation, industry relationships, and entrepreneurial opportunities. Innovative ways to improve the performance of the tech transfer process are critical in these times of challenging research funding. A few years ago, the University of North Carolina at Chapel Hill created the Carolina Express License Agreement, the goal of which is to speed up the start-up process for technology by standardizing the licensing agreement [16].

Finally, one area of increased scrutiny within research (and elsewhere) has been management of individual and institutional conflict of interest, particularly when discoveries move towards commercialization. Asking how self-interest and integrity of research are impacted is more important than ever, given that support for research and the academic enterprise in general will become more dependent on sources other than the NIH.

REINVENTING MEDICAL EDUCATION AND TRAINING

The third classic mission of AHCs, medical education and training, has in most institutions been expanded to include a broad array of degree granting, certification, and continuing education programs, and in fields ranging from nursing and pharmacy to public health and informatics. This educational portion of the enterprise must also be reinvented to adapt to the changing health care paradigm.

AHCs should consider this as an opportunity to reexamine and challenge traditional beliefs and approaches to medical education in particular. Why is medical education so expensive, and is it possible to reduce its cost? Should one consider the use of an instructor core for the basic sciences component of medical school as well as a greater utilization of technology for teaching? What about the use of online teaching, simulation, and gaming?

The Duke-NUS (National University of Singapore) Graduate Medical School in Singapore, and now the Duke University School of Medicine, utilizes an innovative approach called TeamLEAD (learn, engage, apply, develop), where students learn most of the material before class via online lectures (as well as reading assignments), and devote in-person classroom time to team-based application of that material. This approach reduces classroom teaching, increases the use of electronic media and simulation, and places team-based problem solving at the center of the curriculum.

Another idea that could reduce the overall cost of medical education is "fast-tracking." For example, one can imagine creating a focused medical school primary care curriculum that takes three years to complete (1 year didactic; 2 years clinical); coupled to a 2-year (if the overlap with the second clinical year can be leveraged) or traditional 3-year residency. By reducing the overall training period by 1–2 years, the relative output over time of primary care physicians for AHC communities could increase.

Of course, reinvention is not only happening at medical schools. Programs for new categories of health care professionals are being created as certain fields become more significant to health care. For example, health coaches are becoming increasingly important in patient care. Accordingly, there is a real need for training programs for these health professionals. In addition, important opportunities exist for the development of a Doctor of Nursing Practice degree, long-distance learning programs in nursing, and the training of nurse practitioners in population health. For example, Duke University School of Nursing is working with an insurance company Blue Cross and Blue Shield of New Jersey (BCBSNJ) to train their nurse practitioners in population health. Single-year fellowship-type programs can cross-train mid-level providers in specialized areas of high need, such as behavioral health. New programs in emerging areas, such as population health management and clinical informatics management, can bring in new revenues while enhancing AHC missions.

Increasing the opportunities for interprofessional education will also be key, given that medicine is evolving from an individual provider basis to a team-based approach to care [17]. Interprofessional education can range from activities, such as case conferences, that bring together students from medical, nursing, and other health professional schools and programs to standing working groups and core faculty. Team training is also an important approach; the US Department of Health and Human Services has funded initiatives that provide interprofessional team training and culture change around patient safety [18].

In particular, leadership and management has been a critical area in medical education and training. Leading AHCs require alignment of strategy and management with care delivery. However, management and care provision are often disconnected and live in two separate worlds. Physicians, as well as other providers, are not engaged in strategy, policy, or planning—important tasks that are usually left to non-clinicians with business and health administration degrees. Providers need to be engaged in administration. Given this is not an area of expertise for most, they need training and background to lead in health care administration [19].

Providers who have made their way into administrative leadership positions within health care organizations have often moved into these roles in a fashion that can be fairly described as "accidental"—usually without formal training in key areas such as finance, strategy, human resources, and operations. Increasingly, more and more aspiring physicians are starting to acquire these skills early in their training as medical students in combined MD/MBA or MD/MHA programs, but do not necessarily have the opportunity to apply these skills until after completing training and achieving a requisite amount of seniority. Recognizing the opportunity to develop this talent pool more thoughtfully, Duke Medicine created the Duke Medicine Management and Leadership Pathway for Residents for individuals who enter residency with a graduate degree in management or equivalent experience [20]. It is the first residency program of its kind: an eighteen-month rotational, project-based experience in applied health care and school of medicine management in conjunction and interspersed with a traditional clinical residency.

There is also a real need for training mid-career physicians in leadership, management, and finance and undertaking management projects mentored by senior leadership. Nonclinical scientists starting to lead and manage research laboratories should also be provided with leadership and management training. Training in leadership and management skills across the human capital pipeline is critical to the future of AHCs.

GOING GLOBAL

With health inequalities, emerging infections, and a growing global burden of chronic disease, the transformation of AHCs will require a global perspective in all three academic medicine missions. Opportunities in globalization include addressing global health disparities, globalizing all three missions (e.g., expansion of clinical services, networks

in clinical research, overseas educational programs), and capacity building.

Many AHCs have taken advantage of these types of opportunities to establish a global footprint [21]. They are developing mission-based initiatives in clinical and translational research, global health, health care management and medical and health professional education with partners in China, Singapore, India, the Middle East, and elsewhere. For example, Johns Hopkins, Cleveland Clinic, and Partners Healthcare have developed significant clinical programs in the United Arab Emirate, and AHCs such as Duke University, Imperial College London, and Johns Hopkins have partnered with local institutions to bring their education and research expertise to Southeast Asia. Such global initiatives not only enhance the academic missions, but can also provide new sources of revenue.

Entrepreneurial activities with overseas pharmaceutical companies, hospital management, capacity development, and clinical research programs are examples of AHC global activities that are generating revenues while extending their missions. For example, the Cleveland Clinic is developing and managing a 360 bed hospital in Abu Dhabi through a partnership with Mubadala Healthcare. President and CEO of Cleveland Clinic, Delos Cosgrove, has said that the success of the operation is putting "Cleveland Clinic DNA" into this hospital—i.e., taking Cleveland Clinic workforce and using those individuals to initially drive the operations of the hospital [22].

AHCs should leverage their pedagogical expertise to create educational opportunities. Among the many examples, two initiatives are particularly noteworthy—Weil Cornell Medical School in Qatar and Duke NUS Graduate School of Medicine in Singapore. These are ambitious projects that have long-lasting impact on education and health care in their respective regions and beyond. With the new electronic age, AHCs can take their educational mission globally via the internet and potentially reach large numbers of students through distance learning. A specific example is the collaboration between the Duke Clinical Research Institute and Kaplan EduNeering, Inc that recently launched a global online certificate program called Clinical Research Education and Training via e Learning (CREATe) for training new clinical research investigators and research personnel, particularly in countries such as India that are increasingly becoming the focus for large clinical trials.

A FOCUS ON INNOVATION

A common theme across the discussions above regarding necessary changes in clinical care, research, and education is the need for transformative innovation. Challenging times require bold thinking and action that expand upon traditional assumptions without compromising on the mission and vision of the organization. Changes to the status quo inherently require risk, which must be fostered, supported, and managed across the institution. Accordingly, we must not only develop innovative models of care that leverage improvements in processes, technologies, financing, organizational structures, and utilization of human capital—but accelerate the scale and sustainability of these models. As Hwang and Christensen note, challenges in health care cannot be overcome by technology and product innovation alone; there must also be "business model innovation," the deployment of fundamentally new models of care that are fit for the evolving dynamics in health care [23]. New models for care delivery should adopt systematic process improvements from other industries, such as automobile manufacturing, as well as new organizational structures and financing mechanisms. Technological innovation should include the development of effective information systems, electronic and digital technologies, and mobile applications and devices that can monitor and manage patients seamlessly. And, in addition to promoting transformative innovation in care delivery and disruptive technology, we must develop new approaches to training and developing the health care workforce to prepare them for the changing health care delivery landscape.

The development of institutional programs and infrastructure to support efforts of innovation across the academic health center spectrum can help to catalyze these changes. Indeed, several institutions have developed major initiatives in health care delivery research and innovations—such as UK Imperial College's Institute of Global Health Innovation, Dartmouth's Center for Healthcare Delivery Science, and the University of Michigan's Institute for Healthcare Policy and Innovation. At UK Imperial College, Lord Ara Darzi has led the efforts in creating the Institute of Global Health Innovation. This Institute seeks to bring together engineers, medical researchers, physicians, and leaders in business and health policy to reduce health disparities and improve health globally [24]. A similar approach has been taken by the University of Michigan in creating the Institute for Healthcare Policy and Innovation, where researchers are engaged in innovative models and changes to health care policy [25].

At Duke Medicine, a new Duke Institute for Health Innovation (DIHI) was launched to lead the way in developing new and disruptive solutions, with a focus on four domains of care delivery: implementation sciences, health technology innovation, health leadership and workforce development, and applied health policy. These four domains of activity will serve as "centers" within DIHI. The goals of these centers are to integrate implementation and health services research with the clinical enterprise, leading to a living laboratory that can pilot and evaluate new models of care and rapidly scale those that are cost-effective, while also developing future leaders of health and health care through novel education and training programs. The development of incubator-type support for disruptive technologies and services, with an emphasis on informatics and medical technologies, will also help support novel approaches while driving commercialization. The Duke Institute for Health Innovation also aims to foster a culture of innovation and entrepreneurship across the organization, which is critical to undertaking the significant changes necessary to maintain the sustainability of AHCs.

LOOKING AHEAD

There will always be a need for AHCs to serve as the nation's leaders and integrators of health care missions. But given the changing health care environment and unique challenges facing them, AHCs will have to reinvent themselves in all the ways described above, evolving into true AHSSs and leading the transformation of medicine—or, alternatively, become boutique providers of fewer highly-specialized clinical, research, and training programs. More specifically, this change will enable them to achieve higher quality and greater efficiency in care—strengthening their preeminence in areas of distinction, developing more effective partnerships, reinventing drug discovery, producing more translation, and increasing commercialization of (and returns from) the fruits of their research—and to train more innovatively and more quickly the health care providers that we need.

THE NEED FOR ENTERPRISE-WIDE PLANNING AND MANAGEMENT

But how do AHCs prepare for this uncertain future? A PricewaterhouseCoopers report highlights AHCs' decentralized structures as one potential inhibitor to their ability to respond to these and other challenges [9]. However, such decentralization—traditionally along focused disciplinary lines—has also permitted much of the innovations for which AHCs are known. It is clear that AHCs need to unite and behave more like single integrated organizations, i.e., their clinical faculty need to be better integrated with the hospitals and affiliated clinics, and the health delivery systems need to be better aligned with the schools of medicine, nursing, and other health professions—as well as the research enterprise.

Change will require more accurate, mission-based accounting. Currently in many AHCs, the clinical, research, and educational components function to varying degrees like separate organizations. However, the individuals and units—including centers, institutes, departments, and divisions—carrying out these missions often overlap. As a result, the allocation of resources and costs among the different missions is often redundant, unclear, and confusing at best. At worst, this contributes to: a lack of transparency regarding how decisions are being made and resources are being used; the absence of a common financial model underlying all three missions; and reduced clarity and accountability needed to identify and support shared priorities across the missions.

To address these problems, AHCs should undertake an integrated approach to planning—"Enterprise-Wide Planning." Representatives from across the enterprise—including senior health system and school leaders, department chairs, program and center directors, and other key stakeholders, including faculty and students—engage collectively in a coordinated planning process to: affirm, align, and prioritize missions; develop clarity in decision rights and accountability; and recommend critical changes that will enable long-term sustainability and success [26].

Therefore, AHCs must first understand the true cost of each mission through a transparent accounting system and then determine what the current and future finances can support. This step enables AHCs to take the next difficult but critical step of addressing the priorities, balance, and tradeoffs of the missions in a transparent and participatory manner.

Going forward, AHCs must become AHSSs that function as an integrated organization with alignment among all missions, clarity in distribution of decision rights, and financial accountability across the enterprise. Such a system should ideally include a vertically integrated care delivery system, a research model that achieves a seamless "discovery to care continuum," and education programs that produce an effective health care workforce for the future, all working as a team for the same goals—to transform medicine.

Finally, as this evolution from singular AHCs into networked AHSSs continues, we must remember that optimizing patient and population health is the goal, and the AHSSs of the future are ultimately accountable to society for what they do to promote, preserve, and restore the health of all peoples—locally and globally.

ACKNOWLEDGMENT

The authors would like to thank Alex Cho, M.D., M.B.A., Ziggy Yoediono, M.D., M.B.A., Devdutta Sangvai, M.D., M.B.A., Bimal Shah, M.D., M.B.A., and David Zaas, M.D., M.B.A. for their contributions and editing assistance. This paper is based on the original work published by the authors in the New England Journal of Medicine. Transforming Academic Health Centers for an Uncertain Future. Dzau, Victor J; Cho, Alex; ElLaissi, William; Yoediono, Ziggy; Sangvai, Devdutta; Shah, Bimal; Zaas, David; Udayakumar, Krishna. The New England Journal of Medicine. Sep 2013, 369; 11, p991–993.

REFERENCES

[1] Social Security Advisory Board. The unsustainable cost of health-care: social security advisory report. Washington, DC: Social Security Advisory Board; 2009.

[2] Hearing before the committee on ways and means, one hundred ninth congress. In: T-EH, editor. Sector. 2005 [Vol First Session].

[3] Academy Health. The impact of the affordable care act on the safety net academy health policy brief. Washington, DC: Academy Health; 2011.

[4] http://nexus.od.nih.gov/all/2011/10/17/how-do-you-think-we-should-manage-science-in-fiscally-challenging-times/ [accessed 13.06.12].

[5] Mann S. Proposed federal budget cuts GME, freezes NIH funding. AMC Reporter. http://www.aamc.org/newsroom/reporter/march2012/276736/budget.html; 2012.

[6] Panel IECE. Core competencies for interprofessional collaborative practice: report of an expert panel. Washington, DC: Interprofessional Education Collaborative; 2011.

[7] Association of American Medical Colleges. AAMC data services. AAMC data book: statistical information related to medical schools and teaching hospitals. Washington, DC: AAMC.

[8] [a] Association of Academic Health Centers. How much does research cost? http://www.aahcdc.org/Resources/BenchmarksandMetrics/The-CostofResearch.aspx; 2014. [Last accessed 29.10.14].
[b] National Academies. Research Universities and the Future of America: Ten Breakthrough Actions Vital to Our Nation's Prosperity and Security. Washington, DC: National Academies Press; 2012.

[9] PricewaterhouseCoopers. The future of the academic medical center: strategies to avoid a margin meltdown; 2012.

[10] Shortell SM, Casalino LP, Fisher ES. How the center for medicare and medicaid innovation should test accountable care organizations. Health Aff (Millwood) 2010;29(7):1293–8.

[11] Every NR, Hochman J, Becke R, Kopecky S, Cannon CP. Critical pathways: a review. Committee on Acute Cardiac Care, Council on Clinical Cardiology, American Heart Association. Circulation 2000;101(4):461–5.

[12] Christensen CM, Grossman JH, Hwang J. The innovator's prescription: a disruptive solution for health care. New York: McGraw-Hill; 2009.

[13] International partnership for innovative healthcare delivery website. http://www.ipihd.org/.

[14] Dzau VJ, Ackerly DC, Sutton-Wallace P, et al. The role of academic health science systems in the transformation of medicine. Lancet 2010;375(9718):949–53.

[15] http://www.ucsf.edu/news/2011/01/8370/sanofi-aventis-enters-two-research-development-collaborations-ucsf [accessed 25.06.12].

[16] Foundation EMK. Facilitating the commercialization of university innovation: the Carolina express license agreement, 2010.

[17] Gawande A. Commencement address: Harvard medical school. http://www.newyorker.com/online/blogs/newsdesk/2011/05/atul-gawande-harvard-medical-school-commencement-address.html; 2011.

[18] Henriksen K. United States Agency for Healthcare Research and Quality. Advances in patient safety: new directions and alternative approaches, vol. 4. Rockville, MD: AHRQ publication; 2008. Agency for Healthcare Research and Quality.

[19] Fairchild DG, Benjamin EM, Gifford DR, Huot SJ. Physician leadership: enhancing the career development of academic physician administrators and leaders. Acad Med 2004;79(3):214–8.

[20] Ackerly DC, Sangvai DG, Udayakumar K, shah BR, Kalman NS, Cho AH, et al. Training the next generation of physician-executives: an innovative residency pathway in management and leadership. Acad Med 2011;86(5):575–9.

[21] Ackerly DC, Udayakumar K, Taber R, Merson MH, Dzau VJ. Perspective: global medicine: opportunities and challenges for academic health science systems. Acad Med 2011;86(9):1093–9.

[22] Cosgrove DM. Cleveland clinic's Delos M. Cosgrove: 'We are in Abu Dhabi to help a country shape its healthcare delivery system'. Knowledge @ Wharton. http://knowledge.wharton.upenn.edu/arabic/article.cfm?articleid=2644; 2011.

[23] Hwang J, Christensen CM. Disruptive innovation in health care delivery: a framework for business-model innovation. Health Aff (Millwood) 2008;27(5):1329–35.

[24] Lord Darzi to lead institute taking innovative approach to global health. London: Imperial College; October 04, 2010. http://www3. imperial.ac.uk/newsandeventspggrp/imperialcollege/newssummary/news_4-10-2010-11-57-6.

[25] University of Michigan Institute for healthcare policy and innovation. http://ihpi.umich.edu/; 2010.

[26] Inside Duke Medicine. Duke medicine launches unified planning process for continued Success. http://www.insidedukemedicine.org/news/duke-medicine-launches-unified-planning-process-for-continued-success/; 2012.

ABOUT THE AUTHORS

Victor J. Dzau, MD is the 8th President of the Institute of Medicine (IOM). He is Chancellor Emeritus and James B. Duke Professor of Medicine at Duke University and the past President and CEO of the Duke University Health System. Previously, he was the Chair of Medicine at Harvard Medical School's Brigham and Women's Hospital, and Chair of Medicine at Stanford University. Dr. Dzau also served as Chair of the Board of Directors of the Association of Academic Health Centers.

William F. ElLaissi, MBA, MHA currently leads partnership development for the Duke Institute for Health Innovation. Prior to this position, Mr. ElLaissi was the Special Assistant to the Chancellor for Health Affairs at Duke University.

Krishna Udayakumar, MD, MBA is Associate Professor of Global Health and Medicine at Duke University and leads multiple initiatives in the rapidly evolving fields of health innovation and globalization of healthcare. Dr. Udayakumar has also led the International Partnership for Innovative Healthcare Delivery (IPIHD), a non-profit co-founded by Duke Medicine, McKinsey & Company, and the World Economic Forum, since its inception in 2011.

Educating the Future Health Workforce

Chapter 7

Disruptive Technologies Affecting Education and Their Implications for Curricular Redesign

C. Donald Combs and Bertalan Meskó

Universities are the only institutions that are set up to transcend the limits of time, location, and immediate circumstance that constrain just about all workplaces.

Nicholas Lemann

In a recent essay in *The Chronicle of Higher Education*, Lemann made this observation about the innate flexibility of universities. His observation stands in marked contrast to the broader public perception of universities as large, expensive, and mostly inflexible organizations more interested in surplus generation and service to their faculty and staff than effective service to their students and public and private funders. One of the pressing questions in higher education generally is to what extent contemporary university practices will be supplanted or radically transformed by applications of digital technologies [1].

This chapter discusses the impact of disruptive technologies on education generally and on health care practice and health professions education specifically. It also focuses ultimately on the more important question of disruptive innovation in education, both present and projected. The narrative proceeds through definitions, to descriptions of emerging digital technologies, the role those technologies are playing in education and in the practice of health care, the implications for educational programs in the medical and health professions, some examples of responsive educational change, and concluding thoughts.

The basic premises are that rapid and pervasive technologic changes are affecting education and health care and that both sets of changes will disrupt the way that medical and health professions education programs are conducted. That disruption can be negative—it can drive the development of new business models that undermine and eventually replace existing business models and institutions that do not change; and it can be positive—better educational outcomes can be achieved, and new business lines can be developed through the effective incorporation of emerging technologies.

DISRUPTIVE TECHNOLOGIES AND DISRUPTIVE INNOVATION

Clayton Christensen and his colleagues have been the seminal researchers addressing disruptive change since Christensen's 1997 book, *The Innovators Dilemma*, which built on a decade of research on organizational decision making. Their research began by reviewing the effects of new technologies on various businesses, and they discovered a pattern wherein tools are introduced into industry, leading to new or simpler ways of achieving results that had been difficult or inefficient to attain. This pattern of introduction of new technologies that fundamentally change the business model of an industry is well illustrated by a few examples: email/postal mail; telephones/telegraphy; smartphones/telephones; ultrasound/radiography; and digital photography/chemical photography. In each instance, the new technology, over time, substantially and negatively impacted companies with business models based on the older technology.

Indeed, this finding led to an important insight by the researchers: Christensen replaced the term *disruptive technology* with *disruptive innovation* because he recognized that few technologies are intrinsically disruptive; rather, it is the business model enabled by new technology that creates the disruptive impact [2,3] (March and September). This evolution from a technological focus to a business modeling focus is central to understanding the evolution of business at the industry level. That is, it is not the technology per se that causes disruption; rather, it is the incorporation of technology into new business models that is disruptive.

Generally, disruptive innovations were technologically straight forward, consisting of off-the-shelf components put together in a product architecture that was often simpler than prior approaches. They offered less of what customers in established markets wanted and so could rarely be initially employed there. They offered a different package of attributes valued only in emerging markets remote from, and unimportant to, the mainstream [4].

The digital technologies described in the following narrative are not, in and of themselves, disruptive. They are, however, beginning to be deployed in a dizzying array of business plans that fit what Christensen has called "new market disruptions" that occur when an existing market segment is not being well served by the incumbents in the industry. For example, the higher education and health care industries currently share several attributes that make them targets for disruptive business models—there is limited access, they are expensive, service is often impersonal, the quality of the outcomes is not linked to the cost, and there are inexplicable variations in outcomes [5].

The Educational Context

In the 1970s, William Bowen, then provost and later president of Princeton University, lectured on the cost of a university education. His main point: there is little productivity gain in a university; therefore, educational costs can be expected to outpace inflation by roughly the rate of productivity increase in the overall economy, amounting to 1–2% per year on average.

In fact, cost increases have been more dramatic: universities outperform hospitals, physicians, and the pharmaceutical industry when it comes to increasing prices. The cost of higher education in the US rose almost 8% annually in the 40 years since Bowen's observation. In comparison, health care costs rose about 6% and aggregate consumer prices increased about 2% annually [6]. The public tolerance for that cost differential is, however, unlikely to continue because the dominant business model of higher education is antiquated. In general, the current model involves:

- A residential, facilities-based approach with continually increasing operational and fixed costs and which requires students to reside away from home and, often, job opportunities.
- Parallel and uncoordinated development of curricula across 3000 universities and 140 medical schools.
- Delivering roughly the same curriculum face-to-face on each campus and, only reluctantly and in a relatively limited manner, incorporating cost-effective digital technologies into the delivery system.
- Roughly equal allocation of resources (faculty, staff, and facilities) for programs that lead to degrees which are economically in-demand as well as those that are not. Fewer than 10% of US bachelor-level graduates are in the high-demand science, technology, engineering, and mathematics fields.

- Cross-subsidization across the various university businesses—undergraduate, graduate, professional, and continuing education programs; clinical care programs; research programs; and various auxiliary programs such as housing and sports.
- Focus on degrees (whether undergraduate or professional) as the essential credential, rather than on metrics of knowledge, skills, and competence achieved by learners.
- Little investment by universities in the career success of graduates. University graduates often lack basic skills such as how to make presentations, use spreadsheets, communicate effectively, and understand the requirements and obligations of being an employee.

In attempting to apply Christensen's research to online learning, there is early evidence of both disruption and wishful thinking. Enrollments in online programs totaled 937,000 students in 2004, 3.9 million in 2007, and 7.1 million in 2013, amounting to about 30% of the total student enrollment in degree-granting institutions. It is not possible to estimate how many courses use online learning to improve student knowledge or enhance the quality of on-campus classes, but the figure is probably high. These enrollment figures might also indicate that online learning is tapping into a new market of students previously unable to enroll, but the number also surely includes currently enrolled students who are augmenting their on-campus courses with online courses. So far, online learning has not reinvented the higher education marketplace despite enrollment growth rates averaging 20% per year. Nonetheless, this growth and the widespread use of online materials in on-campus coursework might be early signs of disruption [7].

The US Department of Education has compiled studies comparing online and face-to-face education and found that students performed better in online courses than in face-to-face courses, while courses that blended online and face-to-face instruction yielded the largest gains [8], due in part to:

- Students spending more time on-task;
- Students being given the opportunity for more control of their learning; and
- Students being provided more opportunities for reflection.

What qualities make online learning disruptive, if it is found to be so? Christensen, Michael Horn, and Curtis Johnson recently applied the theory to K–12 education [5,9]. They concluded that the insertion of computers in K–12 classrooms was an unlikely disruptor because it did not change the usual instructional models in common use; but using computers to provide student-centric education could do so. For students in schools that do not offer certain courses, online options allow them to follow their interests. The authors documented the rise in enrollments in Apex Advanced Placement (AP) courses, totaling over a million student enrollments in more than 4000 school districts. Virtual high schools, virtual chemistry labs, and other innovations contribute to the growth in online K–12 education, which will continue to increase for four reasons:

- Continuing improvement
- The ability for students, faculty, and parents to select a learning pathway that suits individual learners
- Teacher shortages
- Decreasing cost of digital technologies

Customized learning also results from the explosion of tools that allow students to produce their own content and seek help, when needed, from tutoring software [10–12].

Christensen's research suggests three central questions to assess the disruptive potential of technology [5]:

1. Does the innovation target customers who have been limited by lack of money or skills to perform the task themselves?
2. Is the innovation aimed at customers who will be delighted to have a simple product?
3. Will the innovation help customers achieve more easily and effectively what they are already trying to do?

We will return to these questions in the concluding section of this chapter.

Emerging Educational Technologies

Learnnovators [13], an Indian consulting firm focused on e-learning, recently identified 11 technologies that have the potential to transform education. They are summarized below (and include some technologies that overlap with those Meskó identified in health care, *see* p. 62).

- *Big data*: Big Data refers to the large amounts of unstructured data flowing through numerous sources in our digital world every second. In the e-learning scenario, Big Data is the data produced by learners interacting with the learning content and collected through learning management systems (LMS), content management systems (CMS), and other media, including social networks that enable learners to interact with learning programs (see also Ref. [14]).
- *APIs*: Application Programming Interfaces or APIs are one of the most promising technological innovations for education and learning. An API is a software interface specification that allows applications to easily interact, providing improved or added functionality to users. There are a number of highly popular application mashups available on the Internet today, including ones that cater to learning and education. Some of the most popular APIs related to education are applied in After the Deadline, Khan Academy, Blackboard Collaborate, and Knewton.
- *Tin can*: This particular API is a new and evolving learning technology specification; and may prove to be an important successor to the Sharable Content Object Reference Model (SCORM). SCORM integrates a set of related technical standards, specifications, and guidelines designed to ensure accessible, interoperable, durable, and reusable content and systems. Tin Can addresses many of SCORM's limitations: SCORM compliant tools work via a web browser, require a constant Internet connection, require an LMS to initiate learning, track only formal learning activities, and do not support platform transitions (such as computer to mobile device).
- *HTML5*: Version 5 is an updated version of the Hyper Text Markup Language (HTML) developed by the Web Hypertext Application Technology Working Group and the World Wide Web Consortium. The most important features of this disruptive technology are its support for mobile devices and platforms that do not support Flash, more accurate formatting, more interactive experiences, greater flexibility (than Flash), better performance, and the use of less processing power (hence longer battery life).
- Responsive e-Learning Design (*RED*): In the emerging multidevice-learning world, current learners spend most of their time in front of multiple screens—a desktop computer, laptop, smart phone, or tablet. They move between these different types of devices in sequence (called "Sequential Screening") or use them simultaneously (called "Simultaneous Screening"). This scenario makes it very challenging for designers to ensure that the learning solutions they design can work on devices with different screen sizes, shapes, resolutions, or operating systems. RED is the technology that assists in the design and delivery of learning solutions which can work with a broad range of devices.
- *Wearable computing*: This technology will lead to wearable learning. Wearable digital devices with electronic performance support systems that can provide continuous learning and support for mobile workers.
- *Videos*: Video clips are not only going viral in chronicling the lives of the rich and famous. They are one of the best media for effective learning because they offer better learner engagement. User-generated videos are already going viral and clearly indicate the potential for using videos in innovative ways for learning and education.
- *Digital textbooks*: Many schools and universities have started using this new technology as an effective learning tool because of the instructiveness, engagement, and cost-effectiveness that it offers. Digital textbooks, wired for Big Data Analytics, will revolutionize the learning landscape through personalization for the learners.
- *3D printing*: This technology turns a digital design (a three-dimensional file) into a three-dimensional physical product.
- *Mobile learning*: Platform independent applications offer the flexibility to learn across contexts using a wide range of digital devices. Smart Phones and tablets will assume an even greater role in just-in-time learning and performance support.
- *Cloud computing*: Cloud learning refers to learning in a virtual world that is interconnected through the web or

mobile networks. Many universities and higher education institutions are moving toward cloud campuses—virtual campuses via which they deliver learning programs in a highly efficient way. The massive open online course (MOOC) is a recent development powered by cloud computing technology. MOOC promoters (such as Coursera and Udacity) offer a wide range of online educational programs from leading universities.

Whether these emerging technologies do, in fact, prove to be transformative or disruptive depends almost entirely on whether the leaders of today's academic institutions understand the imminent challenges they pose and whether the leaders are committed to creating and sustaining the required changes to existing business plans. If current leadership is unable or unwilling to implement fundamental change on an ongoing basis, then other leaders and organizations will seize the opportunities and erode the foundations on which the existing business models are based.

Michell Zappa, founder of the Envisioning Technology Research Foundation in Brazil, and his colleagues have reviewed educational technology, both current and projected, and concluded that:

- Educational technology is generally centralized today, either in computer labs or classrooms.
- Comparatively, the world outside of the classroom makes ubiquitous use of mobile, digital technologies.
- As classrooms digitize, students will be free to collaborate with peers globally.
- Classrooms, as physical teaching models, will tend to be replaced by learning studios and virtual teaching.
- Instruction and assessment will increasingly become project, performance, and portfolio based.
- Over time, education will evolve into a continuous, interconnected effort, allowing learners to adapt to a perpetually changing world.

The infographic shown in Figure 1 was designed by Zappa and, as he notes, "attempts to organize a series of emerging technologies that are likely to influence education in the upcoming decades. Despite its inherently speculative nature, the driving trends behind the technologies can already be observed, meaning it is a matter of time before these scenarios start panning out in learning environments around the world" [15].

One key insight from this visualization is the shift from a model that is primarily classroom-based to a model based on studios and virtual environments. In the context of this visualization, "Classroom" refers to a learning environment with fixed, centralized technology paralleled with a fixed-location teaching and learning style. This is the prevailing paradigm. One teacher addresses students in a unidirectional method of teaching and assessment.

The term "Studio" refers to a shift in teaching and learning styles that resembles those work environments where peer-to-peer learning is prevalent and groups gather to discuss, share, learn, and solve problems together with the teacher acting as a facilitator. This educational model has several implications for the design of the learning environment, as it no longer requires individual desks in rows facing forward, but various settings supporting peer-to-peer discussions and project areas.

"Virtual" refers to a disembodied learning environment where many aspects of student learning experiences—including discussion, knowledge sharing, project creation, and assessment—occur regardless of geography and time. Examples of this learning environment are emerging with online courses, video lessons, enhanced educational applications for mobile devices, open courseware, and e-books.

In addition to these three learning environments, Zappa envisions six categories of technological development that help to organize the infographic. These are digitized classrooms, opening of information, gamification, disintermediation, tangible computing, and virtual/physical studios. The essential components of those categories follow.

- *Digitized classrooms*: Rather than considering information technology (IT) as a standalone tool or skill, digital technologies tend to disperse throughout every aspect of the classroom. Examples include:

  Tablets
  Electronic paper screens
  Interactive whiteboards
  Data projectors
  Eye tracking
  Attention tracking
  Object-embedded intelligence
  Desk-sized screens
  Wall-sized screens
  Cascading knowledge maps
  Classroom performance dashboards

- *Opening of information*: Dissemination of information outside the physical silos of schools and classrooms provides content, feedback, and assessment to students virtually. Examples include:

  Online school communities
  Video lessons
  Education app stores
  Open courseware
  Digitization of books
  Portable academic histories
  Flipped classrooms
  Interschool teaching platforms

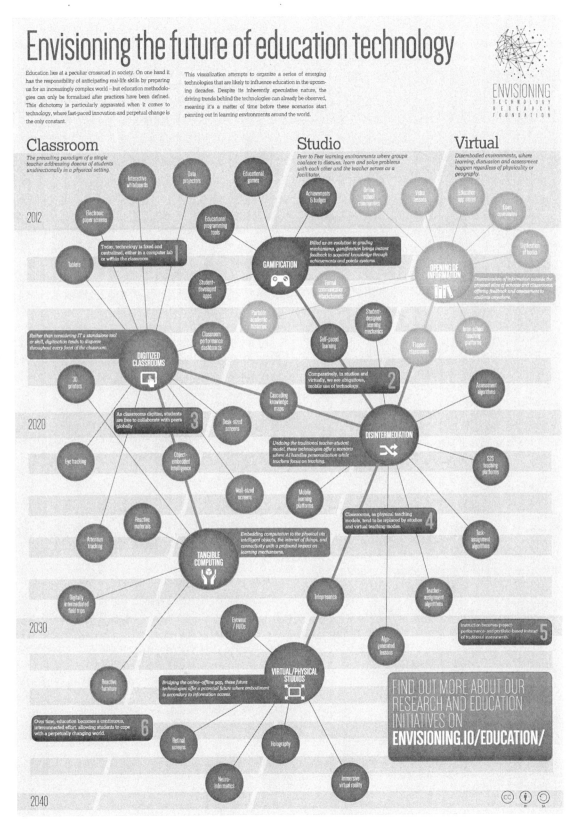

FIGURE 1 Envisioning the future of education technology by Michell Zappa.

- *Gamification*: Transforming content and assessment of mastery through gamification brings instant feedback on the level of acquired knowledge and skills through competency-based achievement and assessment systems. Examples include:

  Achievements and badges
  Educational games
  Educational programming tools
  Student-developed apps
  Self-paced learning

- *Disintermediation*: Improving the traditional teacher-student knowledge transfer model, these technologies create a scenario where artificial intelligence handles personalization and teachers focus on managing the educational process. Examples include:

  Student-designed learning mechanisms
  Assessment algorithms
  Student to Student teaching platforms
  Task-assessment algorithms
  Teacher-assessment algorithms
  Algorithm-generated lessons
  Telepresence
  Mobile learning platforms

- *Tangible computing*: Embedding computation to the physical via intelligent objects, the internet of things, and connectivity will have a profound impact on learning. Examples include:

  Reactive materials
  Reactive furniture
  Digitally intermediated field trips

- *Virtual/physical studios*: Bridging the online–offline gap, these future technologies offer a potential future where embodiment is secondary to information access. Examples include:

  Eyewear/Heads Up Display
  Retinal screens
  Neuroinformatics
  Holography
  Immersive virtual reality

Underlying these emerging educational platforms is an explosion of data about knowledge acquisition, skills competency, and pedagogic strategies and their interrelationships. Instructional designers who understand that current learners are, in all likelihood, not like them and who are open to building new learning environments and employing data-based pedagogical strategies will lead the transformation, or possibly the disruption, of current educational business models.

Potentially Disruptive Technologies in Health Care

Changes are adapted more readily in health care practice than in health professions education. In large part, these changes are driven by the provider's need to respond to the incentives of the health care marketplace, where the financial and legal incentives are frequently more substantial and time-limited than in the higher-education marketplace. The following description of technologies that have the potential to become disruptive innovations is not exhaustive, but provides a sense of the breadth of technologies that are under development.

There are six major categories of digital advances that have begun to disrupt the model of health care delivery developed in the twentieth century—the cell phone, personal computers, the internet, smart digital devices, gene sequencing, and social networks. There is, within these categories, an amazing array of technologies that are beginning to disrupt health care practice and must be addressed in health professions education programs [16,17].

One of this chapter's authors (Meskó) categorized approximately 40 of these current and emerging technologies in his online monograph, *The Guide to the Future of Medicine* [18]. In Figure 2, Meskó breaks technologies into four categories—prepare and prevent, data input and diagnostics, therapy and follow-up, and outcomes and consequences—and displays those categories in three dimensions: those that are already available, those in progress, and those still needing time to mature. A brief description of some of these technologies follows.

Prepare and Prevent

- *Digital literacy*: This refers to the evolving digital and social media applications, including, but not limited to, search engines, medical blogs, Twitter, YouTube, Facebook, Wikipedia, and other social media channels.
- *Virtual dissection*: Virtual dissection tables and screen-based applications are complementing and beginning to replace human cadavers, allowing students to perform dissections that achieve educational objectives virtually. Examples include Anatomage, ImageVis3D, and 4DAnatomy.
- *Full physiological simulations*: There are a number of international efforts (Virtual Physiological Human and HumMod) that seek to integrate research findings to create a functional digital patient for use in patient care, teaching, and research.
- *Curated online information*: The vast information available through the Internet is being assessed by experts and combined into reliable and valid repositories that will be available to practitioners, patients, and learners (Wikis and Webicina.com are two examples).

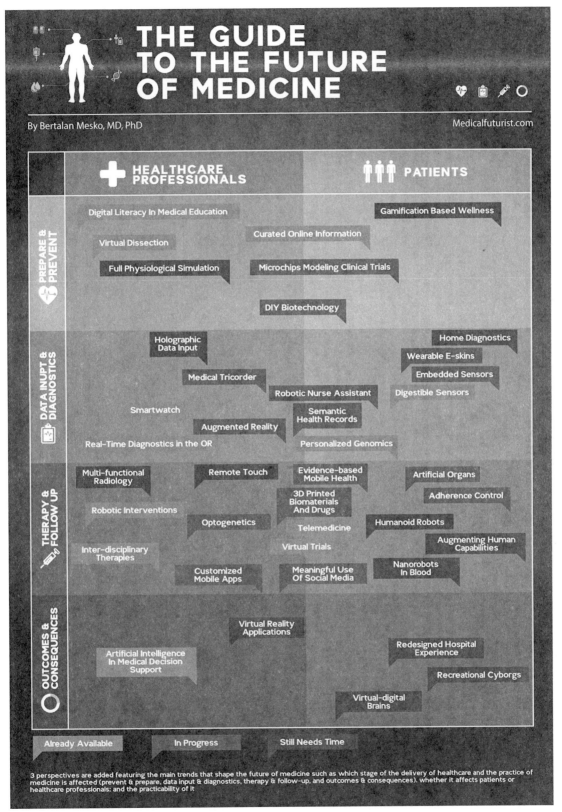

FIGURE 2   *The Guide to the Future of Medicine* by Dr. Bertalan Meskó, MD, Ph.D.

- *Microchip modeling*: Organs-on-Chips technology is an example of using human data rather than humans to conduct clinical trials. This technology provides a range of chips that model different organs, e.g., the liver, kidney, and heart. Researchers plan to connect these chips and create the first connected organs-on-chips human model by 2016.
- *DIY (do it yourself) biotechnology*: Methods and materials in biotechnology have become more broadly available (iGEM, DIY division), allowing interested parties to conduct experiments outside of the traditional laboratory setting.
- *Gamification of patient care*: Applications that function like games are being developed and used to achieve better health outcomes (Shine, FitBit, and Lumosity). Physical activities, cognitive functions, and other skills can be acquired, monitored, and improved with online gaming solutions.

Data Input and Diagnostics

- *Holographic data input*: Entering and retrieving data from a pervasive health information network will be possible on any surface, thus making clinical data available everywhere.
- *Medical tricorder*: Development of portable, wireless devices that can be customized to monitor and diagnose a variety of diseases is underway.
- *Smartwatches*: The early prototypes are being used for consultations, making calls, sending messages, and scheduling visits.
- *Augmented reality*: Tools such as Google Glass and soon-to-be-developed digital contact lenses are beginning to supplement and store clinical information in real time. Surgeons have already streamed their operations live from their own perspectives with Google Glass, making it simpler for students to learn new skills they could only learn previously by trying to look over the shoulder of the surgeon.
- *Semantic health records*: Making the narrative portion of health records as easily searchable as lab and radiologic results will improve the understanding and implementation of effective therapies for both providers and patients.
- *Personalized DNA sequencing*: DNA analysis, combined with improved prediction models, will offer providers and patients better prediction and treatment models.
- *Home diagnostics*: Technologies that can sense across the skin barrier or process drops of blood and urine provided by patients at their homes will shorten the diagnostic and therapeutic time frame.
- *Wearable e-skins, embedded sensors, and digestible sensors*: These sensors measure the most important health parameters, including vital signs, blood biomarkers, and neurological symptoms 24/7. As an addition to digestible

and wearable sensors, tooth-embedded sensors can recognize jaw movements, coughing, speaking, and even smoking. Digital devices and tiny sensors for gathering/storing data and transmitting body temperature, heart, and respiration rates to an external device are extensions of digitized personal health monitoring (Proteus, Digital Health, Sano, and Equivital).

Therapy and Follow-Up

- *Robotics*: Robots can be used in remote surgery, surgical rehearsal in preoperative planning, intraoperative navigation, and simulation and training. One of the best examples is still the Da Vinci system, but other robots in the fields of emergency response or radiosurgery are also available.
- *Customized mobile applications*: The number of general medical mobile apps has been increasing for years. In the near term, apps that can be personalized based on provider treatment preferences and individual patient characteristics will become more prevalent.
- *3D printed biomaterials*: The biotechnology industry is booming with 3D printing of living tissues, cells/drugs, bionic ears, and potentially transplantable human organs.
- *Artificial organs*: Surgeons are already able to implant artificial skin, cartilage, synthetic windpipes, and artificial blood vessels.
- *Adherence control*: Compliance is a crucial issue in improving patient health and decreasing the cost of delivering health care. New technological solutions, such as a pill bottle that glows blue when a medication should be taken and red when a dose is missed, are emerging.
- *Augmenting human capabilities*: The replacement of nonfunctioning organs, or enabling certain functions of the human body, is beginning to occur. The recent use of exoskeletons as a solution for paralysis or loss of extremities is one example.
- *Nanorobots*: Special function nanorobots will keep patient tissues safely oxygenated for up to 4 hours after a heart attack, serve as white blood cells, and remove platelets or repair damaged cells.

Outcomes and Consequences

- *Artificial intelligence in medical decision support*: IBM's Watson supercomputer can process over 200 million pages in a second and has been tested in oncology centers to help oncologists select the most effective cancer treatments. This type of intervention, not limited to Watson, will provide important decision-support to practitioners.
- *Virtual reality applications*: Virtual reality will be used more broadly. For example, in psychotherapy, patients will be able to experience things virtually that they would never be able to experience in real life as part

of their therapy. Patients, for example, might choose to experience an upcoming operation step-by-step as part of preparation for surgery.

This list of technologies that currently impact health care delivery or that will do so in the future is not intended to be exhaustive. Rather, the intent is to illustrate the breadth of technological changes affecting health care delivery and to foreshadow the need to address those changes in the medical and health professions curricula, while simultaneously incorporating emerging educational technologies into the underlying business model.

Incorporating the Disruptive Technologies into Health and Medical Education

There are, of course, many examples where medical and health professions educational programs are responding to these new technologies and making serious efforts to improve and update existing curricula. Five examples, ranging from the relatively simple to the more complex, serve to illustrate the beginnings of effective curricular response to the pervasive changes in technology.

Using Video Clips to Flip Classrooms

Medical accrediting agencies are demanding more active learning in the medical education curricula, including activities such as team-based learning and simulations. To take full advantage of active learning formats, however, students must come prepared with the underlying conceptual knowledge. A growing number of medical schools are using recorded lectures to supplement student learning in the first and second years of medical school [19]. Recently, medical schools have begun to supplement traditionally structured education with short, segmented videos followed by group discussion (flipped learning). Students view lecture material as homework and then solve problems in the classroom. For example, Stanford University teamed with Khan Academy to produce and distribute short instructional videos to medical students, who then use the information learned in the videos to prepare for active, team-based learning sessions [20].

It is now easier for subject-matter experts to produce and distribute video clips. With inexpensive full-motion screen-capture software installed on portable devices, subject-matter experts can create annotated instruction in the comfort of their own office, lab, or home. In the past, creation of instructional videos required a full production studio.

Ensuring Digital Literacy in Medical Education

In 2013, Semmelweis University's Medical School launched a new course dedicated to preparing students for the future of pervasive medical technologies. The course, organized by the scientific vice rector, Professor Mária Judit Molnár, and Bertalan Meskó, provides lectures by expert speakers covering topics such as bionics, robotics, augmented reality, home diagnostics, personalized genomics, social media, mobile health, telemedicine, and information overload.

Incorporating innovative lectures full of real-life examples, the long-term goal of the curriculum is to provide medical students with the understanding and skills needed to assess and implement new technologies in their professional careers. The course covers skills including: dealing with information overload through practical methods; analyzing the quality of the research underlying the various technologies; understanding emerging concepts of the future of medicine; and assessing the quality and usability of new technologies, online resources, and companies.

Students answer questions about the topics covered during the lectures on the course's Facebook page, where related news about these issues is also updated on a regular basis, providing students with a constant and current curated channel of educational resources.

The knowledge and skills students gathered during the course are assessed through a written test at the end of the semester. In 2015, the organizers will add an online curriculum MOOC to allow students worldwide to take the course [18].

Social Networking and Crowdsourcing through Osmosis

Medical education requires a broad knowledge of the biomedical sciences as a precursor to clinical practice. However, the knowledge obtained from each course often decays rapidly following course completion, thus negatively affecting performance during clerkships, licensing exams, and clinical practice. Formative assessment has been shown to improve medical student performance and retention, but many learners lack access to formative assessments because faculty members have limited time to create such resources and these commercial review banks are expensive.

Some students at Johns Hopkins medical school developed an innovative mobile- and web-based software, called Osmosis, which aims to improve long-term retention by uniquely combining three evidence-based concepts in education: (1) adaptive spacing through quick and periodic reviews; (2) quiz-based learning to improve absorption of material through practice questions; and (3) gamification through social network-enabled peer-to-peer learning.

Osmosis (www.osmosis.org) provides students free access to hundreds of high-yield practice questions and explanations developed by faculty and peers and sorted by course module. Question quality is enhanced through a novel rating and commenting feature. The Learning Mode allows students to cycle through questions and encourages understanding by retiring a question once it is answered correctly twice. The Retention Mode facilitates long-term retention by periodically e-mailing selected questions with high-yield clinical concepts to students. Osmosis records data (e.g., student usage and answers) on a secure server.

Initial usage statistics and student and faculty reception are promising, indicating that the platform is flexible, fits well into busy schedules, and helps participants self-assess knowledge gaps. Since Osmosis was launched at the Johns Hopkins School of Medicine in January 2012, approximately 250 students in the first and second year classes have spent 2400 hours answering more than 5000 questions over half-a-million times (~2000 questions answered/student). In addition, over 1500 Creative Commons-licensed images and YouTube videos have been shared. In August 2013, a free iOS app was released through the iTunes store. The app applies the testing effect and spaced repetition by sending medical students push notifications with short clinical vignettes as well as factoids. So far, more than 2500 medical students worldwide have downloaded the Osmosis Med app [21].

Using Artificial Intelligence (AI) to Improve Learning

Much instruction is designed by intuition, drawing on the experience and reflection of instructional designers and subject-matter experts. However, conscious access to one's own knowledge is quite limited—estimated to be only about 30% of what is known. The techniques of cognitive task analysis (CTA) (e.g., conducting structured interviews of experts) can reveal such hidden knowledge. Furthermore, course redesign based on CTA analysis has been shown to improve student learning. Greater levels of automation in CTA can be achieved by "mining" the log data from users of educational technology. By employing AI and statistical analysis, better cognitive models have been discovered across multiple subject matter domains. One benefit of this data-driven approach to CTA is that it supplements human qualitative judgment with automated quantitative metrics that rigorously test purported cognitive model improvements. A critical next step is to the "close the loop" by using the improved cognitive models to redesign instruction and then to compare, in a controlled experimental study, whether the redesign produces better student learning than the original [22].

Past experiments testing the benefits for student learning in CTA-based course re-designs have had impressive results, but have typically taken a broad-strokes approach to redesign. The redesigned "treatment" course usually differs from the original "control" course in many ways, not all of which are clearly attributable to cognitive model improvements or to the insights obtained from CTA. The phrase "how we categorize student activities" is another way of saying "cognitive model." Students learn the elements (the knowledge components) of the cognitive model they spend time practicing. However, the structure of that model is not obvious. Knowledge components are not directly observable and

most are not open to conscious reflection despite our strong sense of self-awareness regarding our own cognition. They can be inferred and discovered from student performance data across multiple tasks via a statistical comparison of alternative categorizations, that is, of alternative cognitive models.

There is an emerging opportunity for using AI in education, not only to mine the data to discover better cognitive models, but also to redesign systems based on the resulting insights and test them for improved student learning.

Introducing Ultrasound Use Early in Medical Education

Touted as the "stethoscope of the future" for its ability to provide real-time diagnostic information, ultrasound also offers benefits of increased portability and affordability that are making it ever more practical. The American Medical Association (AMA) acknowledged the trend, recently affirming for the first time in its House of Delegates resolution H-480.950 that "ultrasound imaging is a safe, effective and efficient tool when utilized by, or under the direction of, appropriately trained physicians."

Alfred Z. Abuhamad, MD, Professor and Chair of Obstetrics and Gynecology (OB/GYN) at Eastern Virginia Medical School (EVMS), and a long-time ultrasound pioneer, says the AMA resolution is a major step forward for a technology whose potential is only now being realized. "Ultrasound has some important advantages over other imaging tools," says Dr. Abuhamad, who is also national president of the American Institute of Ultrasound in Medicine, a sponsor of the AMA resolution. "From a safety perspective, ultrasound uses no ionizing radiation, so it is safer. From a cost perspective, it is less expensive than MRI (magnetic resonance imaging) or CT (computed tomography). And, from a clinical perspective, it is at least as effective as other imaging modalities" [23].

Ultrasound's versatility, safety, portability, and affordability have helped bring it to such remote locales as the Mt. Everest base camp, the battlefields of the Middle East, and the international space station. As units become even more portable and cost-effective, ultrasound is becoming ubiquitous in hospitals, clinics, and at patient bedsides. In the emergency department, ultrasound can be used to diagnose bleeding in internal organs in trauma victims as well as check the status of vital organs, such as the heart and aorta. It is also valuable in diagnosing problems in numerous other areas of the body, e.g., the eyes, muscles, and bones.

EVMS is taking a leadership role in advancing the everyday use of the technology. That effort includes incorporating the use of ultrasound throughout the school's medical curriculum. The course involves faculty members from anatomy, physiology, Obstetrics and Gynecology, internal medicine, and emergency medicine, and uses anatomical specimens, standardized patients, and hand-held ultrasound devices in a coordinated curriculum spanning the 4 years of medical education. It covers the basic physics of ultrasound, the knobology of the machines, and their use across the body, and concludes with both written and objective structured clinical exams.

Students are learning its value early in medical school by using advanced new equipment—including 30 units approximately the footprint of a smart-phone—provided through a grant from General Electric (GE). Their experience with the devices, both during medical school and later in clinical practice, has become the subject of a research study in collaboration with GE. "We are one of the first adopters of what is going to become a trend in medical education," Dr. Abuhamad says. "Our hope is that students will think about ultrasound when they need an imaging technology in clinical practice. They will understand its benefits and limitations and will know how to order it appropriately" [23].

THE IMPLICATIONS OF EMERGING TECHNOLOGY FOR CURRICULA

This chapter began with a quotation about the capacity of universities to transcend the limits of time, location, and immediate circumstance. The arguments in support of that statement are manifold; there are daily announcements about courses and degrees with new content and new delivery methods. Yet, as the magnitude of the current and future technologic onslaught becomes apparent to those in academic leadership positions, there are many reasons to be concerned. Christensen's conditions for disruptive innovation are pervasive in higher education generally and in health professions education specifically.

The aging model of higher education appears ripe for disruption. It is residential, expensive, and degree oriented; uses institution-specific curricula; and shows little concern for equipping students with the basic skills of employment. Disruptive potential is shown, for example, by the rapid growth of online certificate programs that use competency-based metrics. These certificates target potential learners who have not been able to find the educational programs they need; who seek simple, focused, educational programs; and who want easier access to demonstrably effective programs. New business models based on educational processes that make full use of emerging technologies and on outcomes measures that are competency, rather than time or degree, based are required.

Medical and health professions educational programs must, by their very nature, be closely linked to professional practice. Successful professional practice begins with skills competency that is supported by an appropriately curated knowledge base. Professional practice prospers when it also includes competence in communication and teamwork and a willingness to incorporate new technologies that meet the quality standards required by science and by regulation.

That, in a nutshell, is the overriding challenge that these emerging technologies pose to existing curricula—how can they educate initially competent professionals capable of maintaining their competence in a rapidly changing practice environment and do so in a manner that is flexible, that fulfills universities' transcendent potential, and that is much more cost-effective and competency based.

The characteristics of curricula that are successful in the future will, we believe, be based on business models that embrace the participatory, democratic culture emerging from ubiquitous digital devices and social networks; that take advantage of big data and cloud computing to increase the customization and personalization of educational programs; and that achieve balance between the constant connectivity afforded by digital devices and the need for offline reflection on the abiding philosophical questions of what is real, how do we know, and what difference does it make?

The characteristics of academic leaders who can successfully shepherd the evolution of their current curricular business models and avoid or mitigate disruption are likely to be those who create an organizational context that follows Rogers' concepts about the diffusion of innovation [24]. That is, they help colleagues understand the relative advantages of implementing new technologies, their compatibility with the values of the academy, their relative simplicity and ease of use, and their dependence on experimentation and revision based on observable results.

Adaptive academic leaders can fulfill Lemann's notion of universities as the only institutions "set up to transcend the limits of time, location, and immediate circumstance" and ensure that the risk of disruptive innovation aptly described by Christensen is managed.

REFERENCES

[1] Lemann N. The soul of the research university. The Chronicle of Higher Education; May 2, 2014.

[2] Christensen CM, Overdorf M. Meeting the challenge of disruptive change. Harvard Business Review. Boston (MA): Harvard Business Publishing; March–April 2000.

[3] Christensen CM, Bohmer R, Kenagy J. Will disruptive innovations cure health care? Harvard Business Review. Boston (MA): Harvard Business Publishing; September 2000.

[4] Christensen CM. The innovator's dilemma: when new technologies cause great firms to fail. Boston (MA): Harvard Business Review Press; 1997.

[5] Christensen CM. Disruptive innovation. 2014. Retrieved March 2014, from Clayton Christensen: http://www.claytonchristensen.com/key-concepts/.

[6] Bowen WG. Higher education in the digital age. Princeton University Press; 2013.

[7] Porto S. The EvoLLLution: illuminating the LiveLong learning movement. 2013. Retrieved March 2014, from The Impact of Disruptive Technology-Based Innovations in Higher Education: http://www.evolllution.com/opinions/impact-disruptive-technology-based-innovations-higher-education/.

[8] Garrison D, Kanuka H. Blended learning: uncovering its transformative potential in higher education. Internet High Educ 2004;7(2):95–105.

[9] Christensen CM, Horn MB, Johnson CW. Disrupting class: how disruptive innovation will change the way the world learns. New York: McGraw-Hill; 2008.

[10] Briggs S. Innovation excellence. 2013. Retrieved March 2014, from 10 Emerging Educational Technologies & How They Are Being Used Across the Globe: http://www.innovationexcellence.com/blog/2013/07/29/10-emerging-educational-technologies-how-they-are-being-used-across-the-globe/.

[11] Blankenship M. How social media can and should impact higher education. Hispanic Outlook; November 29, 2010.

[12] Conole G, de Laat M, Dillon T, Darby J. 'Disruptive technologies', 'pedagogical innovation': what's new? Findings from an in-depth study of students' use and perception of technology. Comput Educ 2008;50(2):511–24.

[13] Kumar S. Learnnovators: bringing innovation to learning. (F. Magazine, Producer). 2013. Retrieved March 2014, from Top 11 Disruptive E-Learning Technologies For 2013: http://learnnovators.com/top-11-disruptive-e-learning-technologies-for-2013/.

[14] Manyika J, Chui M, Bughin J, Dobbs R, Bisson P, Marrs A. Disruptive technologies: advances that will transform life, business, and the global economy. McKinsey Global Institute. McKinsey & Company; 2013.

[15] Zappa M. Envisioning the future of education technology. 2014. Retrieved April 2014, from Envisioning: http://envisioning.io/education/.

[16] ebizMBA Guide. eBizMBA Guide. April 2014. Retrieved April 2014, from Top 15 Most Popular Social Networking Sites: http://www.ebizmba.com/articles/social-networking-websites.

[17] MacQuarrie A. Learning Liftoff. 2011. Retrieved March 2014, from Transforming the Way We Learn: Engaging the Millennial Generation: http://www.learningliftoff.com/transforming-way-learn-engaging-millennial-generation/#.U2E8D9y4mFI.

[18] Meskó B. 2013. http://scienceroll.files.wordpress.com/2013/10/the-guide-to-the-future-of-medicine-white-paper.pdf.

[19] Association of American Medical Colleges. 2013. Organization of learner representatives. Lecture video recording and attendance. Retrieved from: https://www.aamc.org/download/268838/data/lecture_video_recording_and_attendance.pdf.

[20] Prober CG, Khan S. Medical education reimagined: a call to action. Acad Med 2013;88(10):1407–10.

[21] Gagiani S, Hayes MR. Learning through osmosis: an online platform and mobile app for medical education. 2013. In: A poster presented at the 2013 AAMC Annual Meeting MedEdPORTAL poster session on "Excelling in Health Education Assessment."

[22] Koedinger KR, Stamper JC, McLaughlin EA, Nixon T. Using data-driven discovery of better student models to improve student learning. 2013. Retrieved from: http://pact.cs.cmu.edu/pubs/Koedinger%20Stamper,%20McLaughlin%20&%20Nixon.pdf.

[23] EVMS Magazine. 2013. Issue 5.1. "Is the stethoscope about to become extinct? Medical ultrasound comes of age."

[24] Rogers EM. Diffusion of innovation. 5th ed. New York: Free Press; 2003.

ABOUT THE AUTHORS

C. Donald Combs, PhD is Vice President and Dean, School of Health Professions, at the Eastern Virginia Medical School (EVMS). Dr Combs holds several faculty appointments and currently teaches graduate courses in strategic planning, medical simulation, and interdisciplinary leadership.

Bertalan Meskó, MD, PhD is a medical futurist, the author of The Guide to the Future of Medicine, and the founder of Webicina.com.

Chapter 8

Diversity in the Academic Health Center: Progress and Opportunities

Sonja Haywood, Ron Berkman, Jeanne C. Sinkford, Louis Sullivan, Jeffrey L. Susman and Jay Gershen

INTRODUCTION

Despite decades of efforts, the United States has an enduring shortage of racial and ethnic minorities in the health professions—the result of segregation, discrimination, and unequal opportunities. According to the 2010 US Census, African Americans, Hispanic Americans, Asian Americans, and Native Americans or Alaska Native Americans make up >38% of the US population [1]. The Census Bureau projects that by the year 2042 there will no longer be a white majority in the United States [2]. However, <15% of the nation's physicians, nurses, dentists, and other health professionals are minorities [3]. By the year 2050, these ethnic groups are projected to comprise only 14% of the nation's physicians, 13% of the nation's dentists, and 11% of the nation's nurses [4]. These figures illustrate a daunting diversity gap in our health care workforce that contributes to poorer health status, shorter lifespan, and higher burden of disease and disability.

Fortunately, research demonstrates that health outcomes can be improved by health professionals who reflect the populations they serve, and that minority clinicians are more likely to be stewards for diverse communities. For example, black and Latino physicians are three to five times more likely to establish their practices in low-income neighborhoods or the barrio [5].

Efforts to educate a more racially and ethnically diverse health professions workforce have included initiatives with academic health centers (AHCs), health professions schools, colleges, high schools, and health professions associations that address barriers ranging from limited educational opportunities to the high cost of health professions education, from conscious and unconscious bias in evaluation and admission procedures to a lack of role models and mentors. Strategies to address this array of challenges were presented in the Sullivan Commission's report in 2004, *Missing Persons: Minorities in the Health Professions* [6]. Among the 37 recommendations was the formation of educational and research collaborations between AHCs and institutions of higher learning with a significant number of students from racial and ethnic minorities background.

In January 2005, The Sullivan Alliance was formed to support, monitor, and implement these recommendations through the formation of state alliances—including coalitions in Virginia, Nebraska, Florida, North Carolina, Maryland, Ohio, and Alabama. Crucial elements for a successful state alliance have been proven to be (a) leadership within the state, (b) local financial support, and (c) on-going interinstitutional communication.

It is clear that there is no easy answer to addressing the diversity of our health profession workforce and the health disparities of minority communities. This chapter provides a critical framework for this work, successful model programs, and approaches that can enhance direct care to diverse communities. No chapter can do justice to the innovative and creative interventions that AHCs are implementing. It is hoped that this overview provides encouragement, guidance, and strength to initiate and perpetuate successful work in this arena.

BEYOND DIVERSITY TO INCLUSION

The traditional view of the United States is as a nation fueled by immigrants, a "melting pot" that aspired to assimilate diverse populations into one harmonious culture. More recently, as this utopian vision of assimilation and loss of cultural identity was dismissed not only as unrealistic but undesirable, other metaphors have evolved. The "salad bowl" concept suggests distinct cultures that retain their unique qualities while joining together—where the whole is greater than the sum of its parts. Likewise, the "mosaic" or the "tapestry" views suggest distinct images melding to create a cohesive work of art.

Today, the term "inclusion" has emerged as an even more powerful metaphor. Diversity has been likened to inviting people of different cultures and backgrounds over to your home for a party. Inclusion is the deliberate act of making them feel happy that they came. This distinction between diversity and inclusion provides an insight into both the success and failures of efforts to diversify our health care workforce. There has never been more attention given to

The Transformation of Academic Health Centers. http://dx.doi.org/10.1016/B978-0-12-800762-4.00008-6

inviting students from diverse backgrounds to embark upon a health professional career, and these pipeline efforts are having a positive effect on encouraging students to pursue the health professions [3]. Nonetheless, the consensus is that this work has been inadequate in sustaining and nurturing these students. An alarmingly high percentage of these students fail to graduate, and despite three decades of effort there were 100 fewer African-American males enrolled in medical school in 2013 than in 1980 [7].

Inclusion goes beyond the initial recruitment efforts and scholarships to assuring an affirming atmosphere for students, embracing and celebrating their differences, and providing support to overcome barriers to success. In other words, both diversity and inclusion are required. The house party metaphor offers an insight into this distinction. It is one thing to invite people to the party, open the door, and then have them find their own way; it is quite another to be greeted warmly, welcomed inside, and shown around. Inclusion in health care workforce development means that the host institution is equipped to affirm the new student immediately as someone who belongs. This is especially important for students who enter an environment where they are a distinctive minority or where conscious and unconscious biases exist. Peer mentors, affinity groups, and intrusive advising all contribute to this aspect of inclusion, which is counter to higher education's traditional "sink or swim" culture.

A second component of inclusion mandates that the host provide ongoing attention to the guest's needs. The warm greeting quickly dissipates if it is followed by indifference. First-year programs are only good for the first year, while inclusion must be managed and sustained. There is no panacea, but rather a longitudinal and developmentally appropriate cycle of services, including role modeling, advising, and mentoring.

Finally, inclusion requires that the guest not just be served, but eventually is empowered to give back. It is the transition that happens when the visitor feels comfortable enough to take the initiative to wash dishes after dinner, advancing from acquaintance to friend. The evidence of inclusion, ultimately, is not so much being a recipient, but being a contributor, using one's gifts and talents to make the institution better for everyone. Inclusion creates space for leadership within the institution. Thus, just opening the door to diverse students is necessary—but insufficient. Diversifying the health care professions requires acts of inclusion.

THE VALUE OF DIVERSITY

If man is to survive, he will have learned to take a delight in the essential differences between men and between cultures. He will learn that differences in ideas and attitudes are a delight, part of life's exciting variety, not something to fear.

—Gene Roddenberry

The face of America is changing rapidly (see the Introduction), and the value of this diversity may be found in many forms: enhanced creativity and innovation, a richness of culture and tradition, the unleashing of human potential, and community economic success. Embracing difference strengthens society by providing innovative approaches, ideas, and solutions to complex problems and challenges. A diverse health care workforce contributes to enhancing the care delivery to, and health outcomes in, all members of the community.

One lens to understand diversity and its impact on health care is to consider the meaning of culture: the beliefs, values, behaviors, language, and customs that are common to and followed by people of a particular group, race, or country. A person's culture—a vital aspect of an individual's identity—influences health illness beliefs and health behavior ranging from nutritional habits to sexual practices. When health care practitioners are ignorant of a patient's cultural background, it becomes challenging to successfully engage the patient in their health care and compromises the development of a trusting patient–practitioner relationship. These challenges make it critical for health care to be provided by a qualified, culturally competent workforce that is reflective of the population it serves.

Increasing diversity has profound implications for enhancing the cultural competency of the health care system at all levels. The more a health care practitioner understands a patient's culture, the better the health care practitioner will connect and effectively communicate with the patient. A diverse workforce increases the opportunities for patients to encounter providers with shared experience, a common background, and a better understanding of cultural nuances—including cultural beliefs, communication style, diet, familial hierarchy, and past experiences of discrimination that can be critical in ensuring adherence to treatment plans. Moreover, a diverse health care workforce increases health care access by providing patients the option to choose racial concordant physician–patient encounters, which build trust, increase satisfaction, and improve adherence with the health care system [8].

A diverse workforce also has powerful economic benefits. Improved health status of individuals and their communities promotes community prosperity and economic development by enchaining employment through increasing employee well being. Moreover, health care professionals from the community are more likely to understand how to build on local assets and develop sustainable economic growth.

Diversity 3.0 and Beyond: A Historical Perspective

The twenty-first century is a defining moment for the US health care system and for health professional education. Medical education admission has been influenced by a race and class system where only the majority class had access.

Before the late 1960s, underrepresented minorities (URMs) made up only 2% of medical school students (CHBAS) and were educated predominately at historical black colleges and universities. This exclusionary practice continued until the civil rights era when the race and class system began to unravel.

Federal actions—including *Brown v. Board of Education*, the Civil Rights Act of 1964, and the Elementary and Secondary Education Act of 1965—prompted the enrollment of minority students. In the early 1970s, many medical schools established minority affairs offices, and the enrollment of underrepresented minority students in medicine (URMs) increased to 8%. But by the middle of this decade, progress stalled. Not until renewed efforts in 1990, with the Association of American Medical Colleges (AAMC) Project 3000 by 2000, did URM admissions regain momentum. Enrollment peaked at of 2340 (8%) for URM students in 1996 (AAMC Data Book 21), and declined to 1922 students nationwide in 2001. While programs promoted students' academic success and inclusion, progress was hindered again by inadequate mentoring and support services, underperforming primary and secondary school systems, and uncoordinated pipeline efforts. Since 2001, enrollment of URMs has remained stagnant at around 8%, while enrollment of African-American males has fallen to the lowest level since the 1970s [7].

Today's AHCs offer curricula focused on cultural competence, public health, health care disparities, and health care equity while promoting student and faculty diversity and inclusion. As a result, a deeper, more nuanced view of diversity's role in the health professions has emerged. In this paradigm, the value of diversity has become redefined. AHCs now aspire to leverage the rich spectrum of human talents necessary to effectively serve a diverse population. A new understanding of the policies required to recruit, retain, and support the success of URMs in the health professions has emerged.

Affirmative Action and the Evolving Legal Environment

While many AHCs have prioritized student diversity, they face the vexing legal challenge of implementing race-conscious admissions programs that will pass legal muster. The US Supreme Court first addressed this issue in the 1978 case *Regents of the University of California v. Bakke* [9]. Bakke challenged the admission program of the Medical School of the University of California at Davis, which was designed to ensure the admission of a specified number of students from certain minority groups. Bakke argued that the program violated his constitutional rights under the Equal Protection Clause because he was denied admission on the basis of race. The Court applied the most rigorous standard of review utilized for constitutional challenges and conducted a "strict scrutiny" review of the University's admissions program.

The *Bakke* Court was deeply divided; the nine justices issued a total of six opinions. A majority of the Court agreed that racial quotas were impermissible; four justices agreed that they violated the Civil Rights Act of 1964, while Justice Powell held that they violated the Equal Protection Clause. Five justices did agree that the consideration of an applicant's race was acceptable as one element to be considered in the medical school's criteria for admission. Justice Powell, writing for the Court, acknowledged that, while attempts to remedy historical patterns of discrimination in a particular profession, or in society in general, could not qualify as a compelling governmental interest, a university's attempt to attain a diverse study body could. Given the fractured opinion of the court, there was ambiguity on how to legally achieve a diverse study body when educating health care professionals.

In 2003, the high court faced its next challenges to race-conscious admission programs in two University of Michigan cases, *Grutter v. Bollinger* [10] and *Gratz v. Bollinger* [11]. The court again applied the strict scrutiny standard to the University of Michigan's Law School admission program (in *Grutter*) and its admission program for the College of Literature, Science, and the Arts (in *Gratz*). Under the law school's program, the school sought to enroll a "critical mass" of minority students by accepting students "who without this commitment might not be represented in (their) student body in meaningful numbers." The Court's *Grutter* decision concluded that postsecondary institutions have a compelling interest in the benefits that flow from achieving a diverse student body. The court noted that it was equally important to ensure that enough members of URMs were in the classroom in order to promote crossracial understanding, break down stereotypes, and encourage students to participate in a meaningful way without feeling isolated or as if they were acting as spokespersons for their race.

The law school had narrowly tailored its program after considering and rejecting other race neutral alternatives that it felt would sacrifice diversity and negatively affect academic quality. The Court embraced the *Bakke* concept that race and ethnicity were only part of an individualistic and holistic review of each applicant. Recognizing that the benefits of a diverse student body are "substantial as well as important and laudable," the Court focused on the highly individualized holistic review, where race was not the only possible basis for a diversity admission. It was, as in *Bakke*, a "plus factor" in the overall consideration of the applicant's credentials. The decision also introduced the concept that race conscious admission programs should be time limited and subject to periodic review. Unlike the *Grutter* case, in *Gratz*, the Court reviewed and rejected the "point system" in place for applicants to the College of Literature, Science, and the Arts, whereby every URM was automatically awarded 20 points out of 100. The court held that the "point system" was not sufficiently flexible because the factor of race was decisive in the admissions decision and there was no individualized review.

The issue of race conscious admissions came before the Supreme Court again in 2013, in *Fisher v. University of Texas at Austin* [12]. Here, an applicant challenged the university's admission program arguing that its continued use of race as a factor in admissions was discriminatory and that the admission program had achieved its goal of accepting a "critical mass" of racially diverse students through its "top 10%" program already in place. The *Fisher* Court applied a strict scrutiny review, but noted that the district court had incorrectly deferred to the University of Texas' judgment as to whether the program was narrowly tailored to achieve the interest of diversity. The Court ruled that this was an analysis and judgment that is fundamentally a part of the district court's responsibility. Accordingly, the Supreme Court remanded the case to the district court for further consideration of that critical issue.

After the *Fisher* decision, the US Department of Justice and the US Office of Civil Rights published a set of questions and answers about the *Fisher* case as a supplement to their existing guidance document on this topic (the guidance was established in 2008 and later revised in 2011) [13]. The 2011 guidance is broken into four sections that describe a legal framework for considering race as a compelling interest in achieving diversity. It sets forth considerations for postsecondary institutions that voluntarily seek to use race to achieve diversity in their admissions program. It summarizes key steps for institutions seeking to achieve diversity, and provides acceptable examples for ways to achieve this compelling interest. In essence, the guidance suggests that institutions document the institution's rationale for attaining a diverse student body; evaluate race-neutral approaches to achieving diversity within the student body; conduct a periodic review of the program to determine the continuing need to evaluate race as a component of the admission program; determine the extent to which race is used to assure that it is used to the least extent possible to affect admission; and maintain documentation that describes the compelling interest, the process followed to arrive at decisions, the contours of the admissions program, and ways in which the chosen approach does, in fact, achieve diversity.

In 2014, the Supreme Court issued a 6-2 ruling in *Schuette v Coalition to Defend Affirmative Action* that upheld a Michigan state constitutional amendment that bans the use of race and ethnicity in admissions to the state's public universities. The decision effectively endorsed similar voter-backed initiatives in Arizona, California, Nebraska, Oklahoma, and Washington. The majority left intact the 2013 *Fisher v. University Texas* decision [14]. Writing for the controlling opinion, Justice Anthony M. Kennedy stated, "This case is not about how the debate about racial preferences should be resolved. It is about who may resolve

it." In her dissent, Justice Sonia Sotomayor evoked Chief Justice Robert's famous line, "The way to stop discrimination on the basis of race is to stop discriminating on the basis of race" with a twist: "The way to stop discrimination on the basis of race is to speak openly and candidly on the subject of race, and to apply the Constitution with eyes open to the unfortunate effects of centuries of racial discrimination." Justice Stephen G. Breyer, in voting with the majority, opined, "The Constitution foresees the ballot box, not the courts, as the normal instrument for resolving differences and debates about the merits of these programs." Clearly, this decision reflected a deeply divided court.

This brief summary of the historical legal landscape of achieving diversity in higher education reflects the evolving, sometimes contentious, and often complex nature of institutional measures that support diversity and inclusion in the admission process.

DEVELOPING A CONTEMPORARY CONCEPTUAL FRAMEWORK

Despite diversity efforts in the health professions for >50 years, disparities in educational attainment, career development, and health persist. Two early comprehensive programs stand as seminal achievements in the field of health professions diversity programs. The Dental Pipeline Program in dentistry [15] and the Health Professions Partnership Initiative in medicine [16] have demonstrated appreciable progress in diversifying the health professions. These programs promoted several core components:

- Specialized K-12 and high-school community-based programs, and baccalaureate prehealth profession programs
- Holistic recruitment and admission processes
- Longitudinal curriculum support and enrichment during health profession training
- Professional development efforts to diversify faculty and staff
- Advising, mentorship, and role modeling opportunities

Despite the promise of these programs, both national efforts are defunct, emphasizing the importance of long-term institutionalization, sustainability, and political support.

The rise and fall of these efforts also teaches us that supporting diversity efforts must extend beyond arguments about removal of barriers and improving access or benefits to the training and practice environment. The argument must also include diversity as a key aspect of how excellence is defined within organizations and systems [17]. There is now an emerging wave of efforts that address sustainability by operating on the principles of an asset-based paradigm. Here, local communities, their resources, and their challenges are placed at the center of the initiative. Both existing and new community resources are utilized for capacity building to create naturally sustainable programs. This change is critical as we shift the perspective and approach from an academic focus of "mine or yours" to a community focus of "all of us."

CASE STUDY: THE NEOMED–CLEVELAND STATE UNIVERSITY PARTNERSHIP FOR URBAN HEALTH

Northeast Ohio Medical University (NEOMED) and Cleveland State University (CSU) established the Partnership for Urban Health (the Partnership) to: (1) increase the number of students underrepresented in medicine matriculating into medical school; (2) increase the number of medical students opting for practice in urban primary care; and (3) improve access, decrease disparities in health, and improve the economic vitality of underserved communities.

The NEOMED and CSU were presented with a unique opportunity when they began discussions regarding a potential new partnership. NEOMED was founded as a community-based medical school to develop a primary care workforce for Northeast Ohio. At the same time, CSU, located in the heart of one of the most impoverished cities in the country, has the mission of offering equal access to high-quality education and focusing on the needs of a city in the process of reinventing itself.

It quickly became apparent that these missions were consistent with national concerns about health and health care disparities. Among the identified causes of health disparities were a lack of access to quality care for disadvantaged communities, a limited number of providers from these communities, a paucity of health promotion and prevention services, and a lack of opportunity for residents to obtain the education that would lift them out of poverty and encourage service to their communities. Important factors in implementing the partnership were support for the expansion of NEOMED's class size, local philanthropic support, and developing a community advisory board of distinguished and connected members of the community.

The Partnership is supported by four pillars of an Education for Service Initiative reflective of the critical components of successful intervention strategies: pipeline initiatives, innovative instructional design, community partnerships, and financial incentives.

Pipeline Initiatives emphasize a pathway of longitudinal, articulated, and developmentally appropriate activities running from middle school to health practice. One approach, "The Health Professions Affinity Community" (HPAC), builds on the context of engaging, educating, and partnering with communities to address health and educational disparities by making students meaningful contributors and architects of their future direction. A second approach is the establishment of an STEM + M(edicine) high school, namely, the Bio-Med Science Academy, on the NEOMED campus, as

well as partnering with existing STEM institutions—such as John Hay Science Academy and MC^2 in Cleveland. These pathways include (but are not limited to) a substantial effort early in students' education to provide academic support, encouragement, enrichment, and career identity information that nurtures students through the transition to university and early college years [18].

Innovative instructional design links either a baccalaureate or postbaccalaureate education at CSU with a medical education at NEOMED that emphasizes urban primary care and interprofessional practice. Hallmarks of this component include: (1) holistic review process that selects for academic achievement, personal and interpersonal proclivity for caring for the underserved, and a clear commitment to urban primary care; (2) academic and social support throughout the curriculum accompanied by a carefully crafted set of opportunities to explore urban communities and urban primary care; (3) mentoring and advising to support the process of higher education and transition to medical school; (4) a focus on interprofessional education engaging students in pharmacy (PharmD), occupational therapy, nursing, physician assistant, social work, and speech and hearing; (5) guidance in the residency selection process and job selection; and (6) financial support throughout the educational process. This model repositions diversity and inclusion as an aspect of excellence that supports service to a diversified patient population [17].

Community partnerships focus on recruitment, education, and practice embedded in medically underserved communities with the goal of establishing a diverse resource base to sustain the program. The model asserts that communities must be involved in the educational endeavor, and students must have opportunities to build close relationships that will support the community, sustain their own interest, and enhance their skills and connection to the community.

The final pillar is financial incentives in the form of scholarships, work study, and fellowships that reduce, if not eliminate, the costs of higher education in exchange for student service in their local communities. These instruments are particularly designed to reduce financial barriers for students from economically underserved and underrepresented backgrounds in an effort to diversify the health professions and to meet the needs of the Affordable Care Act—which promises to enroll a disproportionate share of persons from medically underserved communities.

Even with such sustained and comprehensive efforts, the pipeline from middle school to practice is long. It is important to frame success not just as producing primary care physicians who practice in underserved communities, but as a resource that enhances job skills beyond the health professions, contributes to economic vitality, and sustainably links the university, community, and local and state governments to solve challenging issues of poverty and disparities.

Early results from the Partnership have been encouraging, including successful recruitment of minority students, mentors, and educational partners; substantial support from local foundations and health systems; and promising interprofessional educational offerings. Critical learnings from this experience include the importance of academic and service support, the importance of mentoring and peer support, and the challenges of negotiating realistic community expectations.

CASE STUDY: THE UNIVERSITY OF ILLINOIS AT CHICAGO, COLLEGE OF PHARMACY: URBAN PHARMACY HIGH SCHOOL PIPELINE PROGRAM

Most colleges of pharmacy, except those that are minority-serving colleges, still grapple with the challenge of how to increase the number of URM students in their respective colleges. Of the 99,821 applicants to the first professional Doctor of Pharmacy (PharmD.) program by US colleges of pharmacy in 2011–2012, only 15.7% were from URMs, of which African Americans/blacks accounted for 10.2%, Hispanics/Latino 5.3%, and American Indian/Native Americans 0.2% [19]. The total number of first professional PharmD degrees conferred in 2011–2012 was 12,719. Of this number, White Americans received 58.1%, African Americans/blacks 6.5%, Hispanics/Latino 4.2%, and American Indians/Native Americans 0.5% [20]. These low percentages are equally reflected in URM faculty in pharmacy schools [19].

The demographic shifts of URMs becoming the majority population in the United States will result in increased racial and linguistic diversity, presenting a complex health care patient population with different cultural values and beliefs. A 2000 report by the Department of Health and Human Services stated that, while there may be a shortage of pharmacists in general, the shortage was more severe for URMs and concluded that, without fundamental changes in pharmacy practice and education, the shortage of URMs will persist [21]. Fourteen years after the report, increasing the number of African Americans/blacks, Hispanics/Latino, and Native Americans students and pharmacists is still a challenge.

Despite the University of Illinois Chicago College of Pharmacy (UIC-COP) location in the southwest side of Chicago—a heavily African-American and Hispanic community—the number of URM's enrolled remained low. In 2005, UIC-COP developed and implemented the UIC-COP Urban Pipeline Program (UPP), targeting Chicago Public School (CPS) students in collaboration with CVS/CareMark, the Chicago Board of Education's Office of College and Career Preparation, and the Coalition for United Community Labor Force, a community grassroots agency.

The theoretical and conceptual framework used for the UPP is the socialization framework, defined by Merton,

Reader, and Kendall as "the processes through which a person developed a sense of professional self, with characteristic values, attitudes, knowledge, and skills...which governed his/her behavior in a wide variety of professional and extraprofessional situations." [22] This need-based, sociopsychological model of career choice served to deepen the understanding of the complexities involved in the career decision processes of URM high school students and provided another framework for the program structure (e.g., work motivation and expectations) and cultural–environmental factors (e.g., socialization and structure of opportunity), which were incorporated into expectations about work, career choice, and work behavior [23]. Embedded in the UPP structure were the Pathways to College Network concept of high expectations—which incorporates inclusive leadership, collaborative partnerships, flexible resources, and professional development. Based on this, a comprehensive theoretical and conceptual framework Model for Success in Pathways to Pharmacy High School Program was developed [24]. The uniqueness of this model from all the other pathways to pharmacy programs is its grounding in an epistemological framework with adaptability and applicability for use in other high school pathways to career programs.

Starting in 2005, officials from the Chicago Board of Education sent invitational letters to all CPS principals and science teachers asking their help in identifying students from their respective schools who met the criteria for participation (≥3.0 grade point average, high aptitude in math and science, 98% school attendance record, and junior standing). All CPS students, regardless of race/ethnicity, are eligible to apply to the program. This UPP was structured as two three-week experiences. The first 3 weeks were spent at the college of pharmacy and the last 3 weeks at a CVS/pharmacy.

During the first 3 weeks at the COP, students were assigned to work with clinical faculty mentors observing firsthand the day-to-day work of pharmacists in an academic institution. To further expose students to clinical pharmacy experience, an "Introduction to Pharmacy Practice" lecture series and hands-on workshops were presented by UIC-COP faculty members and by fourth-year pharmacy students. The curriculum integrated academic and social activities, including opportunities for students' professional and personal development, faculty–student interactions, and mentoring; exposure to basic science faculty members and researchers; activities to foster team building and student collaboration and interactions; communication/public speaking activities; and, required daily reflective writing exercises.

Problem-based learning cases focused on diseases common to minority communities, such as diabetes. Student volunteers from the college of pharmacy tutored program participants in pharmacy technician calculations and customer service etiquette and offered mentoring and advice.

Presentations were made by representatives from the UIC Honors College, Financial Aid Office, Office of Admissions and Records, and other campus resources and support programs helping to promote the successful transition from high school to college. The program was driven by an apprenticeship paradigm with multilevel socialization processes. At the end of the first 3-week session at the UIC-COP, student groups presented their case study capstone projects.

During the second session block, students worked as interns/technicians at a CVS/CareMark pharmacy store for four days a week. Illinois law allows high school students who are 16 or older with a valid pharmacy technician license to work as pharmacy technicians/interns. Mentoring and personal development activities occurred on Fridays.

A comprehensive learner and program assessment was developed. Of the 120 participants in the program from 2005 to 2008, 88 (75%) responded on a posttest that they would pursue a prepharmacy curriculum in college and pursue pharmacy as a career, compared to 40 (33%) students on the pretest. All 120 participants in the UPP enrolled in college after graduating from high school, and 88 (75%) enrolled in the prepharmacy curriculum. Of the 25 students from the 2005 group, 19 applied to pharmacy schools, and the remaining students decided to pursue teacher education in math and science and graduate studies. To date, the UIC-COP has 10 of the UPP students currently in college. From 2005 to 2014, a total of 270 students are in the pharmacy pipeline, of whom 58% are African Americans, 30% Hispanics/Latino, and 12% are either White Americans or Asian Americans. Thus, an intensive pipeline program in partnership with a college of pharmacy, employers, and members of the community can successfully influence the diversity of the future pharmacy workforce.

CASE STUDY: NEW MEXICO COMMUNITY HEALTH WORKERS AND HEALTH EXTENSION REGIONAL OFFICES

The US health care system impacts 10–15% of a community's health, yet community health consumes >90% of the US health care dollar. The social determinants of health, on the other hand—including education attainment, food security, safe housing, social integration, good jobs, and access to transportation—greatly impact health, but receive far less attention and resources from our society than they deserve. The challenges associating with caring for New Mexico's population are a clear reflection of how the determinants impact overall health.

New Mexico is a minority–majority state—45% Hispanic/Latino, 10% Native American, and 2% African American—with a sizeable immigrant population, a mix of rural and frontier counties, and a small number of urban cities. For providers in the New Mexico health care system to address social determinants, these health care workers must be well versed in cultural and linguistic competence. During 2004–2014 to "upstream" the impact of social determinants, the University of New Mexico Office for Community Health has worked with the Health Extension Regional Offices (HEROs) to develop two new members of the health workforce team.

HEROs were modeled after the popular and successful agricultural Cooperative Extension Service [25]. As with the Cooperative Extension, Health Extension agents are community based, community selected, and focus their time on linking community health priorities with university resources in all mission areas—education, clinical service, research, and health policy. They can be considered "outposted" agents of the university—bringing rich resources to a community's front door as selected by communities based on their priority needs. HERO agents are culturally and linguistically competent, know their communities well, and are seen as a trusted source of information. They are ethnically diverse, reflecting the ethnic diversity of the state. Their ultimate goal is to improve community capacity to address local needs.

Community health workers (CHWs) are known as Promotores de Salud on the US–Mexican Border or Navigators, a term used more commonly among those employed by health systems. These individuals—who share the same culture, language, lifestyle, and ideas as the majority of their clients—have a long tradition of service to New Mexico's communities. CHWs are trained in social and health issues affecting their communities and serve as conveyers of basic health information and navigators of social services. Their presence, leadership, and knowledge of people and resources within the community facilitates and strengthens their role in broadening access to and coverage of health services in remote, rural areas and urban settings—especially for marginalized, disadvantaged, or underrepresented populations.

The Office for Community Health applied this rich CHW resource to the emerging Medicaid managed care market. With implementation of the Affordable Care Act and the shift of millions of uninsured to Medicaid, it became critical for insurance companies to manage the cost of the expanding number of enrollees who often experience the highest rates of adverse social determinants.

CARE New Mexico, a program created by the Office for Community Health, mostly employs minority CHWs through funding by the New Mexico Medicaid Managed Care program [26]. The CHWs intervene "upstream" in areas such as health literacy, access to food stamps, transportation, or accompanying enrollees to their doctor's appointments. This program has resulted in significant savings in providing care to high-cost Medicaid enrollees. Other examples of the services provided by HEROs and CHWs include outreach to vulnerable elders to enhance medication adherence, interventions that improve behavioral health delivery, and improved linkages between families and schools. Integrating the work of the CHW with the HEROs has fostered community capacity building, preventive medicine, and value-driven care.

LESSONS FROM THE PRIVATE SECTOR: KAISER PERMANENTE

Kaiser Permanente (KP) grew in the mid-1940s on a foundation of equal access to health care for all populations, beginning with the diverse dockworkers who had been refused treatment through traditional avenues. Since then KP has recruited and retained a diverse workforce that reflects the patient they served. In November 2011, KP recognized that newly-insured members under the Affordable Care Act will come from the growing population of Latinos, who would also bring low health literacy, language, and cultural barriers and would challenge the inclusiveness of institutional processes. An example of a program to enhance health equity in this population is KP's efforts in colorectal cancer screening among Latino patients.

A 2012 Center for Disease Control and Prevention survey found that 66% of non-Latino whites were adequately screened, compared to 53% of Latinos. In early 2012, KP San Francisco's interdisciplinary Spanish bilingual module in the Department of Adult and Family Medicine looked at the processes to improve fecal immunochemical testing (FIT) for its Latino membership. KP's *Diversity and Inclusion* and *Hospitals, Quality and Care Delivery* functions partnered with unit-based teams (UBTs) of Spanish bilingual physicians, nurses, and medical assistants (MAs). Goals were developed collaboratively among this group and with patients.

UBTs modified existing instructions for appropriate health literacy levels, an "FIT Room" was created so that MAs could show patients the correct FIT process with a model toilet, and "how-to" videos were created in Spanish and English featuring bilingual physicians and a powerful patient testimonial. MAs kept track of every kit that was sent out and personally followed up with a phone call. Some patients were hesitant to send a stool sample through mail, so the charge nurse was available on site for patients to drop off samples. While originally targeting Latino patients, substantial increases in screening were also achieved with all patient groups.

In the first quarter of 2012, the colorectal cancer-screening rate among Latino members was 71% programwide at KP. By the first quarter of 2013, it increased to 75%.

Hence, the Spanish Language Module model of care provides a patient-centered medical home for Latinos and is one element of a comprehensive language access strategy. It provides an ideal environment for creation and implementation of performance improvement projects that are applicable to other populations and can be replicated at other sites. Alignment of goals promotes collaboration among bilingual team members and diverse parts of the organization. Utilizing culturally and linguistically responsive multidisciplinary teams of Spanish bilingual health professionals can achieve delivery of excellent service as well as quality outcomes—and thus decrease health disparities and long-term cost expenditures. Such interventions suggest the necessity of a diverse heath professions work force when implementing options for innovative population-based care in AHCs.

CASE STUDY: THE MINORITY DENTAL FACULTY DEVELOPMENT PROGRAM (MDFDP) INITIATIVE

Academic health systems are frequently partners with specialty societies and health professional organizations. In this example, the impact on diversity of one such association—the American Dental Education Association (ADEA)—is explored. Woven into the mission of ADEA is a commitment to address the dental faculty shortage and the shortage of URMs in US dental schools. A special task force in August 1999 issued the report "Future of Dental School Faculty." The task force concluded that "dental education is now in a crisis" due to the shortage of dental faculty. URMs comprise 13% of the total full-time dental school faculty, 12% of postdoctoral trainees, and 13% of the entering dental school class. The lack of diversity in the dental academic pipeline is being addressed by ADEA/W.K. Kellogg Foundation Minority Dental Faculty Development grants [27].

Ten US dental schools and one advanced dental education institution were awarded grants over a six-year time period, 2004–2010. A key strategy of the grantees was a "grow your own" philosophy for the recruitment and advancement of a diverse faculty. Program objectives included: establishing academic and community partnerships, mentoring, faculty development, and pipeline activities designed to increase the number of URMs interested in dental careers. The grants were also used to promote health system changes that focused on primary care and public health concepts through service experiential learning and patient treatment in community outreach facilities. Examples of these activities include

- Formal mentoring programs developed for undergraduate and postgraduate dental students and mentoring strategies expanded to include informal peer-to-peer mentoring using online technology;
- Community partnerships that involve a broad range of clinical services and settings including school clinics, mobile units, nursing homes, Indian Health Service clinics, migrant workers, and a health disparities center;
- Portfolios (narrative descriptive logs) as a useful tool for: mentoring, career guidance, tracking, decision making, and evaluation;

- Acknowledging the importance of leadership styles and effectiveness—obtained through surveys of deans and trainees, institutional site visits, and interviews with the higher education leadership—to the institutional climate and to diversity in its three dimensions: compositional (demographic), curricular, and institutional; and
- Recruitment that included partnerships with local schools (K-12) for student recruitment; mentoring; dental treatment; career days; visits to the dental schools for hands-on experiences; as well as health fairs with collaborating community partners, teacher, parents, and local dentists [28].

Lessons Learned from the MDFD Program

1. The Dean's leadership is critical to success.
2. A clearly articulated diversity policy statement is a major driver for resource support.
3. Diversity is not simply numbers (compositional), but reflects instructional design, institutional culture, and other elements of inclusion.
4. Mentoring programs are needed for predoctoral and advanced dental education trainees and for faculty.
5. Academic climate change requires supportive resources and opportunities for social and inclusive interactions.
6. A diverse executive leadership pipeline is missing.
7. Cultural competency concepts require the inclusion of social determinants.
8. Logic modeling, GAP analysis, and other tools of strategic planning are critical to long-term success.
9. Innovative ideas should be considered in interprofessional education, collaboration, resource sharing, and leadership development opportunities.
10. The value of short-term strategies to achieve long-term goals is being realized at MDFD grantee institutions.

These lessons learned are being incorporated into the latest MDFD grants focusing on allied dental leadership and faculty development, academic/community partnerships, and team-based care [29]. The new dental school accreditation Standard 1–4 (adopted August 2010) states that dental schools must have policies and practices to achieve diversity among its students, faculty, and staff [30]. This Standard is expected to serve as a stimulus for diversity and inclusion in all dimensions of dental academic programming. ADEA sponsors faculty and leadership development through a variety of experiential programs, especially those at the ADEA Leadership Institute.[1]

ADEA's dental student recruitment fairs and workshops are complemented by free and accessible online resources: ADEA GoDental[2] and ADEA Explore Health Careers.[3]

---

[1] http://www.adea.org/LeadershipInstitute/
[2] http://www.adea.org/GoDental/Pages/default.aspx
[3] www.explorehealthcareers.org.

LESSONS LEARNED AND OPPORTUNITIES FOR THE FUTURE

Diversifying the health care profession workforce is essential to the health of the American public. Leadership at AHCs recognize that diversity is not only providing access or promoting policies of inclusion but also promoting sustainable partnerships with the community. Organizations, such as The Sullivan Alliance, demonstrate the power of coalitions to develop, replicate, and adopt innovations nationally. Sustainable efforts in Northeast Ohio and Chicago demonstrate the power of comprehensive, evidence-based programs with strong underlying models of change. Meanwhile, Kaiser is bringing its tradition of innovation to develop culturally competent programs to its diverse communities of patients. In New Mexico, health extension regional offices and community health workers are melding efforts in community health and medicine. Finally, the ADEA shows how a national association uses best practices to enhance workforce diversity.

These innovative programs exemplify the understanding of the "multiplier effect" of diversity in health care. By crafting a pathway to success paved with support from minority communities, strong mentorship, and solid financial support, AHCs are playing an important, facilitative role. No longer trying to be the sole actor in this complex process, AHCs can lead in accessing financial support for students while convening families, neighborhoods, business and faith-based organizations, and government.

At a time of radical change in health care—with the implementation of the Affordable Care Act, development of accountable care organizations, the move from inpatient- to outpatient-centered care, and the growing importance of interprofessional health care teams—AHCs can lead the country in diversifying the health professions. Further, there is a growing opportunity to meld community-building efforts, public health and wellness initiatives that marry social determinants of health, cultural aspects of care, community empowerment, and healthy lifestyle initiatives.

Comprising both educators and health care providers, AHCs can move beyond their traditional role in leading quaternary care to fostering community economic prosperity and health equity. Such efforts must begin at the undergraduate level and in health professions education, where interprofessional teamwork, service learning, and community-based education become the foundations for training. AHCs are in a unique position to create fluid, developmentally appropriate paths from the community to practice that take advantage of shared curricula and community strengths, offer opportunities to deliver education more efficiently, and promote the elimination of health disparities. This linkage between the twenty-first century AHC and their stakeholders—especially their local neighborhoods—suggests new priorities for the future. Community advisory councils, community health workers, and health extension services take on equal importance with transplant and cardiovascular centers. Most importantly, AHCs transform into champions of health workforce diversity, eliminating health disparities, and enriching community economic development.

ACKNOWLEDGMENT

The authors wish to thank the following contributors to this chapter, who substantively enriched its content and provided their invaluable expertise.

Clara Awe, PhD, EdD, University of Illinois, Chicago;
Bette Bonder, PhD, Cleveland State University;
Andre Burton, JD, Northeast Ohio Medical University;
Ronald L Copeland, MD, Kaiser Permanente;
Arthur Kaufman, MD, University of New Mexico;
Eric Porfeli, PhD, Northeast Ohio Medical University;
Franciso J. Ronquillo, PA, University of New Mexico;
Maria Schimer, JD, Northeast Ohio Medical University;
Richard W. Valachovic, DDS, MPH, American Dental Education Association;
Gina Weisblatt, PhD, Northeast Ohio Medical University;
Joseph F. West, ScD, Sinai Urban Health Institute;
Byron P. White, EdD Cleveland State University.

REFERENCES

[1] Sullivan LW, Mittman IA. The state of diversity in the health professions a century after Flexner. Acad Med 2010;85:246–53.

[2] U.S. Department of Commerce. Economics and Statistics Administration, US Census Bureau. 2010. In: Population of the United States, by race and ethnicity.

[3] U.S. Department of Health and Human Services. Pipeline programs to improve racial and ethnic diversity in the health professions: an inventory of federal programs. 2009. In: Assessment of evaluation approaches, and critical review of the research literature.

[4] Smedley BD, Stith AY, Nelson AR. Unequal treatment: confronting racial and ethnic disparities in health care. 2006. In: Institute of Medicine of the National Academies.

[5] Mitchell DA, Lassiter SL. Addressing health care disparities and increasing workforce diversity: the next step for dental, medical, and public health professions. Am J Public Health 2006;96(12):2093–7.

[6] Sullivan Commission on Diversity in the Health Workforce. September 2004. Missing persons: minorities in the health professions. Washington, DC.

[7] Diverse Issues in Higher Education. Hefty debt dissuades blacks from attending med school. November 4, 2013. http://diverseeducation.com/article/57246/ [last accessed 30.10.14].

[8] LaViest TA, Carroll T. Is doctor–patient race concordance associated with greater satisfaction with care? J Health Soc Behav 2002;43(3):296–306.

[9] 438 US 265. The regents of the university of California v. Bakke. 1978.

[10] 539 US 306, 321. Grutter v. Bollinger. 2003.

[11] 539 US 244. Gratz v. Bollinger. 2003.

[12] et al, 539 US. Fisher v. University of Texas. 2013.

[13] http://www.ed.gov/about/offices/list/ocr/doccs/dcl-qa-201309.html. Last viewed 2.20.14.

[14] Liptak A. Court backs Michigan on affirmative action. N. Y Times. April 22, 2014.

[15] Bailit HL, Formicola AJ, D'Abreu KC, Bau I, Zamora G, Stavisky JS. The dental pipeline program: the national program office perspective. J Dent Educ 2009;73(2 Suppl.):S15–22.

[16] Cavazos LP. The health professions partnership initiative and minority education: lessons learned from the health professions partnership initiative (HPPI), 1996–2005. Acad Med 2006;81(6):S58–9.

[17] Nivet MA. Commentary: diversity 3.0: a necessary systems upgrade. Acad Med 2011;86(12):1487–9.

[18] Porfeli EJ, Lee B, Vondracek FW, Weigold IK. A multi-dimensional measure of vocational identity status. J Adolesc 2011;34(5):853–71.

[19] AACP Institutional Research Report Series. Profile of pharmacy faculty, 2011. Alexandria, VA: American Colleges of Pharmacy; 2011.

[20] AACP Institutional Research Report Series. Profile of pharmacy students, fall 2012. Alexandria, VA: American Association of Colleges of Pharmacy; 2012. http://www.aacp.org/resources/research/institutionalresearch/Pages/StudentApplications,Enrollmentsand DegreesConferred.aspx [accessed 20.02.14].

[21] Department of Health and Human Services. The adequacy of pharmacist Supply: 2004–2030. Department of Health and Human Services Health Resources and Services Administration, Bureau of Health Professions; December 2008. Report ftp://ftp.hrsa.gov/bhpr/workforce/pharmacy.pdf [accessed 20.02.14].

[22] Merton RK, Reader GR, Kendall PL. The student physician. Cambridge: Harvard University; 1957.

[23] Astin AW. Student involvement: a developmental theory for higher education. J Coll Stud Pers 1984;40(5):518–29.

[24] Awé C, Bauman J. The theoretical and conceptual framework model for success in pathways to pharmacy high school program. Am J Pharm Educ 2010;74(8):149.

[25] Kaufman A, Powell W, Alfero C, Pacheco M, Silverblatt H, Anastasoff J, et al. Health extension in New Mexico: an academic health Center and the social determinants of disease. Ann Fam Med 2010;8(1):73–81.

[26] Johnson D, Saavedra P, Sun E, Stageman A, Grovet D, Alfero C, et al. Community health workers and medicaid managed care in New Mexico. J Comm Health September 28, 2011 (published online).

[27] Sinkford JC, West JF, Weaver RG, Valachovic RW. Modeling mentoring: early lessons from the W.K. Kellogg/ADEA minority dental faculty development program. J Dent Educ 2009;73(6):753–63.

[28] Brunson WD, Jackson DL, Sinkford JC, Valachovic RW. Components of effective outreach and recruitment programs for underrepresented minority and low-income dental students. J Dent Educ 2010;74 (10 Supp.)S74–86.

[29] Sinkford JC, Valachovic R. Growing our own. The ADEA minority dental faculty development program. A manual for institutional leadership in diversity. In: Project support from the W.K. Kellogg Foundation. Washington, DC: American Dental Education Association; 2011.

[30] Commission on Dental Accreditation. Accreditation standards for dental education programs. 2014. www.ada.org/sections/educationAndCareers/pdfs/predoc.pdf [accessed 10.03.14].

ABOUT THE AUTHORS

Ron Berkman, PhD is the President of Cleveland State University (CSU), a public research institution in Ohio, where he has served since 2009. Prior to his presidency at CSU, Dr Berkman was Provost, Executive Vice President, and Chief Operating Officer at Florida International University in Miami.

Jay Gershen, DDS, PhD became the sixth president of the Northeast Ohio Medical University (NEOMED) in January 2010. His prior positions include Executive Vice Chancellor at the University of Colorado Health Sciences Center, where he played a leadership role to create the University of Colorado's twenty-first century health sciences center at the former Fitzsimons Medical Army Garrison.

Sonja Haywood, MD is the Director of the Northeast Ohio Medical University (NEOMED)–Cleveland State University (CSU) Partnership for Urban Health, created to yield primary care physicians specifically trained to address underserved urban communities. Prior to her position at NEOMED, she was at Case Western Reserve University and University Hospitals Case Medical Center in Cleveland as an assistant professor of family medicine, as well as a medical staff physician and preceptor.

Jeanne C. Sinkford, DDS, PhD is the Senior Scholar-in-Residence in ADEA's Office of the President and CEO. Dr Sinkford previously served as Director, then as Associate Executive Director, of the Association's former Center for Equity and Diversity. Currently Professor and Dean Emeritus of Howard University College of Dentistry, Dr. Sinkford served as the school's Dean from 1975 to 1991.

Louis Sullivan, MD is the Chair of the board of the National Health Museum in Atlanta, Georgia, and is the Chairman of the Washington, D.C.-based Sullivan Alliance to Transform America's Health Professions. He served as the Chair of the President's Commission on Historically Black Colleges and Universities from 2002–2009, and was co-Chair of the President's Commission on HIV and AIDS from 2001–2006. He also served as Secretary of U.S. Department of Health and Human Services (HHS) from 1989 to 1993.

Jeffrey L. Susman, MD is the Dean of the Northeast Ohio Medical University College of Medicine and professor of family medicine. Dr Susman currently serves on the National Board of Medical Examiners, Computer-based Case Simulation Committee, the National Quality Forum Standing Committee on Behavioral Health, and the Boards of the Aultman College of Health Professions, Menorah Park, William Goldman Foundation, and Canton Medical Education Foundation.

Chapter 9

The Growing Integration of Health Professions Education

Frank B. Cerra and Barbara F. Brandt

THE CHALLENGE

Mrs. Anderson is an elderly female diagnosed with hypertension, obesity, coronary artery disease with mild congestive heart failure, and metastatic breast cancer on chemotherapy. Different doctors and clinics treat her for these conditions. Each doctor prescribes multiple medications; she does not know what the drugs are or what each does, and no one had followed up with Mrs. Anderson at her home. She has side effects from the medications and does not understand whom to call. After multiple discussions with the receptionist, she finally gets through to a doctor who is not sure what all the medications are as she did not prescribe many of them. The medications are also dispensed from multiple pharmacies, none of which has a complete list of medications. When Mrs. Anderson does visit a clinic, tests are performed, many of which are repeats of tests just completed at another clinic. The primary care physician who is trying to coordinate the care with the specialists is not on the same electronic health record system as the specialists and has very delayed communications from them regarding Mrs. Anderson. The visits with the primary care physician do not last a sufficient time for her questions to be answered and she is hurriedly ushered out of the exam room only to sit in the waiting room until the receptionist can hand her a printout summary of her visit and schedule her next appointment. Mrs. Anderson is progressively unable to perform her activities of daily living, which no one asks her about. She also lives alone and transportation is a major challenge for her. She would like to arrange for some of her care in her home, but does not know how to arrange it. She also makes frequent visits to the emergency room as her primary care clinic closes at 5 p.m.

THE CONTEXT

This brief case is illustrative of the challenges facing the current health system. While there is general agreement about redesigning the current health system, there are several factors slowing the pace, one of which is the separation of education and clinical practice that has progressively occurred over the last 20–30 years [1,2]. One consequence of this divergence is that, on graduation from health professional schools, care providers, while competent in their core skills, are not well prepared for working in health systems. For instance, newly graduated doctors are skilled in differential diagnosis and complex medical decision making, but are less skilled in the use of electronic health records, systems of care, informatics, population health, and in the competencies and skills of interprofessional work and collaborative practice, particularly when it involves student learners. A lot of time and resources are used by health systems to retrain the providers who work for, in, or with them.

There are a number of forces in the health care marketplace that are stimulating major changes in both practice and education. Some of these include:

- the transition to a focus on outcomes—such as improving the health of populations and the experience of care for patients, families, and communities and reducing the per capita cost of care while adding value to the care process;
- the movement from nonintegrated to integrated care systems with providers as employees;
- an increase in the number of insured people needing access to care;
- the transition from acute care and disease management to prevention, wellness, and nonhospital care; and
- the growing use of teams.

The health marketplace is testing new models of care while the educational system that produces practitioners for the health marketplace has been slower in its redesign. However, recently, some programs are beginning to address the gap.

One approach to improving the preparedness of providers is to reconnect education and clinical practice so that they can learn from each other what is needed to transform the health system. The most common way to accomplish this

The Transformation of Academic Health Centers. http://dx.doi.org/10.1016/B978-0-12-800762-4.00009-8

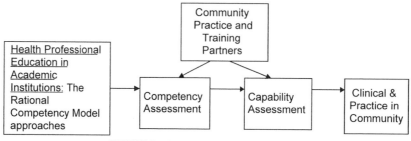

FIGURE 1 The rational competency model.

is to employ interdisciplinary, interprofessional teams [3,4]. This resurgence of interest in team approaches stems from:

- the success of the team approach—generally in the absence of student learners—in quality improvement when applied to patient safety;
- the Patient Protection and Affordable Care Act promotion of team activity;
- the use of collaborative practice and care coordination in primary care;
- new payment systems;
- the interest of accrediting bodies; and
- the success in applying teams in nonhealth areas—such as manufacturing, the airline industry, and athletics.

The application of collaborative practice and the reconnection of education and clinical practice is a concept that has been discussed for a number of years [3,5–7]. The resurgence of interest in the use of this interprofessional education and collaborative practice (IPECP) model stems from the belief that the relinking of education and clinical practice has the potential to achieve improved health outcomes, informing the redesign of the health workforce, and the process of care itself [4,6,7]. And, there is hope that this reconnection will occur in a way that produces progressive and sustainable improvement in the efficiency and effectiveness of both education and clinical practice.

The Rational Competency Model

Conceptually, and as a general principle, the educational system currently uses the Rational Competency Model that is depicted in Figure 1 [1]. This linear model of education relies on the presence of community training sites and community practitioner educators. For example, more than half of the education and training of a physician occurs in the community by community practitioner educators. The model is similar for other provider professions, with little cross-interaction between and among the provider professions and the interprofessional use of experiential training sites. This Rational Competency Model is built on "expertise" and has been codified over the past 40 years by such functions as accreditation, scopes of practice, and board certification. This model has increasingly built the silos that

have made interprofessional education and practice difficult to implement. Interestingly, one area of current interprofessional collaboration that is achieving some success is in improving patient safety and quality. However, this movement in general has not emphasized the inclusion of student learners and is one of incremental change.

Educational performance sites in the practicing community are becoming increasingly difficult for student placement because of the negative impact of their presence in observational roles on the efficiency of clinical practice. The pro bono model of expecting community faculty to volunteer to educate the future workforce is also being increasingly questioned and challenged by health systems and market forces. And, the process of care in the community is yet to embrace the use of the interdisciplinary-team approach with the meaningful inclusion of student learners at the point of care. These logistics have defined a clear need for community–academic interaction and performance feedback from the clinical setting and practicing community faculty.

The Rational Competency Model does, however, provide a platform for educating students in teamwork and team-based care and an opportunity for the academic and practicing community to partner in interprofessional education and practice. Redesigning the education and practice interface has the potential of creating a "nexus" that will meet the demands of this needed change in care delivery, as well as address the challenges of educating and training the next generation of health professionals. Unfortunately, health system transformation is in rapid-cycle change, and the rational competency approach will likely not create the urgency to keep pace with this rapid transformation. Rather, a different approach is needed—referred to as the Nexus and the learning environment it has the potential to create.

The Nexus Learning Model

The transformative redesign of the process of care requires a reconnecting of the education and care delivery systems at the point of care using an interprofessional model. This is defined as a functioning Nexus model and is depicted in Figure 2 [1]. Challenges include the rapid-cycle change in the community, the disarray of health workforce planning, payment systems, and administrative and management

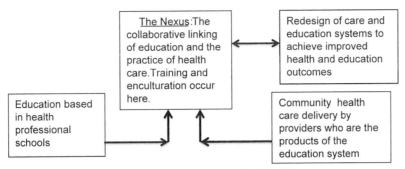

FIGURE 2   The Nexus Learning Model.

approaches to sustainably implement the transformative change. New roles, responsibilities, and scopes of practice are also emerging, initiating a change in the hierarchy of the provider professions.

The Nexus Learning Model, while increasingly operative, is often constrained in its successful implementation due to a number of challenges—such as culture and resources, national policy issues such as graduate medical education funding, competing marketplace forces between community and academic health systems, and general financial and political stress on higher education institutions. However, there is increasing recognition of the need for clinical sciences in the education and training of the next generation of health professionals. The clinical practice, research, scholarship, and professional degrees of academia are essential for producing the health workforce, community outreach and partnerships, financing of the health professional schools education and research programs, and the development of new treatment modalities. Likewise, much of the current process of care needs to be redesigned and current personnel retrained—e.g., interventions and their supporting infrastructure for acute disease requiring invasive or intensive treatments and innovation in the process of care particularly in the outpatient and community settings.

The real need is to form collaborative partnerships of education and practice systems, "the Nexus," in a way that drives:

- Working together to transform education to keep pace with the rapidly transforming processes of care;
- Creating a closed-loop education-practice model for the continuous improvement of the delivery of health care;
- Working collaboratively to achieve the Triple Aim in both health care *and* higher education—cost, quality, and the user experience; and
- Working collaboratively to create a new interface between education and clinical care.

IPECP: AN OVERVIEW OF THE CURRENT STATUS

Most institutions and leaders in the field have adopted the World Health Organization's operational definitions of

what interprofessional education and collaborative practice looks like [8]. Thus:

Interprofessional education "occurs when two or more professions learn about, from, and with each other to enable effective collaboration and (to) improve health outcomes."

Interprofessional (or collaborative) care/practice "occurs when multiple health workers from different professional backgrounds provide comprehensive health services by working with patients, their families, careers (caregivers), and communities to deliver the highest quality of care across settings."

The use of IPECP to improve health outcomes has been an evolving concept. Since the 1970s, there has been consensus that IPECP has the potential to substantively reshape the process of care and health outcomes [3,4,8]. In 2008, the Triple Aim was proposed by Berwick et al. as a way of refocusing the health system on outcomes [9]. The influence in connecting IPECP to improved health outcomes came in 2010 in a World Health Organization report [10] espousing that IPECP [11] should result in more efficient and effective health care services. Hence, the interest in IPECP as a way to achieve the transformative redesign of health care.

This view was also supported by the "borrowing" of successful team models from nonhealth sectors such as industry, the military, and sports. While these applications occurred in vertically integrated decision-making structures, their success has led to a potential transference to health care delivery—a horizontally integrated decision-making model that requires teamwork in a more voluntary way. With the creation of the National Center for Interprofessional Practice and Education [12], the growing acceptance of reconnecting education and clinical practice in new partnerships (nexi) organized to improve both education and clinical outcomes, and the movement of health professional societies and the Interprofessional Education Consortium in this direction with the definition of competencies needed for IPECP [11], the Nexus model is now the center of attention. While definitive information and evidence to make this connection is in development [13], the movement is occurring in early adopter academic–practice partnerships. In addition, the National Center has established a team intervention network that consolidates team intervention data into a National Center Data Repository. This repository is collecting, analyzing, and working to disseminate information and evidence regarding interventions of interdisciplinary teams to improve outcomes. Remaining cognizant of the ecological environment within which these interventions occur, the repository will perform comparative effectiveness analysis as successful models are implemented in different environments.

There is a great demand for the "how do I do it" for IPECP, especially with the presence of student learners. Precisely what the education and training approach should be for student learners and practitioners is, currently and largely, an "I think this works" approach that is fragmented, lacking a coordinated system of communicating results, and in need of tools for assessment of: readiness for interprofessional learning and practice in the presence of student learners; what was learned; team dynamics and interactions; and, ultimately, outcomes achieved [14–17]. Determining exactly who should be on the teams is also in this muddled approach. The major criteria for success mostly remain the outcomes achieved with some assessment of participant satisfaction with the interprofessional experiences—including the patients.

Additional challenges include improving population health and engaging patients, families, and communities [18]. The actuality of these outcomes, how the process of care is to address them, and how those efforts will be supported are in active stages of development with public health agencies and organizations, community organizations, and other essential partners in addressing the social determinants of health, as well as funders interested in supporting these efforts. For instance, there is an active discussion as to what the common knowledge-base of population health should be for all health professionals. The roles of other providers—such as social workers, providers with skills in care of the aging, and community health workers—are currently being defined and developed.

The Interprofessional Education Collaborative (IPEC) has defined competencies for interprofessional work, which have become generally accepted by accrediting bodies and educational institutions [11]. They are: values and ethics, roles and responsibilities, interprofessional communications, and teams and teamwork. As IPECP and care redesign has progressed, other competencies and skills have been further defined and include: population health, informatics, evidence-based patient-centered care, quality improvement technology, an understanding of systems, and cost-effective practice.

While student learners are beginning their interprofessional experiences at clinical sites or in simulated environments, challenges remain—such as coordinated scheduling, integrated curriculum development, faculty education and training, and the presence of clinic and hospital sites that are not welcoming to IPECP, particularly with student learners present.

There are a number of stakeholders involved in the redesign of the health system, e.g., educational institutions, health systems, payers, policy makers, and regulators. These groups are very open and supportive to IPECP and many are experimenting with models of team approaches. However, they are all looking for the evidence of effectiveness in improving outcomes, examples of team models that produce the results and factors that are essential for sustainable change, and a potential for return on investment when implementing IPECP. Each of the stakeholders also face unique challenges for the cultural change to occur.

INTERPROFESSIONAL EDUCATION AND COLLABORATIVE PRACTICE: WHAT DO WE KNOW

One area where there is supporting information regarding the application of IPECP is in the care of adults with multiple chronic conditions (MCCs). Population demographics and trends indicate an association between increased demand for access to health services—including additional considerations resulting from the social determinants of health—and the ability to perform activities of daily living. As an index of the changing demographics, people 65 years and older constitute almost 14% (41.4 million) of the population and are projected to increase to 21% by 2050 [19]. Many chronic diseases—such as diabetes, arthritis, and hypertension—are age prevalent [20]. Patients with MCCs are high users of all types of health services and are influenced by nonclinical factors such as functional ability, informal caregiver support, family support, financial circumstances, and community services availability.

This increased access demand, and the complex care required, is often within a delivery system that is fragmented, incomplete, inefficient, and sometimes ineffective [21]. MCCs increase the demand for more effective care coordination, particularly as geriatric syndromes and transitioning into frailty occur. In addition, other negative consequences of poor care coordination include duplication of services, inappropriate or conflicting care recommendations, medication errors from polypharmacy, patient/caregiver distress, and higher costs of care.

Collaborative care is frequently practiced in the primary care health home model and is thought to optimize care for patients with MCC, particularly when there is transition to geriatric syndromes and frailty. New collaborative-care models are being developed and implemented in patient-centered primary care health homes and accountable care organizations, supported by changes in payment systems that reward improved outcomes in the quality, affordability, and experience of care [22,23]. Collaborative care encompasses the following essential elements:

- Common goals shared by patients and members of the care team and a shared commitment to meeting the goals;
- Care-team member roles that are appropriate to an individual's education and expertise;
- A full understanding of team member roles;
- A mechanism for care-team communication—such as a shared care plan with shared documentation that is accessible and useable; and
- A mechanism for the ongoing monitoring of patient outcomes.

This model of care, when inclusive of educators, practitioners and student learners, is referred to as interprofessional education and collaborative practice. In most primary care settings, the interdisciplinary team consists of physicians, nurses, and pharmacists; more recently, care coordinators have been included. As the complexity of needed care increases, there is increasing realization, and some evidence, that the inclusion of skills in care for the aging and in social services will play an essential role in the care delivered. In the spectrum of care, the more routine care or care of a single disease condition usually does not require an interdisciplinary provider team. However, skills in social services and care for the elderly are still required for planning and quality improvement in the process of care, as well as to improve the health of the population for which the primary care health home is responsible [24–29].

This relationship between the increasing complexity of disease in a person and the need for an interprofessional team to plan and provide care is depicted in Figure 3.

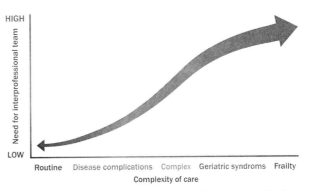

FIGURE 3   Relationship between complexity of care and need for interprofessional team.

Building and Training a Team

Becoming a nexus linked to achieving one or more outcomes using a team approach encompasses developmental milestones or "stair steps" of increasingly complex tasks that can guide team development and its ability to achieve outcomes. The initial team model involves frequently getting to know each other and initiating some "parallel play" as a group. The next advancement occurs when the team members decide they know each other well enough to try designing and completing a simple project together. As trust and experience develop, the team takes on increasingly complex challenges that can fully change a process and achieve more transformative outcomes. This process is depicted in Figure 4.

Much of the success in health care teams has occurred in the area of patient safety and quality, driven by an Institute of Medicine report [23,30,31]. Such technology is now present in most care delivery systems. This linear model has resulted in significant incremental change. In centers where student learners are being added, several have designated contributing roles for the student learners. Some education–practice partnerships are using the quality improvement platform as

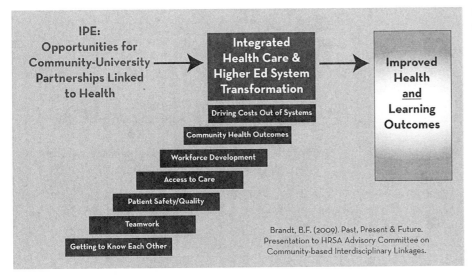

FIGURE 4   Opportunities for community-university partnerships linked to health.

TABLE 1 Lessons Learned About Interprofessional Education (IPE) at the University of Minnesota Academic Health Center

It is about changing a culture.

1. Education must move from teaching to learning: move from talking heads to a learning environment where both students and faculty learn.

2. Changing the culture takes time, commitment, patience, perseverance, passion, leadership, courage and resources.

3. Relationships with policy makers are essential for the culture change to occur.

4. Value-added changes in care delivery will be followed by supportive payment systems.

5. Experiential learning and practice in the clinical setting is a critical success factor.

A compelling vision/case is essential.

1. Students, faculty and administrators need to see how the landing benefits all.

2. Early faculty adopters are essential.

3. Strategy and operational alignment needs to be created among deans, schools and faculty governance.

Resources are needed to achieve the vision.

1. Essential infrastructure needs to be put into place in a service model.

2. Logistical issues must be addressed, e.g. scheduling, technology.

3. Financing needs to be part of the budgeting process.

Faculty reward systems need to be developed and implemented, including academic currency and tenure.

Creating an environment where students and faculty and administrators are willing to take risk needs to be in place.

1. Early wins move the adoption process along.

2. If something is not working it must be addressed through assessment, work with teams, and making unpopular decisions.

3. Success of interprofessional, interdisciplinary research is a great assist in a research-intensive university.

Administrative and faculty leadership is essential.

1. The Chief Executive Officer must visibly champion the vision and the resources.

2. Associate deans for education are essential in moving the agenda ahead, including the leadership selection.

3. Appropriate transition needs to occur to permit the faculty, staff and students lead and do the development.

4. IPE needs teams of leaders at all levels to make team care work.

the point of embarking into higher levels of team performance that then becomes linked to achieving Triple Aim outcomes.

Various forms of team training are being tested in both educational and practice settings. Whether these efforts will move the team "up the steps" to achieve high-functioning status is being anxiously observed, particularly in the current setting of improving health outcomes and the new payment systems designed to support that transition. One challenge is the lack of evidence-based best practices for team composition, training, and evaluation/assessment. A number of investigators and organizations are undertaking research in these areas [17,22,23,30]. A robust literature review of the available assessment tools and their validation and usage indicates that not many are available and those that exist are difficult to apply to the IPECP setting [15]. The National Center is currently assembling a team of experts to undertake the development of tools to evaluate: the state of readiness of clinical sites for IPECP; what is learned by teams who are provided team training of some kind; the status of collaborative practice; and how education and clinical practice fit together to achieve major outcome improvements in education and collaborative practice.

There are a number of efforts to reconnect education and clinical practice using interprofessional teams with student learners [32]. Some examples in place are the University of Utah, the University of Colorado, the University of Arizona, and the University of Minnesota. There is similarity in the lessons learned across these sites, summarized in Table 1.

THE NATIONAL CENTER FOR INTERPROFESSIONAL PRACTICE AND EDUCATION

The need for a national center as a trusted source of information and coordinated leadership in IPECP grew out of the education and clinical practice redesign effort to improve outcomes. A national competition was held and the University of Minnesota was selected to house the National Center for Interprofessional Practice and Education (National Center). The goals of the National Center are to:

1. Provide unbiased, expert guidance regarding IPECP;
2. Collect, analyze, and disseminate information assessing the effectiveness of IPECP;
3. Build a business case for IPECP as an effective education and care delivery model;
4. Evaluate the effectiveness of IPECP for improving population and community health;
5. Identify IPECP environments to serve as exemplar training sites; and
6. Develop new programs and support/enhance existing team-based IPECP programs.

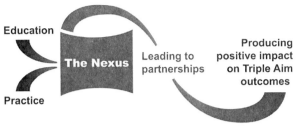

FIGURE 5 The Nexus.

Accomplishing these goals is derived from a vision to reconnect education and collaborative practice, summarized in Figure 5.

The National Center has five core queries regarding IPECP, and is collecting data to:

1. Improve the Triple Aim outcomes on an individual and population level;
2. Create a sustainable, adaptive infrastructure that supports the Triple Aim outcomes of both education and practice;
3. Identify ecological factors essential for achieving Triple Aim outcomes;
4. Identify factors essential for a sustainable, adaptive infrastructure in the transformation of the process of care and education; and
5. Identify changes needed in policy, accreditation, credentialing, and licensing for health care provision and education.

In order to produce this data, the National Center has created two interlocking processes—a National Center Data Repository (NCDR) and an Innovation Incubator Network. Each network site is part of a national network that is implementing an interprofessional team intervention linked to defined educational and clinical outcomes. This information is collected in a site-specific project database and is then transferred (deidentified) to the NCDR. The NCDR is a repository for all the site-specific data, and also contains a series of surveys in which each performance site contributes information. The domains of the data collected include the project-specific data regarding the intervention, the team employed, and the outcomes to be achieved; the costs incurred for the project; ecological/environmental factors affecting the success of the project; information regarding team education and process; and the demographics of the project participants. The NCDR is designed to provide an analysis of return on investment for the IPECP team interventions linked to achieving outcomes. Assessment tools for teams are also available in the NCDR. The incubator sites are actively creating data and the NCDR is now in the log phase of data collection.

This longitudinal approach to data collection is one of real time in the context of becoming a learning organization essential to sustainable redesign. This approach to information and evidenced-based decision making does not deploy the randomized controlled trial (RCT). The RCT will continue to play an important role in efficacy and effectiveness trials, the results of which will continue to inform care delivery, particularly for drugs and devices. However, with the establishment of health and education outcomes and the need to manage toward achieving those outcomes in real time, an outcomes oriented, comparative effectiveness approach using longitudinal databases and informatics tools is required. Also, the various stakeholders require information and evidence, and a demonstration of return on investment, in order to make and sustain their investments in IPECP. The National Center approach to generating information and evidence will inform the various stakeholders as they redesign the process of care, including factors regarding population health, community engagement, and families in the redesign process.

The National Center has developed a number of resources and initiatives to assist education and health system organizations in creating successful nexi and moving ahead to improve outcomes with IPECP teams. A summary of these resources is presented in Table 2.

WHAT COULD THE FUTURE LOOK LIKE?

Amina is a Somali refugee mother with small children who lives in an affordable housing community. She has type I diabetes and uses a mobile phone that keeps her connected with the "care learning team" of clinicians, university faculty, and health professionals-in-training. This system monitors her diabetes. She has just measured her blood sugar; the reading is high and has been so for a day or two. Her glucometer automatically reports her glucose measurements to her local federally qualified health center (FQHC), setting in motion a sequence of events with her care learning team. The FQHC is in partnership with the Somali Community Development Alliance (SCDA) and the university to achieve a new "Nexus" between the process of care and educational systems that provides a team-based, patient-centered health home with the patient and community workers part of the team.

Charged and resourced by the university senior administration, the faculty created a curriculum committee of interprofessional providers, administrators, policy makers, students, and community members to redesign the health curricula to meet the Triple Aim, starting from the day of entrance into a professional school. Students learn together in courses, rotate through communities in interprofessional teams, use state-of-the art learning technologies, and are assessed in their team competency performance throughout the longitudinal educational experience. Students are placed for multiple months in the same health systems to enable them to fully engage in care delivery, complete quality improvement projects, and contribute and add value to the site by participating in authentic experiences in a variety of clinical and community settings (e.g., home health, acute care, ambulatory).

TABLE 2 National Center Initiatives

Initiative	Description
1. NCDR	1. A national repository residing at the University of Minnesota comprised of quantitative and qualitative information and data on educational, practice, costs, outputs, and outcomes of IPECP interventions designed to improve health and education outcomes 2. Has the capacity to causally connect the interprofessional teams to specific health and education outcomes 3. Is constructed in a way that return on investment analyses can be reported out in formats to provide to different stakeholder groups 4. Is linked to a national network of sites performing team interventions designed to improve specific educational and/or health outcomes
2. Innovation incubator network	1. A national network of performance sites that plan and implement an IPECP intervention that is designed to improve health and education outcomes 2. Multiple sites are performing relevant IPECP interventions 3. A network collaborative is operational within which to share information, identify, and solve problems and to interact regarding project specific performance and the aggregate performance reports from the NCDR; and will also look at the intervention portfolio and advise the national center regarding where the gaps are and what interventions need to be implemented
3. Resource exchange	1. This exchange is a multipurpose, open access, web-based site that the National Center uses as a trusted source for information, data, and evidence 2. The exchange receives relevant information on IPECP from any source that wishes to register and input their information 3. A review process is present that identifies information of sufficient quality that its access by users is recommended 4. A user collaborative of all who register on the site participate in. It has regular sessions for information dissemination, dialog, and uses a variety of social media
4. Products and services	1. The National Center has a portfolio of products and services that is available to incubator sites and users who wish to use them 2. The National Center is developing an expanded portfolio of products and services those desiring information regarding the field and how it is developing 3. Nexus Learning System education and training modules that integrate the disciplines of assessment, evaluation, intervention research and informatics in a new way that is permissive of collecting and using date to causally track the linking of education and clinical practice to improve education and health outcomes

5. Developing partnerships and relationships	1. The National Center is implementing a process to create relationships and partnerships to achieve:
	  a. Linkages with sources of information, education and training that it does not have or decides not to develop
	  b. Engagement of the public health community in sharing its knowledge and skills in population health at the level of the practice community and in educational institutions
	  c. Engaging people, families and communities in the redesign process and in being contributing members on the IPECP teams
	  d. Coalitions the National Center informs from its information and evidence
	2. Recruit new performance sites for the incubator network
	3. Recruit more relevant information regarding the IPECP field into the national center resource exchange
	4. To advise and improve the performance of the national center in achieving its goals

6. Developing the business case	1. A business case that is relevant to the various stakeholder groups is a key success factor in the work of the national center

Amina's team of students is integrated into the team providing clinical care serving the SCDA community. A nurse clinician is the preceptor of record. All participants consider themselves learners in this educational and practice environment.

The computer receiving Amina's blood glucose reviews her record, recognizes an increasing deviation from her usual pattern of control, and informs Amina's care learning team by sending an alert to her electronic health record. The team receives the alert and assigns the pharmacist-in-training under the supervision of the nurse clinician to engage the Somali community health worker. The latter speaks Somali and reviews the cultural, diet, family issues, and the current self-care protocol with Amina. A team initiates a conference call with Amina and a care plan is developed. The response time is well under the 1 hour goal established by the clinic's chief medical officer.

Amina walks to the FQHC lab a few blocks away, leaves a urine sample, and has some blood drawn. A few hours later the pharmacy learner discusses the presence of a urinary tract infection and confirms her medication allergy and pharmacy of choice. Amina starts taking her antibiotic within a few hours. Later in the day, the team huddles and reviews the encounters of the day, including a review of Amina's case. The FQHC care goals for diabetes and the care processes and outcomes are reviewed.

Over the next few days, to reinforce her self-care goals, Amina sends text messages via her mobile device to her care learning team updating them on her symptoms, and reviewing the improvement in blood glucose levels. She also receives texted communications regarding her health goals and self-care plan and a reminder to return for her previously scheduled group visit. For Amina, hospital and emergency care, formerly a frequent occurrence, have not been necessary since the new learning and health care delivery partnership became operational.

The student learners update their electronic learning portfolio. At the end of three rotations, each portfolio is assessed by a team of faculty, practicing clinicians, and community members based on previously established profession-specific and interprofessional competency mapping [33].

REFERENCES

[1] Brandt BF, Cerra FB. Community-university collaborative partnership to improve health through interprofessional education Robert Wood Johnson Foundation. February, 2012.

[2] Cerra FB. Reconnecting physician education and clinical practice, vol. 28. Minnesota Physician; 2011. No. 6.

[3] Institute of Medicine. Educating for the health team. Washington, DC: National Academies Press; 1972.

[4] Cerra FB, Brandt BF. Renewed focus in the United States links interprofessional education with redesigning health care. J Interprof Care 2011;25(6):394–396.

[5] Institute of Medicine. Health professions education: a bridge to quality. Washington, DC: The National Academies; 2003.

[6] JH Macy Conference Report. Transforming patient care: aligning interprofessional education with clinical practice redesign January 17–20, 2013. Atlanta, Georgia; 2013.

[7] Institute of Medicine. Interprofessional education for collaboration: learning how to improve health from interprofessional models across the continuum of education to practice: workshop summary. Washington, DC: The National Academies Press; 2013.

[8] World Health Organization. Framework for action on interprofessional education and collaborative practice. Geneva: WHO Press; 2010.

[9] Berwick DM, Nolan TW, Whittington J. The triple aim: care, health and cost. Health Aff 2008;27(3):759–69.

[10] Institute of Medicine. Framework for action on interprofessional education and collaborative practice. Washington DC: The National Academies Press; 2010.

[11] IPEC. Team-based competencies: building a shared foundation for education and clinical practice. In: Conference proceedings. February 16–17, 2011; 2011.

[12] Chen FM, Williams SD, Gardner DD. The case for the national Center for interprofessional practice and education. J Interprof Care 2012;27(5). Accessed at: http://informahealthcare.com/doi/abs/10.3109/13561820.2013.786697?journalCode=jic.

[13] Brandt B, Lutfiyya MN, King JA, Chioreso C. A scoping review of interprofessional collaborative practice and education using the lens of the triple aim. J Interprof Care 2014;28(5).

[14] Canadian Interprofessional Health Collaborative. An inventory of quantitative tools measuring interprofessional education and collaborative practice outcomes. August 2012.

[15] National Center for Interprofessional Practice and Education Team Assessment Tools Resource Exchange. Accessed at: www.nexusipe.org.

[16] Reeves S, Perrier L, Goldman J, Freeth D, Zwarenstein M. Interprofessional education: effects on professional practice and health care outcomes (update). Cochrane Database Syst Rev 2013;3. http://dx.doi.org/10.1002/14651858.CD002213.pub3.

[17] Salas E, DiazGranados D, Klein C, Burke CS, Stagl KC, Goodwin GF, et al. Does team training improve team performance? A meta-analysis human factors. J Hum Factors Ergonomics Soc 2008;50:903.

[18] Garr DR, Margalit R, Jameton A, Cerra FB. Educating the present and future healthcare workforce to provide care to populations. Acad Med 2012;87(9):1159.

[19] US Census Bureau. Health United States. 2012. Table 1.

[20] CDC. Morb Mortal Wkly Rep 2006;55(40):1089–92. http://www.cdc.gov/nchs/data/hus/hus12.pdf#045.

[21] Boult C, Wieland GD. Comprehensive primary care for older patients with multiple chronic conditions: "Nobody rushes you through". JAMA 2010;304(17):1936–43. Nov 3.

[22] Gilman SC, Chokshi DA, Bowen JL, Rugen KW, Cox M. Connecting the dots: interprofessional health education and delivery system redesign at the veterans health administration. Acad Med 2011;89(8).

[23] Milbank Memorial Fund. The medical home's impact on cost and quality. January 2001.

[24] Barnett K, Mercer SW, Norbury M, Watt G, Wyke S, Guthrie B. Epidemiology of multimorbidity and implications for health care, research, and medical education: a cross-sectional study. Lancet Jul 7, 2012;380(9836):37–43.

[25] American Geriatrics Society. Guiding principles for the care of older adults with multimorbidity: an approach for clinicians. American Geriatrics Society Expert Panel on the care of older adults with multimorbidity. J Am Geriatr Soc Oct 2012;60(10):E1–25.

[26] Blacker S, Deveau C. Social work and interprofessional collaboration in palliative care progress. Palliative Care 2010;18(4):237.

[27] Sommers LS, Marton KI, Barbaccia JC, Randolph J, Physician, nurse and social worker collaboration in primary care for chronically Ill seniors. Arch Int Med 2000;160:1825–33.

[28] Barnes D, Carpenter J, Dickinson C. Interprofessional education for community mental health: attitudes to community care and professional stereotypes. Soc Work Educ 2000;19(6):565–83.

[29] Bilodeau A, Dumont S, Hagan L, Paré L, Razmpoosh M, Houle N, et al. Interprofessional education at Laval University: building an integrated curriculum for patient-centred practice. J Interprof Care 2010;24(5):524–35.

[30] Nembhard IM. Learning and improving in quality improvement collaboratives: which collaborative features do participants value most?. Health Serv Res 2009;44(2). [Part I].

[31] Institute of Medicine. Crossing the quality chasm: a new health system for the 21st century. Washington, DC: The National Academies Press; 2001.

[32] The Interprofessional Curriculum Renewal Consortium, Australia. Curriculum renewal for interprofessional education in health. Sydney: Centre for Research in Learning and Change. University of Technology; 2013.

[33] Earnest M, Brandt BF. Adaptation from Josiah Macy Jr. Foundation paper: building a healthcare workforce for the 21st Century healthcare by aligning practice redesign and interprofessional education 2012. Presented January 18, 2013, Atlanta, Georgia.

ABOUT THE AUTHORS

Barbara F. Brandt is the Director of the National Center for Interprofessional Practice and Education, and Associate Vice President for Education, Academic Health Center University of Minnesota. She also leads the Minnesota Area Health Education Center network, an interprofessional workforce development program for underserved rural and urban areas across the state.

Frank B. Cerra is the Senior Advisor and Interim Deputy Director, National Center for Interprofessional Practice and Education, and Professor of Surgery and McKnight Presidential Leadership Chair at the University of Minnesota. He previously served as Senior Vice President for Health Sciences and Dean of the Medical School at the University of Minnesota.

Chapter 10

Advancing Collaborative Global Education Programs

Yang Ke, Qiudan Sun, Lei Zhang, Jianlin Hou and Weiwei Du

INTRODUCTION

Globalization has become the most important feature of the twenty-first century. In this new era, our essential interdependence is increasingly revealed and our communities have become both closer and more diverse [1]. As Austin [2] commented, "transportation and communication technology plus worldwide health, environmental, and security risks and a world economy driven by transnational corporations are connecting us in a new kind of way." Globalization is already an inevitable trend, especially in the field of health.

Globalization greatly impacts human health and health services, posing challenges for both healthcare professionals and the public health [3,4]. Globalization has inevitably changed the determinants of health and means of healthcare delivery. For example, health risks such as SARS and H1N1, and more recently the Ebola outbreak, which occur in one region/country may quickly become a grave concern throughout the world. Physicians are now members of a global community [5]. As Martimianakis and Hafferty noted "The effects of globalization on health are the focus of administrators, educators, policy makers and researchers as they work to consider how best to train and regulate health professionals to practice in a globalized world" [6].

What knowledge and competencies are required to train health professionals in this new era of globalization? Bozorgmehr and Saint et al. have provided one response: "knowledge and competencies in the areas of international migration, cross-cultural understanding, emerging and re-emerging infectious diseases, non-communicable diseases, social and trans-border determinants of health, health inequities and inequalities, global health organizations and governance, human rights, medical peace work, environmental threats and climate change have become increasingly important in our globalized world" [7].

The Commission on Education of Health Professionals for the 21st Century answered this question from another perspective: "all health professionals in all countries are educated to mobilize knowledge, and to engage in critical reasoning and ethical conduct, so that they are competent to participate in patient-centered and population-centered health systems as members of locally responsive and globally connected teams" [8]. These requirements and realities necessitate a new paradigm in global health professions education to prepare future doctors for the new challenges in a globalized society [9]. In this environment, advancing collaborative global programs in health professions education is critically important.

In advancing collaborative global education programs, we must address global health; health problems, issues, and concerns that transcend national boundaries can only be best addressed by cooperative actions and solutions [10]. Among the various definitions of global health, we find that the version proposed by Koplan et al. to be most pertinent: "global health is an area for study, research, and practice that places a priority on improving health and achieving equity in health for all people worldwide. Global health emphasizes transnational health issues, determinants, and solutions; involves many disciplines within and beyond the health sciences and promotes interdisciplinary collaboration; and is a synthesis of population based prevention with individual-level clinical care" [11].

The nature of global health presents powerful opportunities and challenges for health professions education. Exposing students to global health issues, preparing them to work in culturally diverse contexts, and working toward the adoption of a set of essential global competencies are all activities rationalized as preparing physicians to meet these challenges [6]. The increasing trend is for health education institutions around the world to work with one another and establish collaborative education programs through cooperation, communication, information, and resource-sharing.

TYPES OF COLLABORATIVE EDUCATION PROGRAMS

Collaborative global education programs take a variety of forms. Exchange programs between institutions that

The Transformation of Academic Health Centers. http://dx.doi.org/10.1016/B978-0-12-800762-4.00010-4

provide students, lecturers, and researchers with healthcare experiences in other countries is the most popular one [12]. Study-abroad programs have been encouraged to be part of the college experience in most universities worldwide [13]. Medical students are pursuing study abroad in increasing numbers, and there is a continuing growth of globalization in higher education. Universities are responding by developing collaborative programs to address these needs [14]. Partner institutions today increasingly seek mutual benefits. Two-way flow of energies, expertise, and knowledge is greatly encouraged [15].

Among the various forms of collaborative global education programs, we have found the following most popular and commonly adopted throughout the world, as shown in Figure 1.

Degree Programs

Sandwich Program

There are many forms of collaborative degree programs in health professions education. One of the popular models is the "sandwich" model for training, which is usually at the doctoral level. The experiences of Wageningen University from the Netherlands and University of California, San Francisco (UCSF), from the US offer clear examples of such a model.

In the Wageningen University experience, a sandwich PhD student will spend the first 6–8 months of his/her degree at Wageningen University, during which the student will expand his/her research proposal and follow a tailor-made education and training program. The next 3 years are spent conducting research in his/her home country under the supervision of a local supervisor. The student will continue regular contact with his/her supervisor in Wageningen via e-mail. The sandwich PhD candidate often returns to Wageningen for short visits, and a Wageningen University supervisor will sometimes visit the student's home institution. After the student has finished collecting all the data, he/she can return to Wageningen to finish the PhD thesis, which would take approximately 6 months [16].

UCSF runs a joint training program with its partner universities. Trainees, selected jointly by faculty at UCSF and the partner institutions, complete the initial part of their training at their home institutions, and then come to UCSF for advanced training and specific coursework, and then return home to complete their project work and receive their degrees [17].

Similarly, the Graduate University of Chinese Academy of Sciences (GUCAS) conducts a Joint PhD Training Program with European research institutions and universities, in which students from GUCAS typically study in Europe for 12 months and return to China to defend their dissertation before obtaining their doctoral degree [18]. Tsinghua University sends its medical students to the University of Pittsburgh Medical Center (UPMC) for 2 years of research training; these students will then come back to Tsinghua University to complete their MD study [19].

It is generally accepted that sandwich programs are considered to be fruitful and practical and important to the globalization of the research cultures at participating institutions [20]. The students receive first-hand knowledge of the institutions and systems in both countries and they benefit from the expertise of the best scientists in their field in both countries [18].

Dual Degree

Another model is the dual degree program. Dual degree is "a double-degree program, and sometimes is also called a combined degree, conjoint degree or simultaneous degree program" [21]. This program "involves a student's working for two different university degrees in parallel, either at the same institution or at different institutions (sometimes in different countries)" [21]. Usually, the two degrees "might be in the same subject area (especially when the course is split between countries), or in two different subjects"; and students can complete them in less time than it would take to earn them separately [21].

Many universities have dual degree programs in collaboration with their overseas partner universities. The University of Technology, Sydney (UTS)'s collaboration with

FIGURE 1 Classification of Collaborative Education Programs.

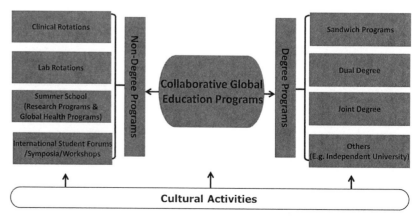

its "Key Technology Partnerships" (including Beijing University of Technology, Huazhong University of Science and Technology, Shanghai University, and Hong Kong Polytechnic University) on the dual doctoral degree program is a good example [22].

Similarly, Doctor of Medicine–Doctor of Philosophy (MD–PhD) programs aim to produce physician scientists who can translate laboratory discoveries into effective treatments for patients. It was created in the US and has been supported by the National Institutes of Health (NIH) since 1964 [23]. In recent years, MD–PhD programs are more often run in cooperation between one university and its overseas partners. For example, Peking University Health Science Center (PUHSC) has established MD–PhD programs with its strategic partners since 2001 [24]. In this collaboration model, MD students enrolled in an 8-year program at PUHSC who have good potential and great interest in research will go to PUHSC's overseas partner universities in their sixth year and spend 3 or 4 years there to pursue their PhD degrees. This MD–PhD program will then transit back to PUHSC where students spend another 3 years to complete their MD education. Under this model, students are trained to gain clinical skills, basic biomedical research capacity, and global perspective, and they are expected to become future physician scientists and leaders in their disciplines [24].

A similar dual degree model is in the training of nurses for the purpose of professional migration. One example is the cooperation between China and Australia, where students who have gained a nursing Associate or Bachelor's degree in China transit to an Australia university for a Bachelor's degree in nursing under cooperation agreements. The students can then work as registered nurses in Australia after graduation.

Joint Degree

There are also partnerships in joint degree programs. For joint degrees, two or more institutions cooperatively conduct programs in which the student will be awarded one certificate with seals from the collaborating institutions upon graduation. The uniqueness of such programs is that a single dissertation will satisfy the thesis requirements of all collaborating institutions [25].

Two good examples include: (1) a collaboration between Peking University, Georgia Institute of Technology, and Emory University in a Biomedical Engineering Joint PhD Program, in which the student will have an advisor on the home campus, a co-advisor on the secondary campus, and will spend at least 1 year taking classes and participating in research in the co-advisor's lab on the secondary campus. This program has the characteristics of "One Degree, Two Countries, Three Universities, Infinite Opportunities" [25]; and (2) the joint MD degree program run by Duke-NUS Graduate Medical School where MD students who

successfully complete the course of study and fulfill all requirements will be awarded a joint MD degree from Duke University and the National University of Singapore [26].

Other Programs

There are also other partnerships for establishing degree programs, such as joint international independent universities. Duke Kunshan University (DKU) is a good example of this kind. DKU is a partnership of Duke University and Wuhan University to create a world-class university offering a range of academic programs for students from China and throughout the world [27].

Nondegree Programs

Nondegree programs usually take the form of student and/or faculty exchange. More and more universities worldwide have realized the importance of exchange programs, and have formed collaborations with their partners in this regard. There are different ways to address exchange programs. We classify the nondegree exchange programs for health professional students under the following categories: clinical rotation, lab rotation, summer school, and international student forums/symposia/workshops.

Clinical Rotation

There is a powerful educational justification for performing clinical rotations abroad (also known as international electives, overseas clinical clerkships/observerships), which is to provide precious opportunities for medical students and residents to gain valuable competencies and skills outside their formal curriculum or training, and raise their awareness of global health [28]. Medical students and residents are increasingly choosing to engage in international electives. For example, in 2006, over 27% of US medical graduates participated in an international health experience [28], compared with only 6% in 1984 [29]. This has already become a trend in medical education and training throughout the world.

Clinical rotations abroad have varied characteristics. They most frequently occur during senior years, when the students may benefit from the program the most, and usually last from 4–12 weeks [30]. There can be exceptions in the program length. For example, Shanghai Jiaotong University (SJU) in China is cooperating with French universities whereby medical students at SJU have the opportunity to take part in a 1-year clinical rotation program in France that increases opportunities for positive experiences for the student [31].

Exchange students have benefited greatly from clinical rotations abroad, and their comments on their overseas experience are usually very positive. For example, an exchange student from PUHSC to the University of Massachusetts Medical School (UMMS) commented: "I always think about my rotation experience in UMMS. I am so lucky and so grateful.

It has opened my mind, and let me re-examine my beliefs and conceptions of the world" [32]. A student from the University of Sydney to PUHSC said: "It is a great opportunity for me to experience different society, different people, and different social mechanism and medical care system" [33].

Lab Rotation

Lab rotations abroad are becoming more popular and have proven to be very beneficial for exchange students. The students can learn new methods, have a chance to expand their professional as well as their personal network, and observe research structures and mechanisms of international institutions [18].

Two cases provide good examples, one from Germany and the other from China. Giessen Graduate Centre for Life Sciences (GGL) at Justus-Liebig-Universitaet Giessen in Germany has established lab rotation programs with universities from the Netherlands, UK, France, Norway, Russia, Australia, and the US, in which students go to their host laboratories with funding from the German Academic Exchange Service (DAAD) for up to 30 days; but, they can stay as long as it takes for them to finish their experiments [18]. PUHSC and SJU both have lab rotation programs with the University of Michigan (UM), which last from 3–8 weeks with the potential for participants to pursue their PhD study at UM.

Summer School

Summer school generally refers to programs in which coursework is done over a vacation period. These programs often include social activities. Usually, the summer school is hosted by one institution with participating students from many institutions (usually the host institutions' partners); alternatively, it can also be jointly hosted by two or more institutions. In terms of health education, we have found the following two models most popular.

Research Programs

These programs usually consist of lectures given by renowned professors, followed by discussions and student presentations, as well as lab experiences, and social activities. Research areas covered may vary depending on the research priority of the organizing institution(s). One unique example is that of PUHSC-Karolinska Institute-Keio University-King's College London Joint Research Summer School. These four institutions take turns hosting the summer research program each year. This is a 4-week program with a 1-week course delivered by professors from all four institutions in a specific medical area, followed by 3-week lab rotations.

Global Health Programs

Global health has become increasingly important in medical education, and there is a growing student interest in matters related to global health, including participation in experiential learning activities abroad [29].

For example, Duke University, in collaboration with its partners in Udaipur, India, and Beijing, China, offers its undergraduate students a program called *Global Semester Abroad: Development, Environment and Global Health in India and China*, which provides opportunities for in-depth exploration of health and development issues in India and China. Students will spend half of the semester in India and the other half in China studying common themes of global health and development [34].

Another typical example of such a program is the International Pediatric Emergency Medicine Elective (IPMEM), which is a novel global health elective that brings together students from Canada and the Middle East for a 4-week living and studying experience, aiming to foster cross-cultural dialogue, networking, and cooperation. The lectures, seminars, and workshops in which students participate are related to pediatric emergency medicine, global health, and research. In this program, students also observe patient care during shifts in a pediatric emergency department and design a research project [28].

International Student Forums/Symposia/ Workshops

International student forums/symposia/workshops can also serve as platforms for promoting joint efforts to increase understanding among health professions students around the world. Many institutions conduct these kinds of forums.

The Graduate University of Chinese Academy of Sciences (GUCAS), the University of Tokyo, and the University of Nebraska Medical Centre are collaborating in hosting an international student forum, which takes place annually in different locations and with participants from different universities. The program usually includes presentation sessions, institution visits, and local tours [21]. Similarly, University of Queensland runs an International Postgraduate Symposium in Biomedical Sciences annually with participants from Oxford University (UK), Utrecht University (the Netherlands), Otago University (New Zealand), Peking University (China), and Sun Yat-sen University (China), and so on [35].

Cultural activities are an indispensable part of all these collaborative global education programs. They can help students not only to learn the local culture, but also to foster better understanding of global health.

CHALLENGES/BARRIERS

Collaborative global education programs have been important for the internationalization of education and play a significant role in enhancing the quality of health professions education. Furthermore, such programs bring numerous benefits to participating students, faculty, and institutions. As research has shown, these benefits include greater cultural understanding and enhanced intercultural competence; stronger motivation to pursue either primary care or future international health work; better understanding of socioeconomic influences on health and illness; broadened medical knowledge; improved clinical skills and foreign language proficiency; deeper appreciation for public health issues and professionalism; and greater recognition of the importance of communication skills [36–38]. More importantly, as Wright noted, students and accompanying faculty "are able to look outside their own comfort zones and thus grow personally and professionally." In a totally new environment, students are likely to be challenged by "new ideas and possibilities, begin thinking beyond what they already know as the usual, and develop personal maturity" [13]. Finally, these programs will also strengthen ties between the home and host institutions and promote collaboration at both the institutional and faculty levels.

The above-mentioned benefits have been well-recognized throughout the world; but how to advance collaborative global education programs is still a challenging task. Just as Clement and colleagues found, the most notable barriers to developing global education programs include limited funding, limited faculty time and logistic support, and low faculty interest [30]. Drain and colleagues also commented that the barriers might "stem from personal considerations and also from a combination of accreditation and administrative issues" [36]. It is important to provide further elaboration on the various challenges and barriers.

Cultural and Language Barriers

Global education programs can pose significant challenges and stress for students. The experience of living and working in an unfamiliar environment and culture is valuable to the learning and growth of students, but could be one of their greatest sources of anxiety. This has been described as a struggle with cultural dissonance and culture shock. In a clinical rotation setting, for example, students may be confronted with health beliefs and care practices which are different from their understanding of "universal" values of health care. This experience may then lead to challenging the students' own beliefs and assumptions. Culture shock or dissonance can occupy a central place in global education programs [39]. As Hawthorne and colleagues found, "delivering western-style medical education to culturally and linguistically disparate groups" is a big challenge [40]. Del and Fins also mentioned in their study that "cross-cultural tensions could emerge when introducing themes from Western medical ethics and humanities into other culture context, and thus bring about pedagogical challenges" [41].

Cultural barriers accompanied with language barriers can make the situation worse, especially when both the students and faculty involved in the programs will have to use English to communicate, rather than their mother tongues. Language barriers may lead to a series of consequences—including incomprehension, misunderstanding, and stress—thus decreasing the benefits of the exchange programs.

Regulatory Differences

Differences in accreditation and licensure requirements in clinical education throughout the world pose substantial challenges for global education programs, such as international clinical electives, residency, and internship. Most medical accreditation organizations neither officially recognize time spent doing clinical training overseas nor count procedures performed abroad toward meeting accreditation requirements, which is an apparent disincentive. But still, 85% of the participants who were resident surgeons in a 2007 survey "reported that they would have pursued such an international rotation even if their overseas surgical cases were not counted toward accreditation," which further illustrates the importance, attractiveness, and great benefits of overseas experiences. Currently, only a few institutions, such as the American Academy of Pediatricians, have established guidelines for overseas clinical rotations [36]. Because medical education curricula and criteria differ among countries, it is difficult to exchange students for a long duration. Faculty members from a college or university usually have concerns that the exchange students will not acquire the necessary preparation for graduation requirements.

Financial Barriers

Financial support is essential in ensuring successful implementation of collaborative education programs. Governmental financial support has been an important source of funding, such as the China Scholarship Council (CSC) in China and the Linnaeus-Palme (L-P) program in Sweden [12]. CSC has supported not only degree programs, but also nondegree exchange programs. The objective of CSC is to "provide financial assistance to the Chinese citizens wishing to study abroad and to the foreign citizens wishing to study in China in order to develop the educational, scientific and technological, and cultural exchanges and economic and trade cooperation between China and other countries, to strengthen the friendship and understanding between Chinese people and the people of all other countries" [42]. Additionally, the Australian government provides significant funding, such as Student Mobility Grants, to help higher education and vocational and educational training (VET) students participate in student exchange and student mobility activities as part of their degree [43]. Other countries, such as the US, UK, and Japan, also have similar funding schemes to support collaborative global education programs.

However, current financial support is still very limited due to the increasing demand for global experiences. Oftentimes costs may restrict excellent students with financial difficulties from studying abroad [44]. Funds from various sources still need to be raised to meet the needs [45].

Unbalanced Development among Countries, Institutions, Disciplines

Generally speaking, students from developed countries have more opportunities for access to study-abroad programs than those from less developed countries. For example, 30–40% of medical students participate in a global elective program in the UK, US, Canada, and Germany [9]. In contrast, the percentage is much smaller for medical students from developing countries. Furthermore, compared with smaller medical schools with limited resources, larger schools with greater access to funding and other resources are able to provide a broad range of opportunities for their students [37].

Unbalance also exists among disciplines. Study-abroad programs have been encouraged to be part of the college experience for medical students around the world. However, these programs have been relatively lacking in other health-related disciplines, such as nursing and public health. For example, as noted by Wright, "despite the encouragement of some nursing leaders that clinical experiences from different parts of the world should be 'sampled and savored', few study abroad programs have been developed and made available to nursing students in the United States" [13].

Institutional Barriers

Leadership

International partnerships take time to establish. To ensure successful partnership, maintaining leadership continuity is critically important [14]. Leadership change may impact existing exchange programs with insufficient support.

Program Design and Management

Program design and management is vitally important to achieve desirable outcomes. Without clear teaching plans, learning objectives, and sound evaluation, it is likely that the program turns into a culture tour. At the same time, poor management may lead to limited faculty time and logistic support, which may cause great inconvenience and unsatisfactory results. It is important for the host institution to pay special attention to this issue.

Students need support before, during, and after the exchange. However, there is no uniformity in prerequisites required for overseas electives or for supervision over participating international electives. In many cases, students take international electives without clear administrative and faculty support or supervision—important factors for quality training and mentoring [13,36,37].

Personal Barriers

For a variety of reasons, not all students are able to participate in exchange programs. Factors such as financial status, scheduling conflicts with their home institutions' curriculum, family responsibilities, concerns for personal safety and health, and extracurricular employment to meet basic needs are key deterrents to undertaking studies abroad [36,45,46]. Moreover, a lack of appropriate skill sets or training for roles while overseas can be an impediment. Some students even display a predominantly vacation mindset, which is certainly another barrier [44].

CONCLUSION

Currently, collaborative global education programs play an important role in health professions education. To further advance these programs, it is necessary to address the challenges and seek solutions. Based on extensive literature review and our experiences, our recommendations are summarized in Table 1.

To overcome cultural and language barriers, predeparture training and/or manuals introducing the host country should be provided for the exchange students. Foreign language proficiency tests are also suggested as a requirement for the exchange student when the instruction media at the host institution is in that language. Furthermore, we also recommend that exchange students have basic-level command of the local language. Additionally, custodial care and support from the host institution are very important, especially during the early phases of the exchange [38]. Orientation courses and student volunteers serving as cultural hosts at host institutions are also considered effective ways to help exchange students overcome culture barriers.

To overcome barriers caused by regulatory differences, standardization in global medical education should be advocated, allowing for increased mobility of medical students, faculty, and practitioners between countries. Some international organizations, such as the Association of Academic Health Centers International (AAHCI), are exploring creation of an accreditation process for its member institutions. This association is now working on an accreditation white paper, making use of the expertise and resources of its members (more than 130 around the world). Accreditation oversight organizations could also help solve this problem. For example, the Accreditation Council for Graduate Medical Education (ACGME) has developed criteria to address this problem [36]. Furthermore, as Duffy and colleagues suggested, "the academic exchange experience should bear credit which is accepted at the home institution. Students should know that the academic work completed during the exchange program is recognized and valued by the home institution" [38].

Global initiatives are needed to address the unbalanced development in collaborative health professional education

TABLE 1 Challenges/Barriers of Collaborative Global Education Programs and Suggested Strategies

Challenges/Barriers	Suggested Strategies
Cultural and language barriers	• Predeparture training and/or manuals on the introduction to the host country • Foreign language proficiency requirement • Certain command of the local language • Custodial care and support from the host institution • Orientation courses upon arrival • Student volunteers from the host institution serving as cultural hosts
Regulatory differences	• Standardization in global medical education • Accreditation white paper • Criteria for acceptable international rotations • Credit acceptance by the home institution
Financial barriers	• Financial support from NGOs and international organizations • More government support • Funds raising from various resources • Virtual exchange projects and international virtual medical schools via the Internet
Unbalanced development among countries, institutions, disciplines	• Global initiatives • Virtual exchange projects and international virtual medical schools via the Internet
Institutional challenges/barriers (leadership, program design and management)	• Sufficient administrative support • Sufficient faculty time • Optimized program structure with clear teaching plans and specific learning objectives • Survey to identify the needs of the home institution • Continued institutional leadership support • MoUs/agreements to guarantee continued leadership support
Personal challenges	• Financial support, scheduling flexibility, and health insurance for the students • Transnational competence (TC) education for five core skills (analytic, emotional, creative, communicative, and functional)

among countries. For the unbalance of exchange programs among disciplines, the home institution should make greater efforts to ensure benefit for all of its students. Overall, financial support is definitely needed at different levels to address this issue.

For student exchange programs, the host institution needs to pay special attention to program design and management, provide sufficient administrative support and faculty time, and structure the program with specific learning objectives to guarantee quality and benefits and adequate education outcomes [28]. Before a program begins, the host institution should conduct a survey to identify the needs of the home institution and the exchange students so as to better design the program. Program coordinators are also needed to organize, coordinate, and manage the program. For the home institution, "a program of orientation to the country prior to departure, support during the study period, and debriefing upon return contribute to a successful experience" [38]. For degree programs, the students will need "knowledgeable on-site supervisors as well as close relationships with their faculty mentors

at home" [36]. For the sustainable development of collaborative education programs, institutional leadership support is a must. In this case, leadership continuity seems to be vitally important. Memorandum of Understanding agreements are a good way to maintain continuous leadership support.

To address the problem of personal barriers, we suggest that the home institution make every effort to provide necessary support—such as financial assistance, scheduling flexibility, and health insurance—to relieve the worries and anxiety of the exchange students. Additionally, the home institution should provide their students with transnational competence (TC) education to prepare them for ethnically and socially discordant encounters and global practice sites. According to Koehn, there are five core skill domains in TC—analytic, emotional, creative, communicative, and functional skills [47].

To conclude, collaborative health education and training is the most appropriate response to the health requirements of the twenty-first century characterized by globalization, migration, and a free market economy [48]. Advancing collaborative global education programs has become an unshakeable responsibility for the stakeholders in health education in general and health professions education institutions in particular.

Looking into the future, with joint efforts, we are very optimistic about the continued development of collaborative education programs. With the help of the Internet and new communication technologies, international virtual exchange programs and medical schools can be an important supplement to direct or traditional exchange and other health professions education programs. And, the advantages of such new methods are inexpensive, quick, and convenient. According to Todhunter and colleagues, "The Virtual Exchange Project is a Web-based application designed to give students a study-elsewhere experience in their own setting. It does not seek to replace the tradition of direct student exchange but opens up alternative dimensions to living and studying in another country" [46].

Facilitated by new learning technologies and pedagogies, the concept and practice of the international virtual medical schools is innovative and revolutionary. It is "a transnational approach in which internationalization is integrated and embedded within a curriculum and involves collaboration between a number of schools in different countries," as noted by Harden [49]. As new collaborative global education programs are developed, new challenges will arise, as well, and we must be prepared in the transformation of global healthcare education programs.

REFERENCES

[1] McElmurry BJ, Kim S, Al GN. Global nursing leadership: a professional imperative. Semin Nurse Manag 2000;8(4):232–8.

[2] Austin W. Nursing ethics in an era of globalization. ANS Adv Nurs Sci 2001;24(2):1–18.

[3] Walker Jr B, Evans C, Mouton C. Workshop on global health trends for health professional education. J Health Care Poor Underserved 2006;17(1):12–6.

[4] Woodward D, Drager N, Beaglehole R, Lipson D. Globalization and health: a framework for analysis and action. Bull World Health Organ 2001;79(9):875–81.

[5] Schwarz MR, Wojtczak A. Global minimum essential requirements: a road towards competence-oriented medical education. Med Teach 2002;24(2):125–9.

[6] Martimianakis MA, Hafferty FW. The world as the new local clinic: a critical analysis of three discourses of global medical competency. Soc Sci Med 2013;87:31–8.

[7] Bozorgmehr K, Saint VA, Tinnemann P. The 'global health' education framework: a conceptual guide for monitoring, evaluation and practice. Global Health 2011;7:8.

[8] Frenk J, Chen L, Bhutta ZA, Cohen J, Crisp N, Evans T, et al. Health professionals for a new century: transforming education to strengthen health systems in an interdependent world. Lancet 2010;376(9756):1923–58.

[9] Rowson M, Smith A, Hughes R, Johnson O, Maini A, Martin S, et al. The evolution of global health teaching in undergraduate medical curricula. Global Health 2012;8:35.

[10] Bozorgmehr K. Rethinking the 'Global' in global health: a dialectic approach. Global Health 2010;6:19.

[11] Koplan JP, Bond TC, Merson MH, Reddy KS, Rodriguez MH, Sewankambo NK, et al. Towards a common definition of global health. Lancet 2009;373(9679):1993–5.

[12] Enskar K, Johansson I, Ljusegren G, Widang I. Lecturers' experiences of participating in an international exchange. Nurse Educ Today 2011;31(6):541–6.

[13] Wright DJ. Planning a study abroad clinical experience. J Nurs Educ 2010;49(5):280–6.

[14] Jaspers MW, Gardner RM, Gatewood LC, Haux R, Schmidt D, Wetter T. The international partnership for health informatics education: lessons learned from six years of experience. Methods Inf Med 2005;44(1):25–31.

[15] Syed SB, Dadwal V, Rutter P, Storr J, Hightower JD, Gooden R, et al. Developed-developing country partnerships: benefits to developed countries? Global Health 2012;8:17.

[16] Wageningen UR. Sandwich Phd candidates and the Sandwich fellowship. http://www.wageningenur.nl/en/Education-Programmes/PhD-Programme/Categories-of-PhD-Candidates/Sandwich-PhD-Candidates.htm; 2014. [accessed 10.04.14].

[17] Macfarlane SB, Agabian N, Novotny TE, Rutherford GW, Stewart CC, Debas HT. Think globally, act locally, and collaborate internationally: global health sciences at the University of California, San Francisco. Acad Med 2008;83(2):173–9.

[18] Giessen Graudate Centre for Life Sciences. Lab rotations abroad. http://www.uni-giessen.de/cms/fbz/zentren/ggl/ggl_inter/lab_rotation; 2014. [accessed 16.04.14].

[19] Tsinghua University. Tsinghua university medical students gets research training in Pittsburgh university. . http://www.tsinghua.edu.cn/publish/news/4207/2011/20110506140757493808326/20110506140757493808326_.html; 2011. [accessed 03.04.14].

[20] Hogstedt C, Ahlbom A, Aragon A, Castillo L, Kautsky N, Liden C, et al. Experiences from long-term research cooperation between Costa Rican, Nicaraguan, and Swedish institutions. Int J Occup Environ Health 2001;7(2):130–5.

[21] Wikipedia. Double degree. http://en.wikipedia.org/wiki/Double_degree; 2014. [accessed 29.04.14].

[22] University of Technology Sydney. Uts: dual-doctoral program student guide. http://www.uts.edu.au/sites/default/files/utsi-dual-doctoral-degree-guide.pdf; 2014. [accessed 10.04.14].

[23] Lee JS. MD-Phd students in the 1990s: finding a niche for the dual degree. JAMA 1995;274(21):1736–7.

[24] Zhang L, Du W-W, Zhou Y-F, Cui S, Li X-J, Sun Q-D. Reflections on MD-Phd dual degree programs. Chin J Med Educ 2012;32(2):289–92.

[25] Department of Biomedical Engineering at Georgia Tech & Emory University. GT/Emory-PKU BME joint venture. http://pku.bme.gatech.edu/; 2014. [accessed 28.03.14].

[26] Duke-NUS Graduate Medical School. The Duke-NUS Story. https://www.duke-nus.edu.sg/about/duke-nus-story; 2015. [accessed 23.01.15].

[27] Duke Kunshan University. About Duke Kunshan. http://dku.edu.cn/en/about/duke-kunshan; 2015. [accessed 23.01.15].

[28] Valani R, Sriharan A, Scolnik D. Integrating CanMEDS competencies into global health electives: an innovative elective program. CJEM 2011;13(1):34–9.

[29] Seymour B, Benzian H, Kalenderian E. Voluntourism and global health: preparing dental students for responsible engagement in international programs. J Dent Educ 2013;77(10):1252–7.

[30] Clement RC, Ha YP, Clagett B, Holt GE, Dormans JP. What is the current status of global health activities and opportunities in us orthopaedic residency programs? Clin Orthop Relat Res 2013;471(11):3689–98.

[31] Tang H-M, Zhang M-J, Zhang Y, Chen G-L, Hu Y-Q. Implementation and evaluation of Sino-French cooperation on medical education. J Shanghai Jiaotong Univ Med Sci 2008;28(z1):64–7.

[32] Office of Education of Peking University Health Science Center and Office of International Cooperation of Peking University Health Science Center. 2013 Short-term oversea student exchange program anthology. Beijing: Peking University Health Science Center; 2013.

[33] Office of International Cooperation of Peking University Health Science Center. 2013 office of International Cooperation of Peking University Health Science Center Annual Report. Beijing: Peking University Health Science Center; 2013.

[34] Elliott CG, Barst RJ, Seeger W, Porres-Aguilar M, Brown LM, Zamanian RT, et al. Worldwide physician education and training in pulmonary hypertension: pulmonary vascular disease: the global perspective. Chest 2010;137(Suppl. 6):85S–94S.

[35] De-Graft AA, Boynton P, Atanga LL, Dosanjh M, Magrin G. Developing effective chronic disease interventions in Africa: insights from Ghana and Cameroon. Global Health 2010;6:6.

[36] Drain PK, Holmes KK, Skeff KM, Hall TL, Gardner P. Global health training and international clinical rotations during residency: current status, needs, and opportunities. Acad Med 2009;84(3):320–5.

[37] Izadnegahdar R, Correia S, Ohata B, Kittler A, ter Kuile S, Vaillancourt S, et al. Global health in Canadian medical education: current practices and opportunities. Acad Med 2008;83(2):192–8.

[38] Duffy ME, Farmer S, Ravert P, Huittinen L. International community health networking project: two year follow-up of graduates. Int Nurs Rev 2005;52(1):24–31.

[39] Bender A, Walker P. The obligation of debriefing in global health education. Med Teach 2013;35(3):e1027–34.

[40] Hawthorne L, Minas IH, Singh B. A case study in the globalization of medical education: assisting overseas-born students at the University of Melbourne. Med Teach 2004;26(2):150–9.

[41] Del PPR, Fins JJ. The globalization of education in medical ethics and humanities: evolving pedagogy at Weill Cornell medical college in Qatar. Acad Med 2005;80(2):135–40.

[42] China Scholarship Council. About Us. http://en.csc.edu.cn/About/c309df7fb3fa40b3a179a7ad93f11988.shtml; 2014. [accessed 30.04.14].

[43] Commonwealth of Australia. Study overseas. http://www.studyoverseas.gov.au/Pages/default.aspx; 2014. [accessed 30.04.14].

[44] Ouma BD, Dimaras H. Views from the global south: exploring how student volunteers from the global North can achieve sustainable impact in global health. Global Health 2013;9(1):32.

[45] Bernstein J. CORR insights (R): what is the current status of global health activities and opportunities in us orthopaedic residency programs? Clin Orthop Relat Res 2013;471(11):3699–700.

[46] Todhunter F, Hallawell B, Pittaway D. Implementing a virtual exchange project for student nurses in Queensland and Nottingham. Nurse Educ Pract 2013;13(5):371–6.

[47] Koehn PH. Globalization, migration health, and educational preparation for transnational medical encounters. Global Health 2006;2:2.

[48] Reitmanova S. Cross-cultural medical education and training. Bratisl Lek Listy 2008;109(2):82–7.

[49] Harden RM. A new vision for distance learning and continuing medical education. J Contin Educ Health Prof 2005;25(1):43–51.

ABOUT THE AUTHORS

Weiwei Du, PhD is a program manager and assistant professor in the Office of International Cooperation, Peking University Health Science Center. He currently works in the area of international collaboration projects, student exchange programs, and MD–PhD co-training programs.

Jianlin Hou, PhD is an associate professor, Institute of Medical Education, Peking University Health Science Center. His research interests include system-based health professional education, students' financial aid and career preference, financing, and costs of medical education.

Yang Ke professor of oncology, is Executive Vice President of Peking University (PKU) and Executive Vice President of Peking University Health Science Center (PUHSC) in charge of the overall administration of PUHSC. Prof. Ke is a foreign associate of the Institute of Medicine of the National Academies in Washington, DC.

Qiudan Sun, MA is Director of the Office of International Cooperation of Peking University Health Science Center (PUHSC). Professor Sun also serves as Vice Chair of the Department of Applied Linguistics at the School of Foundational Education, PUHSC.

Lei Zhang, MD, PhD is an Associate Director of the Office of International Cooperation, Peking University Health Science Center (PUHSC). She was previously a research associate at the Department of Biochemistry and Molecular Biology, Dalhousie University in Canada from 2002–2007.

Chapter 11

The Future of Graduate Medical Education: Is There a Path Forward?

Philip A. Pizzo, Clarence H. Braddock III and Charles G. Prober

HOW DID WE GET TO THE CURRENT STATE OF GRADUATE MEDICAL EDUCATION?

Graduate Medical Education (GME), more specifically internship and residency training, dates back to the first half of the nineteenth century, when the concept of the house pupil was first developed in Cincinnati. As medicine migrated from its proprietary focus of the nineteenth century, to the evolution of the modern academic health center following the 1910 Flexner Report, GME became an integral component of the continuum of medical education.

Interns and residents have been romanticized in a panoply of dynamically changing popular images over the past 100 years. The house pupil, under the tutelage of pathfinding leaders like Sir William Osler or William Halsted, symbolized the long tradition of clinical medicine as an apprenticeship model of education. A series of American theatrical films in the 1930s, becoming a popular television series in the 1960s, depicts a caring, compassionate young Dr. Kildare, guided by a wizened and often challenging senior Dr. Gillespie. Additional real-life portraits are described in Barron Lerner's recent memoir *The Good Doctor. A Father, a Son and the Evolution of Medical Ethics* (2014, Beacon Press, Boston). This theme has continued over the years with portraits of interns and residents ranging from the idealistic (e.g., *St. Elsewhere, ER, Ben Casey*) to parody and dark humor—with *Scrubs* and *Grey's Anatomy* as more recent renderings.

The image of the intern and resident has also been the topic of a remarkable number of books about personal journeys (e.g., Robert Marion's *The Intern Blues: The Timeless Classic About the Making of a Doctor*), the insights of an intern who also became a patient (Sandeep Jauhar's *Intern: A Doctor's Initiation*), and satires (e.g., Stephen Bergman's *The House of God* under the pen name of Sam Schem). The literature is also full of self-help books for everyone from interns to chief residents—offering stories and reflections on what can be both a glamorous and humbling professional journey.

Reflected with various tonalities in works of fiction and fact is the deep personal struggle involved in becoming a doctor, and the indelible mark that internship and residency leaves on physicians-in-training. The life and death issues confronted, along with the other demanding challenges that residents face during a long and arduous period of training, have been a rite of passage unique to medicine. Traditionally, this has often been accompanied by a brute force style of teaching that is too often a humiliating experience for trainees. Coupled with this are long work hours and resulting fatigue that may contribute to personal unwellness and potential medical errors. Symptoms of burnout and depression and a culture of cynicism are far too prevalent during residency training. The sometimes antithetical emotions, reactions, and reflections which occur during residency were captured in a recent *Perspective* article in the New England Journal of Medicine, *Getting Through the Night* (2013;369:2279–2281), in which Dr. Perri Klass contrasts the personal toll of residency against the fond remembrances of having overcome almost unspeakable challenges.

A notable transition in how the training of doctors was perceived and valued occurred after passage of Medicare in 1965, when GME became an important source of revenue for some teaching hospitals. As public and private teaching hospitals became the recipients of GME payments from Medicare, universities and medical schools lost their direct control over GME. With Medicare subsidizing resident salaries as well as providing indirect GME payments to hospitals, the incentives for hospitals to increase the number of residents was irresistible. The resulting increase in the number of residents continued unabated until the Balanced Budget Act of 1997 established caps on the number of incremental resident positions that Medicare would fund.

Workforce trends in GME have tracked in close parallel with the incentives guiding American medicine more generally. Specifically, we have too many trainees seeking to become specialists and too few being trained as primary care physicians. Residency training has fostered and reinforced habits of overutilization of tests

The Transformation of Academic Health Centers. http://dx.doi.org/10.1016/B978-0-12-800762-4.00011-6

and procedures, analogous to the procedure-based, fee-for-service model that has negatively characterized the US health care system. In addition, residency training is associated with an exaggerated focus on disease and hospital-based care rather than preservation of health and attention to ambulatory-based delivery systems. Finally, there is too little attention paid to the quality and safety of patient care and practically no focus on the costs of tests and care being rendered. Unfortunately, many still consider medicine as a group of guilds rather than as a complex interdisciplinary care system.

As detailed in chapter 12, medicine has created a complex array of professional societies and agencies involved in accrediting, certifying, and licensing undergraduate, graduate, and continuing education programs, each addressing their own area of interest. This has resulted in a lack of harmonization, making it difficult to integrate science and evidence-based clinical care that is delivered with high quality and safety in a patient-centered and value-based manner. All aspects of the medical industry complex have a stake in GME, including: hospitals; professional societies; the drug and device industry; allied professional groups (especially nursing, pharmacy, and public health); employers; insurance companies; local, state, and federal government; and patient advocacy groups. Similar to the debate around health care reform, most everyone recognizes the need for changes in GME, but efforts to make substantial changes have been limited. Many physician-led professional organizations and societies project the need to expand the physician workforce to meet the health care needs of the nation, thus promoting the need to increase the number of Medicare-funded GME training positions. However, these projections do not adequately consider the potentially changing role of doctors as a consequence of emerging technologies, the expanded roles of other health care providers, and the shift from disease-focused medicine to the health and wellness of patients and communities.

New devices and technologies are changing the way patients engage their doctors, including where and when medical encounters take place. As workflow changes and becomes more efficient, fewer doctors may be needed to provide medical care. In addition to critically examining the number of physicians needed for the future, we need to improve the balance between primary and specialty care trainees. In our analyses, it also is important to examine how the role of other health professionals will impact health care delivery and physician workforce needs.

While many individuals and organizations have given considerable thought to the future of GME, progress has been mostly reactive. The most notable reform of GME in the last decades relates to work-hour restrictions. The tragic death of Libby Zion in 1984 led to the Bell Commission

Report of 1987[1,2] that triggered a number of dramatic changes in resident on-call schedules and changed the paradigm of resident supervision. The major transition in work-hour rules followed reforms by the Accreditation Council for Graduate Medical Education (ACGME) in 2003 and 2011. These reforms resulted in limitations on the number of consecutive and total hours an intern and resident can work on a shift and total hours per week. There is no doubt that the issues that stimulated these reforms were real, or that reforms of some kind were necessary and important. Nonetheless, the rationale for some specific changes (particularly the 16-hour shift limit for interns) is quixotic, and not supported by data that demonstrate their benefit on either the physical or emotional well-being of residents or in patient safety. Furthermore, there is growing concern that some of these restrictions have negatively impacted resident education.

The balance between learning and service during residency is viewed differently by the trainees, program directors, clinical departments and their faculty, hospital staff, and other leaders. The economics of GME funding impacts how academic health centers and teaching hospitals approach the utilization and time allocation of trainees. The calculus will become even more complex if GME funding from the Centers for Medicare and Medicaid Services (CMS) is substantially altered. The loss of GME funding would negatively impact the balance sheet for teaching hospitals and would impact the clinical work responsibilities and expectations for clinical faculty within academic health centers.

Because GME funding comes directly to teaching hospitals, nearly all residents are designated as hospital employees, even though much of their supervision and oversight comes from faculty employed by medical schools or clinical practice plans. Most clinical faculty are deeply committed to resident education, and the balancing act many faculty face in their time allocations and contributions to education, research, and patient care is substantially buffered by the availability of residents and clinical fellows. These factors contribute to a lack of enthusiasm among faculty to consider changes in GME. Furthermore, the primary anchor for GME in the United States is hospital based and disease focused, whereas the future of medicine needs to optimize ambulatory care and population management. Thus, historical, organizational, and economic issues limit the appetite for transforming GME. But the tectonic shifts impacting medicine overall will change the future of GME; although, the question remains whether these changes will be reactive or proactive. It is notable that the 2014 Institute of Medicine (IOM) report, *Graduate Medical Education That*

1. Philibert I, Friedmann P, Williams WT. New requirements for resident duty hours. JAMA 2002;288:1112–1114.

2. Lerner BH. A case that shook medicine. Washington Post. November 28, 2006.

Meets the Nation's Health Needs,[3] has identified many of these same issues and proposes fundamental reforms in how GME is funded, organized, and modernized. While the IOM acknowledges that the current aggregate amount of funding for GME should be sustained, it recommends that "the current Medicare GME payment system should be phased out." The IOM calls for "building a GME policy and financing structure" within the Office of the Secretary of the US Department of Health and Human Services to provide oversight and develop a strategic plan for GME financing. It also calls for establishing a "GME Center within the Centers for Medicare and Medicaid Services (CMS)" to manage the operational aspects of GME funding and the GME Transformation Fund that "would finance initiatives to develop and evaluate innovative GME programs, to determine and evaluate appropriate GME performance measures, to plot alternative GME payment methods, and to award new Medicare-funded GME training positions in priority disciplines and geographic areas." The IOM calls for fundamental changes in the current Medicare-based payment systems with methodology that better optimizes the outcome, performance, and geographic distribution of GME programs. In the conclusion to the Executive Summary, the 2014 IOM Committee on GME recommends "that Medicare GME funding be leveraged toward the achievement of national health care objectives. Continued federal funding should be delivered by a system that ensures transparency and accountability for producing a workforce suited to the needs of the healthcare system."[4]

IS THERE A PATH FORWARD? TO WHAT DEGREE WILL THIS REQUIRE DISRUPTIVE INNOVATION?

GME stands at the nexus in the education of the physician, linking undergraduate medical school education to a future career in medicine and science. Despite their importance, changes and reformations in resident education have been incremental, not revolutionary. The same can be said of undergraduate medical education and even continuing medical education—each of which have changed much less than might be expected in the past century. Because the role of the physician as the essential provider of health care is being impacted by the expanded roles of nurses and other professionals, it is important to assess the impact of these changes on GME. However, the lack of change is based on tradition and a belief that the ways doctors have been trained in the past are still best for the future. Some of the resistance to change is more perverse and driven by economic and

financial gains and losses to academic health centers. That said, fundamental changes in GME are inevitable.

Defining a path forward for GME begins with how the role of the physician is likely to evolve in the years ahead and whether those changes will impact the number of physicians who need to be trained. If more doctors are needed in the United States to care for an aging population and an increasing prevalence of patients with chronic diseases, it is appropriate to question how the roles of future physicians will differ from those currently in practice. Transformation in resident education is a fundamental step toward preparing contemporary physicians for the future practice of medicine. In addition to the location and scope of GME education, the appropriate duration of residency training is also important to examine.

Does the Path Forward Require an Expansion of GME?

In 1952, the first year of the National Resident Matching Program, approximately 10,400 internship positions were available for 6000 US seniors graduating from allopathic medical schools. Following steady growth, the number of Postgraduate Year 1 (PGY-1) positions reached a record high of 26,678 in 2014. This same year, there were 34,270 applicants, of whom 17,374 (50.6%) were US seniors at allopathic medical schools, 2738 graduates of osteopathic schools, and 12,701 graduates from international medical schools (of whom 5133, or 40%, were US citizens). The number of graduates from US medical schools is expected to increase to 21,349 by 2018. Resulting from an expansion in the number of US medical schools and average class size, an increased number of graduates are projected to meet the putative physician workforce needs—estimated by some to be approximately 90,000 physicians by 2020—divided between primary care and other medical and surgical specialties.[5]

By these accounts, the number of residency positions nationally should grow, lest there be insufficient numbers of training opportunities for the increasing number of US medical school graduates. We have entered an unprecedented period in which a US allopathic MD graduate may not be able to pursue the specialty training he or she wants, or even pursue residency training at all.

The Association of American Medical Colleges (AAMC) has been among the major proponents of the looming physician shortfall projections, estimated to be approximately 90,000 physicians by 2020, driven largely by the rising population of seniors and higher number of insured patients seeking health care once the Affordable Care Act (ACA) is fully implemented. While most academic health center leaders

3. Institute of Medicine. Graduate Medical Education that meets the Nation's health needs. National Academies Press; 2014.

4. Institute of Medicine. Graduate Medical Education that meets the Nation's health needs. National Academies Press; 2014.

5. AAMC 2013. State Physicians Workforce Data Book. https://www.aamc.org/data/#; November 2013.

appear to have accepted the physician shortage as reality, we do not. As noted, the projections for new physicians are based on historical models of care delivery and do not take into account the broader group of nonphysician health care providers—including nurses, nurse practitioners, physician assistants, pharmacists, and others. The projections also assume that care in the future will be delivered much as it has been in the past—which, hopefully, will not be the case. The overuse of technologies and invasive procedures needs to be controlled. We will still need highly-trained physicians and surgeons, but we believe that the scope of their work will change substantially over the coming decades.

The use of nonphysician health care providers will continue to expand. This will necessitate a culture-shift in expectations by the public and by the medical community, but it will ultimately be driven by economic realities. For example, two of the authors of this chapter are pediatricians, and we recognize that many of the children now being cared for by pediatricians can be managed well by other providers. This allows pediatric-trained physicians to focus on children with complex chronic care or other serious disorders. Similarly, the growth and maturation of the primary care medical home holds promise as a delivery model that can meet population health care needs without as many physicians as in traditional primary care delivery models. When the reimbursement system shifts away from fee-for-service, different models of care will be configured—and, with that, the education of physicians during residency also will change.

Does the Path Forward Require a Change in How GME Is Paid and Administered?

To whatever degree GME is expanded, the question of who will pay for the incremental positions will loom large. Currently, most GME funding comes from Medicare, with the exception of those supporting residents in freestanding children's hospital where the Children's Hospital Graduate Medical Education Program (CHGME) is the vehicle for funding.

While Medicare funding for all but children's hospitals began in 1965, CHGME was first enacted in 1999 in recognition that children's hospitals care for few Medicare patients and thus receive minimal Medicare GME funding. When the legislation supporting CHGME was put into place, it overcame this inequity but also made clear that GME funding and Medicare do not have to be linked. In fact, that linkage is historical and harkens back to the view that teaching hospitals required special accommodation because they served as safety-net facilities for uninsured patients. Currently, this amounts to approximately $9.5 billion in Medicare support and $2 billion from Medicaid, making the federal government the highest payer for GME.

Presently, there are approximately 115,000 physicians in residency training programs, which averages to about $100,000 per resident per year. Funds for GME are distributed as direct payments for the salaries of residents and indirect payments to help subsidize hospitals for the incremental expenses accrued in managing a training program. While these expenses had a rationale when Medicare was first enacted, justification for continuing this level of support for GME is being challenged on various levels—including the MedPac Oversight committee as well as budgetary allocation reviews by congressional committees (e.g., Simpson Bowles) and Executive Branch projections. These groups have argued that this current form of payment does not incentivize teaching hospitals to train physicians in a manner that adds value in terms of cost, patient experience, or quality. The more CMS focuses on changing the payment for doctors to include metrics on quality and safety, in addition to volume and procedures, the more GME funding is likely to change.

A reasoned path forward is to address the question of workforce needs and costs more strategically, with a view to right sizing the health care team in a more interdisciplinary team-based manner, which can more effectively address health and care of patients and communities, and help assess the value derived from expanded GME programs for individuals and institutions. Given the broad number of constituencies engaged in health care delivery, the model of the "all payer pool" advocated during the health care reform debates of the Clinton health care era still makes sense, even though it remains unlikely to achieve consensus support. To date, hospitals and physician practices have added financial resources to cover the shortfall in GME positions not covered by Medicare—either because they exceed the resident cap or because they are in specialties not recognized by Medicare. A common reason for hospitals to provide this financial support has largely been based on the assumption that replacement of clinical services currently provided by residents will be more expensive if other providers (e.g., nurse practitioners, hospitalists) are engaged. The reality is that teaching hospitals will need to find a way to be less reliant on the clinical services provided by residents, and that other professionals will have to fill in more of the gaps in clinical service. An important question is whether these changes can be achieved if residents remain hospital employees whose primary expectation (including on the part of their employer) is predominantly inpatient clinical care and management. Certainly, there is great value to medical knowledge and experience acquired in complex medical settings, but there is also the reality that this model of health care delivery has largely over utilized medical services.

As the focus of medical care shifts toward health preservation and community-based management, the locus and focus of residents will also need to change. As noted above, the 2014 IOM Report has called for major reform in how Medicare GME is funded, evaluated, and distributed.

While the reforms called for by the IOM will be disruptive, they represent a well-reasoned strategy for fundamental change that could ultimately enable a better path forward for GME in the future.

Does the Path Forward Require Fundamental Changes in How and Where Residents Work along with What They Are Taught and Expected to Learn?

While the apprenticeship model of GME education and its regulation by the ACGME and various professional organizations has resulted in the training of many outstanding physicians, a critical review might conclude that many GME programs are not adequately training physicians for the work they will be doing in the years ahead. Most of the care models in teaching hospitals are still doctor and hospital oriented rather than patient and family focused. In many ways, technology has proven to be a dual-edged sword. On the one hand, the increasing availability of electronic medical records, along with the emergence of novel hand-held devices and applications, is changing the way doctors and patients interact for the better. Previously unimagined technologies, such as the Google Glass, could offer new opportunities for physicians to access information in ways that overcome some of the barriers that exist with the electronic medical records. At the same time, the increasing use of technology separates physicians and care providers from the patients they serve and can dramatically alter the cherished role of the physician as doctor. Clearly, a balance is needed.

The value of a master clinician who listens to his or her patients carefully, forging a human relationship of shared respect and value, is still very much what most medical providers would seek in their own care. Sadly, over time this style of medicine has become a dwindling resource, and the willingness or ability of physicians to fulfill the more lofty roles and personalized care epitomized by Dr. Welby of the past appears to be a victim of modernity. Today, the term "personalized medicine" is associated with delivering treatments that are guided by one's personal biology and genomics, rather than by the doctor who has a personal relationship with the patient and family. These do not need to be mutually exclusive goals—but they seem to have become so. Most medical students initially approach the doctor–patient relationship with altruism and idealism, only to find their ideals devolve during the period of exposure in the clinical years of training. Rarely do young physicians-in-training encounter senior doctors who sit down upon entering a patients room, provide an introduction, listen to the patient's and family's concerns, and address their questions during a medical encounter. The absence of these behaviors is accompanied by a loss of patient satisfaction and, to varying degrees, a loss of value attributed to physicians and other health care providers.

The art of history-taking and listening to patient concerns must be part of the education and training of residents. This requires time, commitment, and purpose by clinical faculty and teaching institutions. Programs such as C-I-CARE, initially introduced at UCLA and now practiced at many institutions, attempt to incorporate patient service by all members of the health care delivery team in both ambulatory and hospital settings. Seen objectively, the methods and goals are straightforward and obvious, but they need to be embedded in the institutional culture—along with professionalism and respect. Residents play a key role in the success of such programs because of their critical intersections with patients, students, nurses, and other professionals and senior physicians. That said, programs such as C-I-CARE cannot be successful unless they are valued and vigorously supported by institutional leaders (e.g., hospital CEO, medical school dean and clinical department chairs, clinical faculty, chief residents, etc.), and become an integral underpinning in the institutional culture.[6]

While goals around thoughtful patient communication seem obvious, their success requires measurement, evaluation, feedback, and consequences. All providers, including residents, need to be aware of the expectations for patient service and held accountable for their performance through assessment and feedback. The new reality is that physicians and institutions will be compared individually and collectively in publicly available data sources, impacting both where patients, as consumers, go to receive medical care and how medical providers are reimbursed for the services they provide (or fail to deliver). Unless the systems for high-quality and patient-centered care are well defined, clearly articulated, expectant, and accountable, it is not likely that medical care service will improve. It is imperative that these models be incorporated into the education and training of residents such that they become part of lifelong learning.

In addition to listening to patients and using history-taking to formulate diagnosis and care plans, it is also important to resurrect the value of the physical examination. It is true that technology can be used to accurately diagnose many medical conditions and illnesses. And while appropriate technology should be embraced and utilized, the doctor–patient relationship is enhanced when physicians forge a human connection with the individuals they are caring for—a view well articulated by Dr Abraham Verghese.[7]

Over the past decades, bedside rounds and even the value of clinical examination seem to have also become a lost art that deserves reconsideration.[8] Acquiring the skills to perform a physical examination necessitates access to master

6. See: http://medicalstaff.stanfordhospital.org/mss/credentialing/application/md_c-i-care_video.html.

7. Verghese A. Culture shock – patient as icon, icon as patient. N Engl J Med 2008;359:2748–2751.

8. See: McGee S. Bedside teaching rounds reconsidered in JAMA 2014; 311:1971–1972.

clinicians. While some of the aspects of this instruction can be facilitated by immersive learning technologies and virtual reality, there is no substitute for direct patient engagement. To make this effective, learning technologies should be viewed as an adjunct or prelude to direct, patient-centered bedside care and teaching—in inpatient and ambulatory settings. Once again, this can only be achieved if institutional leaders embrace the value of physical examination—as has been done at Stanford with the Educators-4-CARE program used to teach medical students[9] and the "Stanford 25" championed by Abraham Verghese and the Department of Medicine.

In addition to a renewed focus on the art of listening, communication, physical examination, and the appropriate use of technology as a learning tool and patient care modality, attention must be given to improving the quality and safety of patient care, along with its value as measured by evidenced-based outcomes and cost. While a premium has been placed on resident knowledge of both common and esoteric medical facts, too little attention has been placed on whether the care being delivered has proven value, benefit, and safety. Given the dramatic changes that will unfold as medical care delivery shifts from a fee-for-service model to one that gives more attention to evidence-based outcomes and health preservation—as well as the quality, safety, cost, and value of care being delivered—fundamental changes in GME education around these domains will be essential.

While individual institutional programs addressing safety and quality—some engaging residents—are important and valuable, much greater attention to more systemic change is needed. A real problem in achieving these changes is the unfortunate reality that generations of physicians and health care systems have given little credence to these aspects of medicine and are not well equipped to guide or transform it. Without question, the education of residents in professionalism, communication, quality, safety, and value is essential in creating the medical workforce for the future. Accordingly, programs emerging from the Institute for Healthcare Improvement (IHI) and the American Board of Internal Medicine Foundation's "Choosing Wisely" campaign will be essential to lead and implement these changes. To incorporate them into the future of medicine will require new metrics to evaluate physicians and institutions and make their performance publicly available to patients, consumers, and payers. If these changes are to be achieved, the transformation of GME will be pivotal and essential. One can envision that if CMS or other agencies continue to pay for GME, they will make performance standards in patient safety, quality, and value a critical component of their reimbursement for clinical care. With that in mind, the way forward requires

fundamental changes in the education and training of future residents and fellows.

Does the Path Forward Require a Restructuring of Residents' Tracks and Pathways to a Closer and More Vertical Alignment with Undergraduate, Postgraduate, and Continuing Medical Education?

While much has been said about the importance of balancing the workforce pipeline for primary care and specialty physicians, a medical student's decision about which clinical path to pursue still defines the distribution of the physician workforce. This requires a more data-driven assessment of workforce needs, based more on future realities than past practice patterns. This necessitates a local, regional, and national dialog about resident education and a strategic assessment of future needs that guides programmatic expansion.

In the absence of rigorous planning that couples strategic expansion in some areas of medicine and surgery with limits in other specialties, workforce needs will likely be guided more by lobbying than by actual needs. Over the past decade, the choice of residency has also been influenced by lifestyle and economic factors, favoring disciplines such as dermatology and radiology and making them highly competitive for the best students—even if these are not the areas of medicine where the greatest need for services truly exists. Not surprisingly, somewhat perverse incentives have overvalued procedure-based specialties, undervalued those that deliver primary medical and wellness care, and impacted the choice of residency training that medical graduates pursue. While we may have difficulty in expanding the number of CMS-funded GME positions in the aggregate, institutions can rebalance the distribution of these training slots across truly needed services (primary care vs specialty). However, the will to do this is often lacking given the competing pressures and expectations of professional groups—who also align with a hospital's perceived needs—not infrequently influenced by financial incentives.

Attention must also be given to other factors that impact medical education at the undergraduate, graduate, and postgraduate levels. Medical education has become enormously expensive, and the average debt incurred by graduating medical students has important implications for the personal and professional lives of future physicians. While it would be good to have these costs reduced, the likelihood of this happening in our current educational models is unlikely. Also, the prolonged amount of time required for training requires fundamental changes in the highly compartmentalized organization of medicine as we know it today. The challenges are made more difficult because of the separate regulatory boards and organizations

---

9. See: http://med.stanford.edu/e4c/.

responsible for medical education and training. Although this complex array of accrediting, certifying, and licensing boards and organizations is immutable to facile change, it is worth examining ways of creating better integration and coordination around the otherwise disparate strata of medical education. Whether it is essential for all physicians to have an undergraduate baccalaureate degree, versus the path to an MD degree more comparable to many European and Asian programs that begin medical education at the high school level, is worthy of discussion. That current resident and fellow programs require 2–10 or more years of training is also worth critical reexamination. This might depend on the purpose of the education and training and how medical schools organize their programs and expectations.

A path forward might begin with more crisply defined expectations of medical schools and teaching hospitals. The concept that individuals can be proficient in multiple domains (e.g., the triple threat of an academic physician who is an accomplished clinician, researcher, or teacher) or that an institution can be great in research and patient care has been long cherished, but requires critical reexamination if we are to create paths forward. The medical school, residency, and fellowship curriculum and training (in length, scope, and outcomes) are likely highly different for those who will become successful physician-scientists versus those who will pursue careers in clinical medicine as a primary care physician or specialist. Currently, we use role models and even expectations that are not forward-looking or even realistic. Worse is that our culture often offers hierarchical and even parochial views of setting value and excellence. Most institutions and leaders are steeped in creating images of the past and individuals who look like them. Here, too, the path forward requires a fundamental reshaping of values and expectations. This will require major disruptive innovations to break through the status quo.

It is reasonable for medical schools and teaching hospitals to differentiate even more as to which components of the physician workforce they will focus on. Given that there are variegated needs, success depends on institutional resources and expertise. For example, the future workforce requires sustenance and even expansion of successful physician-scientists. While virtually every medical school ascribes to the notion that educating physician-scientists is part of its mission, the reality is that relatively few institutions are able to do this successfully. Further, the path for future physician-scientists may benefit from greater vertical integration, beginning prior to medical school and extending through fellowship training. The current model that results in a physician-scientist achieving his or her first NIH RO1 award at 43 years of age underscores a need for reform in length, scope of training, and outcomes. For individuals interested in pursuing a physician-scientist track, undergraduate courses and research activities might obviate the need for medical school basic science courses. Depending on whether the future physician-scientist is going to focus on research primarily or research coupled with patient care, a different balance of clinical versus research-related experiences could be configured. While this would limit the breadth of education for the student and resident, it would have the advantage of creating more depth in a chosen field of interest and of consolidating course work, clinical rotations, and research experiences. This would require significant collaboration among colleges, medical schools, and residency and fellowship training programs. Ideally, programs would also create flexibility for career development beyond residency and fellowship so that physician-scientists could be more successful and experience less attrition. This is particularly important for women physician-scientists whose attrition has exceeded that of their male counterparts.

Similar vertical integration and innovation is important for programs training primary care physicians as well as medical and surgical specialists. GME programs are at the nexus between college, undergraduate medical education, and postgraduate training. Academic health centers should take the lead in transforming resident and medical education—fostering knowledge development that is broad, deep, and continuing; expanding with technology; and focusing on patient care, service, quality, safety, and value. That will require adding to the knowledge repertoire of practicing teaching physicians, as well as trainees, in tandem with metrics to measure and facilitate the enormous cultural transformation that is now required for medicine in the twenty-first century. Defining the role of the primary and specialty care physician and how these roles interact and intersect with other health professionals in hospital, ambulatory, and community-based care delivery is essential.

THERE IS A PATH FORWARD

The profession of medicine has a rich and enduring history, and the path to becoming a doctor remains an integral and vital force for its future. While the rich traditions of medicine will continue to shape the education and training of tomorrow's physicians, the rapidly evolving depth and breadth of medical science—and the biopsychosocial context in which it is practiced—add new dimensions to the future of medical education. Learning by observation and doing (the classic apprenticeship model) will continue to be a central theme in GME. But today's physician role models will need to embrace new skills and knowledge in order to better define and optimize their effectiveness and success. Clinical knowledge needs to be grounded in a broad understanding of the basic sciences concepts that are fundamental to modern medicine. Added to this is the need for a deeper knowledge in population sciences and, in particular, statistics and informatics. The clinical education of future physicians needs to be grounded in interdisciplinary teams that balance not only the traditional medical and surgical specialties, but include those emanating from other important health professionals, including nurses, pharmacists, psychologists, and others.

The future physician will need to acquire skills that are often less hospital focused and more ambulatory and community-based. That will mean shifting the sites where training takes place along with the individuals who will supervise and educate. Differentiation of the professional paths among future physicians should happen earlier in the journey and foster vertical integration and consolidation of training programs and competency-based expectations and outcomes.

Academic health centers and GME programs also will benefit from differentiation. Opportunities for innovation should be fostered and encouraged. Harmonization of the regulatory oversights for undergraduate, graduate, and continuing medical education must be encouraged to allow new innovations to occur and succeed. The use of technology, including immersive learning tools for learners and new devices for patients, should help shape how knowledge and skills are acquired and how and where medicine is practiced. Such changes in the profile of the medical care team and the tools and technologies employed to deliver care, along with the shifts from the fee-for-service model of health care delivery to one that is patient and population centric, will help define workforce needs more rationally and realistically. Simply adding more doctors to the current mix is not a thoughtful solution to workforce challenges.

Coupled with these changes should be a renewed emphasis on professionalism and service through programs—such as C-I-CARE—along with innovative approaches to teaching quality, safety, and value with programs such as those sponsored by the IHI and the American Board of Internal Medicine. Through integrated efforts that combine scientific and evidence-based medical knowledge with caring and compassionate patient service—delivered with attention to quality, safety, and value—a path forward can be more clearly defined. Building on a foundation created during undergraduate medical education, the education and training during GME can help lead this path forward. While reflecting our opinions, some of the areas that require critical attention to optimize the path forward include:

- **Disruptive innovation in how GME programs are organized and constructed**. While there is no question that residency is a time of exceptional knowledge and skill acquisition in the life of a doctor, there is also little doubt that many of the activities learned have limited relevance to future work actually performed by many physicians. That path forward needs to fundamentally structure the scope and locus of training to meet the future of medicine, rather than the needs of hospitals and academic health centers. For many future physicians this means much greater time spent in ambulatory

settings, whereas for others (e.g., hospitalists) greater inpatient training experiences would be appropriate. For all trainees it means a much greater focus on how to practice medicine that is more patient-centered, evidence-based, quality-driven, service-minded, and value-directed. While residents do need to be conversant in the rapidly changing scientific disciplines—such as genomics and molecular medicine that underpin human biology—they also need to acquire skills in population sciences, critical reasoning, and statistics and clinical informatics. The 2014 IOM report on GME offers recommendations on how to move forward with strategic and more transparent investments.

- **Centers involved in GME need to critically examine their strengths and better delineate their missions**. Modern medicine needs a wide range of talents, from the primary care doctor to the specialist and physician scholar and scientist. GME programs should consider developing areas of focused expertise that optimize training and produce graduates who are truly prepared for the work that they will do. This will require academic health centers to differentiate and, with rare exception, more effectively delineate their training goals. In many cases, this will be abetted by closer alignments with universities or community partners; in nearly all cases, it will necessitate more flexible and creative regulations and policies from accrediting and certifying agencies and societies.

- **There should be much greater harmonization and integration of GME training in the overall preparation of the physician**. Medical education and training remains too compartmentalized and siloed. Academic health centers should form a more visionary and strategic alignment with universities, including their baccalaureate and undergraduate medical education programs, to develop different education paths and opportunities that shorten the duration of training, make it more integrated, and reduce debt burden.

- **The workforce needs for future physicians should be based on more critical strategic planning that fully considers that changing roles of doctors for the future**. Many of the assumptions that have led to increases in medical school class size and that are now calling for increases in Medicare-funded GME positions are based on simplistic suppositions about what different medical professions might want, rather than truly need. Importantly, many of the current projections assume that the role of future physicians will mirror that of those currently in practice, particularly including procedure-based specialties. We believe this is unlikely given the refocusing of medicine on population health and less on disease management and, equally importantly, a focus that recalibrates the scope of medical and surgical practice.

- **GME programs should be better organized to foster team-based medical care that optimizes the contributions of multiple disciplines and providers**, especially including nurses, social workers, pharmacists, and other professionals. The goal should be to test ways of delivering care that is high caliber, evidence-based, quality-driven, service-oriented, and more cost-effective.

GME stands at the crossroads of medical education. It should become a leader of the transformation of medicine and not a servant to the past.

BIBLIOGRAPHY

Adiga K, Buss M, Beasley BW. Perceived, actual, and desired knowledge regarding medicare billing and reimbursement. A national needs assessment survey of internal medicine residents. J Gen Intern Med 2006;21(5):466–70.

Auerbach DI, Chen PG, Friedberg MW, Reid RO, Lau C, Mehrotra A. New approaches for delivering primary care could reduce predicted physician shortage. Santa Monica (CA). http://www.rand.org/pubs/research_briefs/RB9752; 2013. [accessed 21.02.14].

Auerbach DI, Chen PG, Friedberg MW, Reid RO, Lau C, Mehrotra A. Nurse-managed health centers and patient-centered medical homes could mitigate expected primary care physician shortage. Health Aff 2013a;32(11):1933–41.

Auerbach DI, Chen PG, Friedberg MW, Reid RO, Lau C, Mehrotra A. New approaches for delivering primary care could reduce predicted physician shortage. Santa Monica (CA): RAND Health; 2013b.

Bodenheimer TS, Smith MD. Primary care: proposed solutions to the physician shortage without training more physicians. Health Aff 2013;32(11):1881–6.

Bodenheimer T, Chen E, Bennett HD. Confronting the growing Burden of chronic disease: can the U.S. health care workforce do the job? Health Aff 2009;28(1):64–74.

Cassel CK, Reuben DB. Specialization, subspecialization, and sub-subspecialization in internal medicine. N Engl J Med 2011; 364(12):1169–73.

Chandra A, Khullar D, Willensky GR. The economics of graduate medical education. N Engl J Med 2014;370:2357–60.

Chen C, Chen F, Mullan F. Teaching health centers: a new paradigm in graduate medical education. Acad Med 2012;87(12):1752–6.

Cooke M. Cost consciousness in patient care – what is medical Education's responsibility? N Engl J Med 2010;362(14):1253–5.

Cordasco KM, Horta M, Lurie N, Bird CE, Wynn BO. How are residency programs preparing our 21st century internists? A study conducted by staff from RAND Health for the Medicare Payment Advisory Commission. http://www.medpac.gov/documents/Jul09_ResidencyPrograms_CONTRACTOR_CB.pdf; 2009. [accessed 02.04.13].

Crosson FJ, Leu J, Roemer BM, Ross MN. Gaps in residency training should be addressed to better prepare doctors for a twenty-first-century delivery system. Health Aff 2011;11:2142–8.

Dine CJ, Miller J, Fuld A, Bellini LM, Iwashyna TJ. Educating physicians-in-training about resource utilization and their own outcomes of care in the inpatient setting. J Grad Med Educ 2010;2(2):175–80.

Flexner A. Medical education in the United States and Canada: a report to the Carnegie Foundation for the Advancement of Teaching. New York: Carnegie Foundation for the Advancement of Teaching; 1910.

Fuchs VR. The structure of medical education – it's time for a change. Denver (CO): Alan Greg Lecture; November 6, 2011.

Fuchs VR. Current challenges to academic health centers. JAMA 2013;310(10):1021–2.

Ghorob A, Bodenheimer T. Sharing the care to improve access to primary care. N Engl J Med 2012;366(21):1955–7.

Green LV, Savin S, Lu Y. Primary care physician shortages could be eliminated through use of teams, nonphysicians, and electronic communication. Health Aff 2013;32(1):11–9.

Grover A, Niecko-Najjum LM. Building a health care workforce for the future: more physicians, professional reforms, and technological advances. Health Aff 2013;32(11):1922–7.

Iglehart JD. The uncertain future of medicare and graduate medical education. N Engl J Med 2011;365(14):1340–5.

Iglehart JK. Expanding the role of advanced nurse practitioners – risks and rewards. N Engl J Med 2013a;368(20):1935–41.

Iglehart JK. The residency mismatch. N Engl J Med 2013b; 369(4):297–9.

IOM. Best care at lower cost: the path to continuously learning health care in America. Washington, DC: The National Academies Press; 2012a.

IOM. In: Eden J, Berwick D, Wilensky G, editors. Graduate Medical Education that meets the Nation's health needs. Committee on the governance and financing of graduate medical education board on health care services. Washington, DC: The National Academies Press; 2014.

Johns MME, Chair. Ensuring an effective physician workforce for America. In: Macy Jr J, editor. Proceedings of a conference sponsored by the Josiah Macy Jr. Foundation, Atlanta, GA. New York: Macy Foundation; October 24–25, 2010.

Jolly P, Erikson C, Garrison G. U.S. Graduate medical education and physician specialty choice. Acad Med 2013;88(4):468–74.

Kirch DG, Henderson MK, Dill MJ. Physician workforce projections in an era of health care reform. Amer Rev Med 2012;63:435–45.

Korenstein D, Kale M, Levinson W. Teaching value in academic environments: shifting the ivory tower. JAMA 2013;310(16):1671–2.

Ludmerer KM. The history of calls for reform in graduate medical education and why we are still waiting for the right kind of change. Acad Med 2012;87:34–40.

Moriates C, Soni K, Lai A, Ranji S. The value in the evidence: teaching residents to "choose wisely". JAMA 2013;173(4):308–10.

Nasca TJ, Philibert I, Brigham T, Flynn TC. The next GME accreditation system – rationale and benefits. N Engl J Med 2010;366: 1051–6.

Nasca T, Weiss K, Bagian J, Brigham T. The accreditation system after the "Next accreditation system." Acad Med 2014a;89(1):27–9.

Nasca TJ, Weiss KB, Bagian JP. Improving clinical learning environments for tomorrow's physicians. N Engl J Med 2014b;370(11):991–3.

Newhouse JP, Wilensky GR. Paying for graduate medical education: the debate goes on. Health Aff 2001;20(2):136–247.

Post J, Reed D, Halvorsen AJ, Huddleston J, McDonald F. Teaching high-value, cost-conscious care: improving residents' knowledge and attitudes. Am J Med 2013;126(9):838–42.

Quinn KJ, Kane KY, Stevermer JJ, Webb WD, Porter JL, Williamson Jr HA, et al. Influencing residency choice and practice location through a longitudinal rural pipeline program. Acad Med 2011;86(11):1397.

Rosenbaum L, Lamas D. Cents and sensitivity – teaching physicians to think about costs. N Engl J Med 2012;367(2):99–101.

Smith CD. Alliance for academic internal medicine – American college of physicians high value, cost-conscious care curriculum development committee. Teaching high-value, cost-conscious care to residents: the Alliance for academic internal medicine-American college of physicians curriculum. Ann Intern Med 2012;157(4):284–6.

Sommers BD, Desai N, Fiskio J, Licurse A, Thorndike M, Katz JT, et al. An educational intervention to improve cost-effective care among medicine housestaff: a randomized controlled trial. Acad Med 2012;87(6):719–28.

Weinberger SE. Educating trainees about appropriate and cost-conscious diagnostic testing. Acad Med 2011;86(11):1352.

Weinberger SE. Providing high-value, cost-conscious care: a critical seventh general competency for physicians. Ann Intern Med 2011;155(6):386–8.

Weiner JP, Yeh S, Blumenthal D. The impact of health information technology and e-health on the future demand for physician services. Health Aff 2013;32(11):1998–2004.

Weiss KB, Bagian JP, Nasca TJ. The clinical learning environment: the foundation of graduate medical education. JAMA 2013;309(16):1687–8.

ABOUT THE AUTHORS

Clarence H. Braddock III, MD, MPH, MACP has been a national leader in medical education curriculum development and innovation, particularly in bioethics and doctor–patient communication. Vice Dean for Education in the UCLA David Geffen School of Medicine, Dr. Braddock oversees all aspects of medical education, including undergraduate, graduate, and postgraduate medical programs.

Philip A. Pizzo, MD is the David and Susan Heckerman Professor and Founding Director of the Stanford Distinguished Careers Institute. He served as Dean of the Stanford University School of Medicine from April 2001 to December 2012, where he was also the Carl and Elizabeth Naumann Professor.

Charles G. Prober, MD is the Senior Associate Dean for Medical Education at Stanford School of Medicine. He is a Professor of Pediatrics, Microbiology, and Immunology and an expert in pediatric infectious diseases. In his current position, he oversees undergraduate, graduate, and postgraduate medical education at Stanford School of Medicine.

Chapter 12

Guiding the Future of Medical Education through "Enlightened" Accreditation Policy

Deborah M. DeMarco, Terence R. Flotte, Michael D. Kneeland,
Paulette A. Seymour-Route and Michael F. Collins

INTRODUCTION

In the twenty-first century, physicians, nurses, and other health care providers face numerous challenges and opportunities that their predecessors could never have imagined. Those who guide medical education face significant challenges as well, but are presented with an exciting and ever-widening array of new technologies and approaches to adult learning that promise to revolutionize how, where, and when health care practitioners learn. Through this "dynamically wired," "instant digital access" provider community, educators have the opportunity, now more than ever, to directly benefit the health and safety of patients across the globe through innovative new curricula and educational methods.

In this context, those engaged in medical education are more tightly regulated than ever before. Some of the impetus for this increased oversight arises from an appropriate desire to more fully protect patient safety, as well as from a revival of for-profit medical schools, which had historically been abundant prior to the Flexner Report. The more detailed and tighter regulations reflected in accreditation policies present a particular challenge to those engaged in the twenty-first century's digital revolution impacting medical education.

Those crafting accreditation policies are presented with a particular opportunity to lead in this rapidly changing environment. Ideally, accreditation policy could provide the forum for determining which overarching principles and goals should become the strategic guideposts for assessing current and future educational and practice innovations, and which elements are merely tactical methods to achieve those overarching goals. Ideally, this "enlightened" perspective on accreditation policy can provide the "guideposts" within which educational scholars could be creative and innovative in meeting the needs of today's learners as these scholars strive to form more effective interprofessional, patient-centered teams.

In this chapter, we will consider several facets of accreditation policy where particular opportunities exist

for innovation and strategic direction. These include: (1) the coordination and translation of best practice across the continuum of medical education, from entry into medical school through formal training and into practice; (2) the challenge of norming standards among the three pipelines of undergraduate medical education (allopathic US schools, osteopathic US schools, and international schools) as graduates of these institutions enter US graduate medical education (GME); and (3) the convergence of standards imposed by the professions of medicine and nursing as they intersect in the domains of interprofessional education (IPE) and graduate training.

Opportunities for translation of best practice and harmonization of accreditation standards to achieve higher quality and safety levels in patient- and learner-centered environments will be highlighted. The process of accrediting hospitals is extremely complex and is beyond the scope of this chapter.

REGULATION ACROSS THE CONTINUUM OF MEDICAL EDUCATION

A student graduating with a bachelor's degree in 2015 and intending to become a practicing physician in the United States is required to take multiple examinations produced and managed by unrelated organizations and to interface with a myriad of regulatory authorities. A simple list of the entities encountered during this process is shown in Table 1.

Apart from the confusion that can be experienced by physicians-in-training when they encounter this "alphabet soup" of regulatory entities, there is the potential for significant mismatches in the goals and objectives of medical education at each education level. Such mistakes can cause discontinuities to occur during the successive stage-to-stage transitions experienced by a developing physician. Lypson and colleagues [1] showed that graduating medical students do not all enter internship with the same set of skills, as there

The Transformation of Academic Health Centers. http://dx.doi.org/10.1016/B978-0-12-800762-4.00012-8

TABLE 1 Regulatory Entities Encountered through the Continuum of Physician Training

Stage of Training	Process or Examination	Regulatory or Testing Entity
Medical school readiness	MCAT examination	AAMC
Medical school entry	AMCAS application	AAMC
MD and DO programs	Program Accreditation	LCME (AAMC+AMA) AACOM
Readiness for practice	USMLE examination steps 1, 2CS, 2CK, 3; AOA examinations	NBME, AOA, FSMB
Legal authority to practice	Licensure	70 Individual state and territorial medical licensing boards, FSMB
Application to residency or fellowship training	ERAS application	AAMC
Admission to residency or fellowship training	Residency and fellowship matches	NRMP American Osteopathic Association Match
GME programs	Program Accreditation	ACGME and its RRCs; AOA
Completion of specialty and subspecialty training	Board and subboard certifications	Individual specialty boards and subboards under ABMS
Continuing education programs	Program Accreditation	ACCME
Ongoing medical practice	License renewal	70 Individual State and territorial medical licensing boards, FSMB
Ongoing specialty and subspecialty practice	Board and subboard MOC certificates	Individual specialty boards and subboards under ABMS

MCAT=medical college admissions test, AAMC=Association of American Medical Colleges, AMCAS=automated medical college applications system, LCME=Liaison Committee on Medical Education, AMA=American Medical Association, USMLE=U.S. Medical Licensing Examination, AOA=American Osteopathic Association, NBME=National Board of Medical Examiners, FSMB=Federation of State Medical Boards, ERAS=electronic residency application system, NRMP=National Residency Matching Program, ACGME=Accreditation Council for Graduate Medical Education, RRC=Residency Review Committee, ABMS=American Board of Medical Specialties, ACCME=Accreditation Council for Continuing Medical Education.

is often a gap between their skills set and what is expected of them. Framed in a more positive light, enhanced coordination across the education continuum might allow for dissemination of best practices and for congruence among the standards, competencies, and milestones assessed at each level.

A recent study by Angus et al. addresses just this issue [2]. In 2010, the Clerkship Directors in Internal Medicine Subinternship Task Force, in collaboration with the Association of Program Directors in Internal Medicine Survey Committee, surveyed internal medicine program directors to determine which competencies or skills they expected from new medical school graduates. GME competencies and milestones attempt to standardize curriculum and assessment nationally; there is a suggestion that the same could be done for subinternships in internal medicine, and, in fact, the entire fourth year of medical school, with more focus on competency-based teaching and assessment. The qualities that received the highest rankings by the program directors were: working in health care teams (when to seek help and communication with nursing), information management (prioritization), and time management (organization

and efficiency). While there is a lack of uniformity in the internal medicine subinternship, the specialty organizations for family medicine, obstetrics and gynecology, and surgery have developed well-defined recommendations for the fourth-year medical student curriculum for students planning careers in those disciplines. Many of the highly ranked competencies stated above are common across specialties, and development of a standardized, milestone-based curriculum for the fourth year of medical school that focuses on these areas has the potential to enhance medical education across the continuum, improve patient safety, and reduce stress associated with the transition to internship.

This model would align with the Liaison Committee on Medical Education (LCME) standard (ED5 in current standards, Standard 7 in clustered standards) [3] indicating that programs leading to the MD degree must "prepare a student for entry into any residency program...," yet there is little or no mention of several of the six GME competencies within the current LCME curriculum content standards. In particular, the disciplines of quality and patient safety have become central to the modern practice of medicine, yet it is possible for a student to graduate with an MD from an LCME-accredited program without ever learning about the Institute of Medicine's (IOM) six desired characteristics of care delivery, the value of operative checklists, or the execution of a Plan-Do-Study-Act cycle. The proposed approach to establish a framework that focuses on the achievement of medical student competencies might also provide the opportunity for the development of a undergraduate medical education (UME) duty-hour framework with a more thorough competency-based design than the existing LCME duty-hour standard [ED-38 of current standard; Standard 8.8 in Clustered Standards] [3].

To more fully understand the value of such a competency framework, it will be illustrative to review the recent history of the evolving Accreditation Council for Graduate Medical Education (ACGME) standards for GME duty hours and competencies. In 2002, the ACGME adopted new duty-hour standards for residency programs in all specialties, which limited the number of duty hours to 80 per week, required a rest period between "shifts," and placed a 24 hour limit on the number of continuous hours a resident could work, plus an additional 6 hours for transfer of care. These standards reflected the culmination of 20 years of work that addressed resident duty hours in an attempt to avert any action by the federal government due to the perceived link between fatigue and preventable medical errors. The Bell Commission, formed in response to the death of Libby Zion at a New York hospital in 1984, published its report in 1987 and regulations on duty hours became part of the New York State Department of Health Code [4]. Prior to this, residents routinely worked 100 hours or more per week and often 36 hours or more in a row. This schedule was felt to be a necessity for the development of clinical skills. There was little or no attention paid to effects on patient safety. After the Libby Zion case, residents at New York hospitals could work no more than 80 hours per week and no more than 24 hours in a row. These rules were variably enforced, and it was not until the patient safety movement gained momentum and links were established between physician fatigue and clinical performance that the ACGME implemented stricter standards. In addition to the 80 hour work week and the maximum number of continuous hours, residents were no longer allowed to be on call more frequently than one in three nights, and they had to have at least one day off in seven days completely free from any clinical responsibilities.

Many institutions tried to implement the new standards without changing anything else about the learning environment, such as efficiencies and improvement in the handover process, enhanced attending supervision, or the reduction of paper work, and other nonphysician tasks that residents routinely performed. Research into the overall effects of these stricter regulations has been controversial and mixed depending on which variable is studied. While resident well-being may have improved after the 2003 duty hours went into effect, there were no clear beneficial effects on patient safety or clinical outcomes [5]. This is likely due to the increased number of hand-offs necessitated by the

regulations. Additionally, the effect on educational variables was also mixed, but many studies point to adverse effects due to less time available for teaching and attending conferences, as well as fewer cases for surgical residents.

In 2005, the ACGME endorsed four strategic priorities designed to enable emergence of a new accreditation model. In part, this came in response to the Medicare Payment Advisory Commission (MedPAC), the IOM, and the federal government questioning the process of accreditation and the quality of preparation of graduates for the "future" health care delivery system. The strategic priorities of the ACGME were:

1. Foster innovation and improvement in the learning environment;
2. Increase the accreditation emphasis on educational outcomes;
3. Increase efficiency and reduce burden in accreditation; and
4. Improve communication and collaboration with key internal and external stakeholders.

In 2008, five years after the ACGME duty hours went into effect, the IOM issued a report, "Resident Duty Hours: Enhancing Sleep, Supervision and Safety" [6]. This study was requested by the House Committee on Energy and Commerce as part of its investigation into preventable medical errors. It called for hospitals to provide safer conditions for patients and trainees by creating a culture of safety in the learning environment, including direct supervision and standardization of the hand-off process as well as for further modifications of the ACGME duty-hour standards. At the same time, the House of Representatives codified the "New Physician Competencies" [7] in the original health care reform package, and MedPAC called on the Centers for Medicare and Medicaid Services to modulate indirect medical education payments to institutions based on quality of competency outcomes. In response, the ACGME commissioned a task force, initially known as the Duty-Hour Task Force, which ultimately became the Task Force for Quality Care and Professionalism. The group linked duty-hour standard progression with level of training and competency-based milestone performance of trainees and set the stage for introduction of the Next Accreditation System (NAS). The task force report, issued in 2010, led to revised duty-hour requirements effective July 2011. The revised standards include differential rules based on the level of training and expertise. They link duty hours to progressive and conditional independence through requirements for different levels of supervision. They also link duty-hour adherence and reporting to the professional responsibility to patients for safety and quality, establishing the importance of institutional patient safety and quality improvement programs in the education of residents. Although the actual duty hours garnered the most publicity, this report also called for major changes in the learning environment, such as direct and indirect attending supervision, standardization of the handover process, resident involvement in team-based interprofessional care, and meaningful involvement of residents in institutional initiatives around patient safety and quality.

Implementation of the NAS began in July 2012 and is being phased in over two or more years. The intent of the NAS is not to close GME programs but to improve GME. The principles are consistent with recommendations made by the IOM, MedPAC, and Congress. The aims of the NAS are to: [8]

1. Enhance the ability of the peer-review system to prepare physicians for practice in the twenty-first century;
2. Accelerate the ACGME's movement toward accreditation on the basis of education outcomes;
3. Reduce the burden associated with the prior accreditation system, which was structure and process based;
4. Free good programs to innovate;
5. Assist poor programs to improve; and
6. Provide accountability for outcomes to the public.

Key attributes of the NAS are:

1. Specialty-specific educational milestones to be used for trainee assessment;
2. A focus on improvement and self-study for individual training programs and the sponsoring institution;
3. Development of national normative data against which programs can compare themselves as well as individual trainees; and
4. Less prescriptive program requirements.

The previous accreditation process essentially took a snapshot of a resident's and program's performance at a point in time. There was no way to benchmark the individual resident or program performance to national standards, and although each program had to adhere to residency review committee (RRC) requirements, there was little standardization of evaluation or performance metrics. The NAS attempts to address these issues with semiannual reporting on resident performance based on milestones developed nationally for each discipline, and annual reporting by programs on a variety of metrics such as faculty board certification and scholarly activity by faculty and residents. Ultimately, there will be a national data bank housed at the ACGME, so programs and their residents can be compared against national norms. The program assessment process will be much more like an LCME visit with a one-year self-study process in advance of the site visit, which will now occur every 10 years.

In the NAS, there is much more of a focus on quality and patient safety, and a part of the institutional accreditation process will be the Clinical Learning Environment Review (CLER) visit, conducted every 12–18 months. The CLER visit will focus on six key areas: patient safety, health care quality (included here is health care disparities), transitions of care, supervision, duty hours and fatigue management, and professionalism. This process is much like an institutional Joint Commission visit, and one thing the site visitors will evaluate is how well the residents and fellows are integrated into a hospital's system in terms of quality and patient safety.

The milestone and competency-based principles of the NAS are aligned with the Dreyfus model of skill acquisition, with the milestones being the observable, developmental steps moving from novice to expert or master. These milestones are organized under the rubric of the six domains of clinical competency (medical knowledge, patient care, professionalism, interpersonal communication skills, practice-based learning and improvement, and system-based practice) and describe a trajectory of progress from neophyte or novice toward an individual ready for independent practice. Also, they provide a framework and language for discussions across the continuum of professional development and align well with the revised Maintenance of Certification (MOC) process that most specialty boards currently have or will adopt. The new MOC process is designed as a continuous self-assessment practice to promote life-long learning. Therefore, the process of accreditation by the ACGME, at both the individual and program levels, has become much more of a continuous process over time rather than a single snapshot every few years. The process is designed to help individuals be better prepared for unsupervised practice in their chosen field and to ensure that programs are accountable to the public for the trainees they produce.

Since the framework of the Dreyfus model is based on the concept of an individual clinician progressing along a developmental curriculum, it would be natural to ask whether this system could be extended to earlier stages of development, stages appropriate to preclerkship level, clerkship level, and subinternship level students. Such extension across the continuum would require collaboration between the ACGME and the LCME, but would ideally be driven by academic medical educators who set evaluation criteria within their own academic health centers for medical students as well as residents. Medical student clinical experience also involves some aspects of graded supervision and some clinical responsibilities within a more narrowly defined scope of practice. It would only seem logical to integrate UME into this continuous process.

Bringing Osteopathic and Allopathic Training into One Continuum

Effective July 1, 2020, the ACGME will be the single accreditation body for GME. The American Osteopathic Association (AOA) and the American Association of Colleges of Osteopathic Medicine (AACOM) will become ACGME member organizations and will nominate members to the ACGME Board of Directors. Two new RRCs will be created to evaluate and set standards for the osteopathic aspects of GME programs seeking osteopathic recognition. There will be a transition period from 2015 to 2020 for AOA-accredited programs to apply for and receive ACGME recognition and accreditation. This single system will allow MD or DO graduates who have met prerequisite competencies to access any GME program without being required to repeat educational experiences. Also, it will eliminate the need for institutions to sponsor dually accredited programs and will ensure consistent high standards for all GME trainees.

NORMING OF PROGRAM STANDARDS QUALIFYING APPLICANTS FOR GME

Individuals may qualify for GME in allopathic US programs through one of three pathways of doctoral preparation: an MD from an LCME-accredited North American medical school (AMG), a DO from an AOA-accredited school of osteopathic medicine (DO), or an MD from a medical school outside of North America—international medical graduates (IMGs).

These three pathways are fundamentally different in that the AMG and DO pathways are regulated at both the level of the student (who must pass United States Medical Licensing Examinations (USMLE)) and at the level of medical education program or school (by the LCME or AOA). For IMGs, the regulatory standard is only exercised at the level of the student.

IMGs may enter US GME programs through the Educational Commission for Foreign Medical Graduates (ECFMG). A nonprofit organization, the ECFMG notes that its services include certification that IMGs have met certain medical education and examination requirements; evaluation of the credentials of IMGs, their medical knowledge, and command of the English language; information for IMGs planning to enter a US GME program; and sponsorship of applicants for J visas. It issues certificates to IMGs who meet educational requirements. Based in Philadelphia, the ECFMG is governed by a Board of Trustees with up to eight trustees at large. Two nominees from each of the following organizational members serve as Trustees: the American Board of Medical Specialties (ABMS); the American Medical Association; the Association of American Medical Colleges (AAMCs); the Association for Hospital Medical Education, the Federation of State Medical Boards (FSMBs) of the United States, Inc.; and the National Medical Association. The ECFMG's president also serves as a voting member of the Trustees. The ECFMG notes that some 25% of physicians-in-training and practice are graduates of non-US and non-Canadian medical schools.

The ECFMG issued a total of 9642 certificates in 2012. The highest percentage of the total was 1438 certificates from schools in India followed by Grenada with 855 certificates, and Pakistan with 544 certificates. Most medical school graduates apply to GME programs through a process established by the National Resident Matching Program (NRMP), the "match." They reported 20,602 postgraduate year-1 (PGY) positions available through the match in 2002 and 26,392 in 2013, a 28% increase. They also reported that in 2002, IMGs and osteopathic graduates who matched to PGY-1 positions were about one-third the number of US allopathic graduates. However, by 2013, this proportion had grown to nearly 50%. According to the NRMP, in 2013, 93.7% of US allopathic senior medical students were matched to PGY-1 positions. At the same time, 53.1% of US citizens who attended international medical schools were matched, an increase of 49.1% from 2012, while the match rate for non-US citizens from international medical schools increased from 40.6% in 2012 to 47.6% in 2013. Table 2, adapted from NRMP statistics, shows selected applicant match data from 2013 to 2012.

TABLE 2 NRMP Match Results for Students Entering US Allopathic GME Programs through Each of Three Pathways

General Category	Specific Category	Active Applicants 2013	Matched 2013	Active Applicants 2012	Matched 2012
AMG	Senior US allopathic students	17,487	16,390 (93.7%)	16,527	15,712 (95.1%)
	Previous graduates of US allopathic schools	1487	729 (49%)	1317	560 (42.5%)
	Students/graduates of Canadian medical schools	21	14 (66.7%)	17	12 (70.6%)
DO	Students/graduates of osteopathic medical schools	2677	2019 (75.4%)	2360	1764 (74.7%)
IMG	US citizen graduates of international medical schools	5095	2706 (53.1%)	4279	2102 (49.1%)
	Non-US graduates of international medical schools	7568	3601 (47.6%)	6828	2775 (40.6%)

As mentioned above, while AMGs and DOs undergo individual testing and must have a degree from a program with a standardized accreditation, the equivalence of the MD degree held by the IMG is not supported by a single institutional accreditation standard at any level. International medical schools generally do undergo accreditation by bodies within their home country or region, such as the Association of Medical Schools in Africa and the Association of Medical Schools in the Middle East [9]. In turn, regional and national bodies may become members of the World Federation for Medical Education (WFME), which has established certain global standards and procedures for institutional accreditation. While the current norms of WFME do not mandate a single standard for IMGs, they do create an opportunity for improving the consistency of preparedness for IMGs. Ideally, a global set of overarching standards for medical education should be adopted internationally. Until that time, WFME aims to fulfill its mission "to strive for better health care, by scientific, ethical and social standards in the education of medical and related personnel, toward provision of competent medical and health services globally" [10].

Finally, it should be noted that some members of the physician workforce receive their GME training in other countries. ACGME International (ACGME-I) began accrediting GME programs outside the United States in July 2010. The stated mission of ACGME-I is to improve health care by assessing and advancing quality of resident physicians' education through accreditation to benefit the public, protect the interests of residents, and improve the quality of teaching, learning, research, and professional practice. To date, there are three sponsoring institutions accredited by ACGME-I in Singapore, Qatar, and United Arab Emirates with a total of 29 programs. Singapore was the first to receive accreditation and has the majority of the programs, in both core residencies and subspecialty fellowships.

The ACGME-I Institutional Requirements are nearly identical to the institutional requirements that were in place in the United States prior to the implementation of the NAS and set the standards for institutional oversight for training programs in countries outside of the United States. The "Foundational Program Requirements" set general standards and responsibilities for the individual program directors and teaching faculty; resident eligibility; and the educational program, including evaluation and duty hours. The duty-hour standards are similar but not as stringent as the current duty-hour requirements in the United States.

The accreditation process includes site visits to the institution and individual programs, very much like those conducted in the United States prior to the implementation of the NAS. Exceptional candidates of core residency programs accredited by the ACGME-I will be considered for ACGME-accredited fellowship programs in the United States when the Common Program Requirements change in 2016.

INTERPROFESSIONAL EDUCATION: COORDINATION OF REGULATIONS ACROSS PROFESSIONS

Recognition that the prevention of medical errors is critically dependent on teamwork and communication is well supported. The challenge for health professional education programs is to build the distinct professional identity of each learner while understanding the context of team-based care. Complementary to understanding the unique contribution of each discipline is the learner's need to value, respect, and understand the contributions of other health professionals as the foundation for patient-centered care.

The ability to provide patient-centered care depends on an interprofessional, collaborative practice—a learned behavior best developed during deliberate IPE experiences. This presents an opportunity for the strategic convergence and standardization of requirements, imposed by individual health professions schools in the domain of IPE, to improve the collective ability of team-based care.

Zorek and Raehl (2012) published a comparative analysis of IPE standards in the United States and found a lack of consistency among the health professions in addressing IPE. Pharmacy and nursing standards were found to account for the majority of IPE statements and standards. The Interprofessional Education Collaborative (IPEC) recently published a list of core competencies for interprofessional collaborative practice [11]. The sponsors were the American Association of Colleges of Nursing; AACOM; American Association of Colleges of Pharmacy; American Dental Education Association; AAMC; and the Association of Schools of Public Health. The IPEC expert panel identified four core-competency domains: (1) values/ethics for interprofessional practice; (2) roles/responsibilities for collaborative practice; (3) interprofessional communication; and (4) interprofessional teamwork and team-based care.

Currently, each of the health professions has its own unique process for accreditation. Nursing has a complex set of regulations, examinations, and accreditation standards similar to those for medicine. National educational standards are based on practice level, professional nurse (RN), or advanced practice nurse (APN). Additionally, there are national and state practice/licensing regulations. Graduate nursing programs require that an individual be licensed as a professional nurse as a prerequisite for completing their education, certification, and licensure as an APN. The Commission on Collegiate Nursing Education is the accreditation body for the American Association of Colleges of Nursing. Nursing programs develop their curriculum in congruence with the AACN Essentials for the specific Baccalaureate, Masters, or Doctor of Nursing Practice Program. These Essentials build upon each other, and all include reference to IPE and practice outcomes. The specific Essentials can be found on the AACN Website (http://www.aacn.nche.edu/education-resources/essential-series).

The medical education accreditation body, the Liaison Committee on Medical Education (LCME), imparts one IPE standard statement (Standard 7.9 in the Clustered Standards; ED-19 in previous standards [3]): "The core curriculum of a medical education program must prepare medical students to function collaboratively on health care teams that include health professionals from other disciplines as they provide coordinated services to patients. These curricular experiences include practitioners and/or students from other health professions." The ACGME Common Program Requirements reference interprofessional competency expectations in the following standards:

- IV.A.5.f.5: Residents are expected to work in interprofessional teams to enhance patient safety and improve patient care quality.

- IV.A.5.d: Residents must demonstrate interpersonal and communication skills that result in the effective exchange of information and collaboration with patients, their families, and health professionals.
- VI.A.3: The program director must ensure that residents are integrated and actively participate in interdisciplinary clinical quality improvement and patient safety programs.
- VI.F: Residents must care for patients in an environment that maximizes effective communication. This must include the opportunity to work as a member of effective interprofessional teams that are appropriate to the delivery of care in the specialty.

The general competency statements for each IPEC domain and the associated standard from nursing/AACN, medicine/AAMC, and GME are indicated in Table 3.

TABLE 3 Summary of Interprofessional Competencies as Articulated by Nursing and Interprofessional Accreditation Authorities

IPEC Domain	IPEC Competency	AACN BSN Essential	AACN Master's Essential	AACN DNP Essential	AAMC Competency	GME Competency
Values/ Ethics for interprofessional practice	Work with individuals of other professions to maintain a climate of mutual respect and shared values	Essential VI: Interprofessional communication and collaboration for improving patient health outcomes	Essential VII: Interprofessional collaboration for improving patient and population health outcomes	Essential VI: Interprofessional collaboration for improving patient and population health outcomes		
Roles/ responsibilities	Use the knowledge of one's own role and those of other professions to appropriately assess and address the health care needs of the patients and populations served	Essential VI Essential VII: Clinical prevention and population health	Essential VII Essential IX: Master's level nursing practice	Essential VI		
Interprofessional communication	Communicate with patients, families, communities, and other health professionals in a responsive and responsible manner that supports a team approach to the maintenance of health and treatment of disease.	Essential VI	Essential II: Organizational and systems leadership Essential VII	Essential VI		IV.A.5.d: VI.F:
Teams and teamwork	Apply relationship-building values and principles of team dynamics to perform effectively in different team roles to plan and deliver patient/ population-centered care that is safe, efficient, effective, and equitable.	Essential VI Essential II: Basic organizational and systems leadership for quality care and patient safety	Essential II Essential III: Quality improvement and safety Essential VII Essential IX	Essential VI	ED-19	IV.A.5.f VI.A.3: VI. F.

Interest in IPE has grown substantially in recent years. Recognized improvements in quality and patient safety related to team-based care and the financial burden of the current health care system demonstrate the need to reform the way care is delivered. Interprofessional collaborative practice is requisite to changing health care delivery. IPE is now recognized as a deliberate approach to providing patient-centered care versus an optional enrichment experience. The technology available to learners in academic health science centers provides the infrastructure for patient- and learner-centered teaching. Team-based learning using technology and simulating interaction with patients and families in a learning environment have been demonstrated to build confidence and competence in communication and safe patient care. This approach establishes a foundation for life-long learning that is consistent with professional role expectations across medicine and nursing and promotes a culture of safety in the provider work environment. Since the pedagogy and scholarship of teaching related to IPE are developing, accreditation of such activities is now receiving more attention.

IPE accreditation is recognized as a global health profession educational issue. The Lancet Commission Report [12] "puts forward a vision: all health professions in all countries should be educated to mobilize knowledge and to engage in critical reasoning and ethical conduct so that they are competent to participate in patient and population-centered health systems as members of locally responsive and globally connected teams." The report recommends socially accountable accreditation as a responsible stewardship mechanism [12]. The World Health Organization (WHO) published a Framework for Action on Interprofessional Education and Collaborative Practice (2010), which includes actions to support IPE and collaborative practice at the system level and identifies the creation of accreditation standards that include evidence of IPE across health professions programs [13]. The Accreditation of Interprofessional Health Education (AIPHE) initiative published the Interprofessional Health Education Accreditation Standards Guide in 2011 [14]. This publication, funded by Health Canada, identifies common domains and language that could guide the development of standards and language applicable to all health professions interested in embedding IPE into the accreditation process. The guide, accessible on the AIPHE Website (www.aiphe.ca), organizes information into the following categories; context, language, criteria, and examples of evidence relevant to IPE across the school, faculty/academic unit, students, educational program, and relevant resources. The challenge in the academic health science environment is to standardize the approach, criteria, and evaluation methods but to remain flexible enough to integrate those standards within the context of the multiple health professions accreditation bodies.

In summary, an enlightened accreditation perspective includes interprofessional standards and related competencies/outcomes that are applied across the health professions. The recent IPEC publications, the framework developed by the WHO, and the AIPHE guide serve as guardrails for the development of contemporary IPE accreditation standards. The collaboration by the IPEC founders can foster integration into the health professional curriculum and learner evaluations. Ultimately, patients will benefit from the convergence of the collective care and expertise of all health professionals who practice in teams. This will position health professionals to provide focus on delivery models that insure safe, efficient, and effective patient-centered care.

CONCLUSION

The impact of accreditation processes on the quality of medical education is being enhanced by harmonization across the medical education continuum and with other health professions. This harmonization should allow for the propagation of evidence-based best practices, such as those based on the Dreyfus model—which is being implemented first within the GME realm. It may also help the American allopathic medical education system to deal with future challenges, such as workforce planning, the globalization of medical education, and the evolution of interprofessional collaborative practice models. If the committees, councils, boards, and agencies charged with accreditation can fully embrace these opportunities, enlightened accreditation policies could serve as a vehicle for promoting evidence-based educational innovation rather than as an impediment to future creativity.

The bodies empowered to accredit schools and programs and to license and certify health professionals also sit as the gatekeepers for the creation of the health care workforce of the future. The profile of the future health care workforce could have a profound influence on broader issues of health care cost and quality. Could an enlightened group of accrediting bodies potentially help drive those agendas in addition to implementing best educational practices? Finally, as this chapter was in preparation, an IOM Panel released a report on the governance and finance of GME, questioning the entire structure currently in place that finances the $15 billion expense of GME [15]. Will the US Congress take up the charge presented by that report to reshape the funding of GME in the United States? If so, will it trust the current accrediting, licensure, and certification bodies to implement any changes necessitated by their action? These questions remain unanswered. However, how we respond to these issues will likely shape US health care for the remainder of the twenty-first century.

REFERENCES

[1] Lypson ML, Frohna JG, Gruppen LD, Woolliscroft JO. Assessing residents' competencies at baseline: identifying the gaps. Acad Med 2004;79:564–70.

[2] Angus S, Vu TR, Halvorsen AJ, Aiyer M, McKown K, Chmielewski AF, et al. What skills should new internal medicine interns have in July? A national survey of internal medicine residency program directors. Acad Med 2014;89:432–5.

[3] Liaison Committee on Medical Education. Functions and structure of a medical school: standards for accreditation of medical education programs leading to the M.D. degree [Washington, DC]. 2014.

[4] Thorpe KE. House staff supervision and working hours. Implications of regulatory change in New York state. JAMA 1990;263:3177–81.

[5] Fletcher KE, Reed DA, Arora VM. Patient safety, resident education and resident well-being following implementation of the 2003 ACGME duty hour rules. J General Intern Med 2011;26:907–19.

[6] Ullmer C, Wollman DM, Johns MME. Residency duty hours: enhancing sleep, supervision and safety. Washington, DC: Institute of Medicine; 2008.

[7] U.S. Congress. America's affordable health care choices act of 2009. 111th congress, 2nd session ed. 2010. [Washington, DC].

[8] Nasca TJ, Brigham TP. Commentary: the modesto story: back to the future? Acad Med 2012;87:996–8.

[9] World Health Organization. Accreditation of medical education institutions: report of a technical meeting. Copenhagen: Denmark; 2004.

[10] Karle H, Walton H, Lindgren S. The world foundation for medical education: history of the first forty years, 1972–2012. Copenhagen, Denmark: World Foundation for Medical Education; 2013.

[11] Interprofessional Education Collaborative Expert Panel. Core competencies for interprofessional collaborative practice: report of an expert panel. Washington, DC: Interprofessional Education Collaborative; 2011.

[12] Frenk J, Chen L, Bhutta ZA, Cohen J, Crisp N, Evans T, et al. Health professionals for a New Century: transforming education to strengthen health systems in an interdependent world. Lancet 2010;376:1923–58.

[13] World Health Organization. Framework for action on interprofessional education and collaborative practice. WHO department of human resources for health. Geneva, Switzerland: Health Professions Network Nursing and Midwifery Office; 2010.

[14] Canadian Interprofessional Health Collaborative. AIPHE interprofessional health education. Vancouver, British Columbia: Accreditation Standards Guide; 2011.

[15] Committee on the Governance and Financing of Graduate Medical Education BoHCS. Graduate medical education that meets the nation's health needs. Washington, D.C: Institute of Medicine; 2014.

ABOUT THE AUTHORS

Michael F. Collins, MD, FACP was appointed Chancellor of the University of Massachusetts Medical School in 2008, after serving as Interim Chancellor. He also serves as Senior Vice President for Health Sciences at the University of Massachusetts.

Deborah M. DeMarco, MD, FACP is a Professor of Medicine in the Division of Rheumatology and the Senior Associate Dean for Clinical Affairs at UMASS Medical School (UMMS). She has been an active, general Rheumatologist both at UMASS and the VA Health Care System for over 20 years.

Terence R. Flotte, MD is the Celia and Isaac Haidak Professor in Medical Education, Dean of the School of Medicine and Provost and Executive Deputy Chancellor of the University of Massachusetts Medical School (UMMS). In these roles, Dr Flotte serves as chief academic and administrative officer of the School of Medicine.

Michael D. Kneeland, MD, MPH is Associate Dean of Allied Health and Inter-Professional Education at the University of Massachusetts Medical School. Dr Kneeland's responsibilities include interprofessional education initiatives.

Paulette A. Seymour-Route, PhD, RN is Dean and Professor of the Graduate School of Nursing at UMass Medical School. Prior to becoming the Dean, she served as the Chief Nursing Officer for UMass Memorial Health Center and Senior Vice President for UMass Memorial Medical Center.

Chapter 13

The Compelling Need for Education Reform: A Futurist's View of Health Professions Education

Tim Lahey, Greg Ogrinc, Leslie Fall, Susan Reeves, Rich Simons and Wiley "Chip" Souba

INTRODUCTION

Health care financing and delivery are changing with incredible speed, and with them the nature of clinical practice. These historic changes challenge academic health centers (AHCs) to adapt and lead. In this chapter we propose six key ways AHCs should rise to this challenge.

Health care delivery is evolving rapidly from fee-for-service systems centered at tertiary care hospitals into care provided by increasingly larger and complicated accountable care organizations that are geographically distributed and administratively complex [1]. In these new health care delivery environments, provider performance is being more tightly scrutinized, efficiency is king, and pressures to practice at the peak of licensure are mounting. The amount of medical knowledge that providers marshal during clinical care is also expanding, and increasingly accessed online.

Each of these pressures demands that health care providers function effectively in changing professional environments, and often in interprofessional teams that harness technology in the service of efficiency without losing sight of delivering compassionate care to individual patients. This is a tall order and can cause stress among health care providers.

The very structures of AHCs will make it difficult for them to rise to these challenges. Historically, AHCs have been situated as tertiary care centers in urban areas—not in the community practices where an increasing proportion of clinical care is delivered. The educational and research missions of AHCs have been supported by clinical profits, yet clinical profits are now in shorter supply and more explicitly tied by insurers to clinical outcome measures. This means financial support for core activities of AHCs, such as teaching and research, can become underfunded. Traditionally AHCs have been led by physicians rather than the interprofessional teams that increasingly deliver health care today. This creates a potential disconnect between leadership and frontline clinical staff. Each of these mismatches between the way AHCs do business and the emerging reality of modern medicine suggests AHCs need to change to remain viable.

The traditional approach to health professions education too is being challenged by the rapid expansion of medical knowledge, the decentralization of clinical care, and the need for interprofessional education (IPE). None has been a major strength of health professions education, and yet each will be important to trainee success in the transition to modern practice. Similarly, amid the growing and shifting stresses of modern medical practice, health professions education has stereotypically neglected "soft" topics—such as ethics and skills in stress management, otherwise known as professional resilience—in favor of a focus on the "hard" biomedical sciences. Yet, the stress of clinical work and the added challenges of modern clinical practice demand that trainees acquire effective skills in managing stress and change while maintaining the quality of compassionate care they deliver.

Challenges like these present AHCs with an opportunity to innovate and lead, characteristics that align with the historical identity of AHCs. In this chapter, we will focus on six ways AHCs should be positioned for success in the modern health care environment:

1. Development of centers of excellence in education and scholarship in the health care delivery sciences;
2. Reorientation of health professions education with the changing geographically distributed locus of modern health care;
3. Increased commitment to IPE and leadership;
4. Intensified investment in clinical and education technology to support efficient, collaborative, and geographically distributed care;
5. Enhanced focus on the development of leadership skills; and,
6. Improved training in professionalism and personal resilience.

The Transformation of Academic Health Centers. http://dx.doi.org/10.1016/B978-0-12-800762-4.00013-X

These six transformative strategies will be a challenge for AHCs to implement, in part because they require change in the culture of health professions education. Yet implementation of these strategies themselves will alter the culture of health professions education for the better. As former IBM CEO Lou Gerstner wrote, "You can't talk a culture into changing. You can't just exhort people to be different. You've got to point to fundamental strategic changes you're going to implement in a company and then drive the execution of that strategy. And it is in the execution of the strategy that the culture begins to change [2]."

DEVELOPMENT OF CENTERS OF EXCELLENCE IN EDUCATION AND SCHOLARSHIP IN THE HEALTH CARE DELIVERY SCIENCES

AHCs have long been the cornerstone of building knowledge and advancing science. Building knowledge in health care requires combining focused questions with the appropriate methods to answer those questions. In the past, questions focused mostly on the development of new therapies or diagnostic tools. The evaluation and treatment of acute myocardial infarction (AMI) is one example. In 1903, Einthoven developed the electrocardiogram; but this eventually gave way to the use of diagnostic catheterization and balloon angioplasty in the 1960s and 1970s and then to coronary stenting in the 1980s to the present. Further advancements included drug-eluding stents and combinations of stents with appropriate antiplatelet therapy. These diagnostics and therapeutics answer the question, "How can we best treat this patient who is having an AMI?"

This is an important question, and AHCs will continue to pursue innovative diagnostics and treatments for AMI's. Over the past decade, however, a new set of questions have emerged, questions that AHCs are less ready to answer. These questions focus on clinical outcomes (e.g., long-term clinical and functional status after AMI), the design of care delivery (e.g., door to balloon time), and other means to measure the delivery of safe, effective, and reliable care. AHCs are at risk of falling behind in these domains if they cling to more traditional foci of academic work. For example, in 2006 Pronovost and colleagues published a seminal paper on the effectiveness of using a checklist to reduce bloodstream infections. Their analysis found that AHCs were slower to achieve a decrease in median bloodstream infections to zero than nonteaching sites [3].

In 2003, Srinivasan and colleagues from a prominent AHC in the US reported on contaminated bronchoscopes that led to unnecessary exposure of patients to *Pseudomonas*. They identified the problem and took steps to remove the bronchoscopes from use once the increase in infections was noticed [4]. A similar report from a small community hospital was included in the same journal issue. Kirschke

and colleagues, authors of that article, noticed the same contamination problem with the bronchoscopes sooner and promptly removed the bronchoscopes from service [5]. Tellingly, it was the small community hospital that noticed the variance in routine bronchoscopy and more rapidly took action within their health care system. This level of system self-awareness, analysis, and action is needed to keep patients safe and maintain high-quality care. However, the traditional design of AHCs is not as open to this quick system-wide response.

The research mission of AHCs is inextricably tied to the health professions education mission. Unprecedented progress in biomedical science, combined with tremendous pressures on health care, has transformed the nature of health and disease and the practice of medicine. This has led to growing challenges in translating these research findings into actual clinical practice. To meet this challenge, leaders are needed who can think critically, innovate creatively, and work collaboratively to realize evidence-based high-quality health care.

The emerging field of health care delivery science (HCDS) is one way to begin to address these issues. HCDS is the combination of research expertise in health care outcomes with leadership and teamwork, finance, and operations. This field includes knowledge and skills of health care systems, quality improvement (QI), and outcomes research combined with finance, management, and change within systems. HCDS is a set of core knowledge and skills for all learners in all health care professions. It complements the basic and clinical sciences by creating an understanding of the complexity inherent in the delivery of care. HCDS enables the execution of safe, effective, reliable, high-value care in our systems. AHCs must become centers of excellence in HCDS scholarship and teaching for learners to acquire the skills to thrive in and lead complex health care systems.

The Dartmouth Example

The Geisel School of Medicine at Dartmouth has more than a dozen years' experience incorporating elements of QI and systems into its curriculum. This began with introductory lectures in the late 1990s to create student "awareness" about health care quality and systems and has developed into an integrated, 4-year curriculum that imparts a basic understanding of the health care system, interprofessional teamwork, and health outcomes and financing.

The general curriculum includes a primary focus on the scientific foundations of modern medicine in the first year, pathophysiology of disease in the second year, core clinical clerkships and electives in the third year, and advanced clinical training and required capstone courses in the fourth year. In 2006 and again in 2011—after years of introductory material about QI and systems—Geisel's medical education committee sought to increase coverage of this content and add in experiential learning opportunities for second-year medical students. These changes have led to the current integrated 4-year experience.

The first-year core curriculum includes a "Fundamentals of Health Care Delivery Science" seminar course, which provides a broad foundation in HCDS. Five of the nine sessions are interprofessional with medical, nursing, and physician assistant students. Teachers for these sessions come from the Dartmouth Center for Health Care Delivery Sciences, Geisel, the Dartmouth Institute of Health Policy and Clinical Practice, Tuck School of Business, Dartmouth–Hitchcock Medical Center, and the White River Junction Veterans Affairs Hospital. Through student and faculty discussion, the class identifies health care system problems and seeks resolutions for those problems. Topics include basic systems knowledge, measurement and outcomes of care, financing, patient safety, and leadership in health care. The novice students are set to acquire the knowledge and skills needed for the new complexities of care.

In year 2, the HCDS content is integrated into some of the organ-based pathophysiology courses. Students gain the knowledge and skills to describe the people, structures, and processes within a system as well as the principles of variation and measurement, such as using statistical process control charts. They then apply their knowledge to local examples, such as regional variation in care for acute stroke and local performance gaps in the evidence-based practices for patients in the intensive care unit. In the cardiology course, for example, students explore the system-level interventions and measurement tools that are used to assess and improve door-to-balloon time for acute myocardial infarction. The QI and systems curriculum thus becomes a new endpoint to the study of pathophysiology from cells to organs, from organ system to individual patient care, and from individual patient care to health care system improvement. Importantly, these courses focus on outcomes at our AHC and affiliated hospitals and the efforts to close quality gaps at those hospitals.

In year 3, students experience systems and QI during core clinical rotations. For example, students rotate on the inpatient medicine at the White River Junction Veterans Affairs Hospital where QI is integrated into teams' routine work. These teams have redesigned the system of care to increase pneumococcal vaccination rates, deep venous thrombosis prophylaxis, and evidence-based smoking cessation interventions. The Geisel students apply the foundational knowledge that they gained in years 1 and 2 during their clinical clerkships. While these opportunities are not present in every clerkship at every location, we have begun this work so as to provide our medical students new opportunities for collaboration with more senior trainees. In general, academic centers, such as Department of Veterans Affairs hospitals, provide opportunities to focus on HCDS because it is a core component of their mission.

Year 4 brings all students back to campus for required capstone courses that deepen students' HCDS knowledge in the health care system, patient safety, innovation in medicine, and safe prescribing. "Health, Society, and the Physician" is an interprofessional course required for all medical students. There are two components to the course. In the first component, medical students work in teams of about 10 with a system site coach. Over 4 weeks the student teams analyze a system of care delivery, review outcomes data, and present recommendations to the site and their student and faculty colleagues. The second component of this course is interprofessional sessions with nursing and physician assistant students on a weekly basis. In small and large groups, the students discuss cases that focus on ethics, innovation in health care, finance, and policy. These sessions provide an opportunity for all students from several professional schools to explore these important topics together before they enter practice together.

In the coming years, we are preparing for the next iteration of the Geisel curriculum. Our redesigned curriculum will include required 4-year-long core courses in HCDSs for all students. The courses will provide a depth and breadth of knowledge and skills in critical appraisal of health care literature and evidence, design and improvement of systems, leadership in health care, and the contextual factors that impact the delivery of care for patients and populations.

Unprecedented progress in the biomedical sciences and the tremendous pressures to contain the cost of care now require a transformed care delivery system. It is no longer sufficient for physicians to be competent in the biomedical and clinical sciences alone; they must understand the system of care in which they work as well, and they must be able to collaborate with all professionals on the health care team to improve the quality, safety, and value of care. We need leaders who can think critically, innovate creatively, and work collaboratively to realize high-quality evidence-based health care in the US and worldwide. Health profession schools at AHCs must be positioned to prepare students as leaders with the knowledge, skills, and experience necessary to address these challenges.

REORIENTATION OF HEALTH PROFESSIONS EDUCATION WITH THE CHANGING LOCUS OF MODERN HEALTH CARE

The emergence of the teaching hospital in the early 1900s established the locus of clinical education for the next century. The prevalent model of clinical instruction was patterned after the Johns Hopkins Medical School, which had a large, modern, urban, well-equipped hospital. Faculty members at Hopkins also had staff appointments in the hospital and were encouraged to participate in research, in addition to providing clinical care and teaching. The clinical clerkship became the dominant vehicle for delivering clinical education. The inpatient locus for clinical instruction made perfect sense since the dominant conditions for which patients were hospitalized in the early twentieth century were infectious diseases and nutritional disorders. Patients with undifferentiated acute illness stayed in the hospital for several weeks, thus allowing students to observe the evolution of an illness over time, establish relationships with patients, and make contributions to patient care. Bedside rounds were the rule with master clinicians demonstrating the art of history taking, physical examinations, and diagnostic reasoning.

With the advent of Medicare's diagnosis-related group system of hospital prospective payment in the 1980s, the learning environment of the hospital began to change. The length of stay at most AHCs dropped precipitously from an average of 10–12 days to 5 days [6]. Medical care began to shift to the ambulatory setting, particularly for diagnostic workups. The development of less invasive surgical techniques and freestanding surgical centers allowed more surgical procedures to be performed outside of the hospital. Most patients now admitted to the hospital were very ill, with their diagnoses already established or their required complex procedures already planned.

This shift in the makeup of the inpatient population abridged opportunities for students and residents to observe the natural history of an illness. Once patients improved, they were discharged to complete the remainder of the treatment at home. Rounds were now more focused on clinical care, coordination, and efficient discharge planning. For example, at the University of Iowa, faculty members observed a decline in the intellectual quality of hospital teaching rounds due to pressure to discharge patients in a timely fashion [7]. Bedside rounds, a rich tradition at AHCs, was becoming a lost art. Another related contemporary challenge in the hospital learning environment is frequent transitions in the attending physician on service, with many hospitalists serving week-long shifts that left little opportunity for students to develop a longitudinal relationship with faculty members. Mentoring thus can be shallow or nonexistent.

Although the inpatient setting still affords many excellent opportunities for clinical education, much of the medical care provided today occurs in the ambulatory environment. This setting affords the opportunity to teach about chronic disease, health promotion, disease screening, shared decision-making, cost-effective medicine, and population health. Momentum in this direction is growing as most medical schools are including some component of ambulatory training in required clerkships.

Health professions education must align with contemporary patient care by providing opportunities for students to learn in a variety of care settings, including nursing homes, community health centers, and rehabilitation units. The good news is that educational experiences for both students and residents have begun to include learning environments outside of the hospital. Some AHCs are now utilizing longitudinal integrated clerkships (LIC) in which students are assigned to a group of faculty in various specialties with whom they follow patients longitudinally [8]. In such a model, students are able to participate in the comprehensive care of patients in multiple settings (e.g., outpatient clinic, specialty clinic, inpatient medical–surgical ward, nursing home, rehabilitation hospital) while developing longitudinal relationships with both patients and faculty. In addition, such clerkships also offer panel patients in settings such as schools, rehabilitation centers, halfway houses, homeless shelters, and nursing homes. These experiences in the community are consistent with the concept of "community-engaged medical education," with the intent that medical students learn about the community context and how this might impact the patients and their clinical problems.

Similarly, changes are occurring in graduate medical education. Many residency programs in internal medicine, family medicine, and pediatrics are providing ambulatory block time allowing the residents to concentrate on developing their outpatient skills without the distraction of the inpatient ward responsibilities. This represents an opportunity for residents to learn something about practice management, billing, and reimbursement from third-party payers, in addition to traditional ambulatory medicine. Likewise, many internal medicine programs utilize a nursing home experience or palliative care unit for a dedicated geriatric teaching and learning experience. Although surgery is typically a hospital-based specialty, outpatient experience is also critical to surgical training because the practicing surgeon must be proficient in making a surgical diagnosis and formulating the appropriate management plan.

Finally, we need to ensure that our learners are exposed to high-performing health care systems that emphasize QI, patient safety and satisfaction, and appropriate health care resource utilization. The culture and system to which our students and residents are exposed during their training will strongly influence their eventual practice patterns [9].

INCREASED COMMITMENT TO INTERPROFESSIONAL EDUCATION AND LEADERSHIP

Health professions education must also be richly interprofessional to prepare learners to thrive in the new collaborative, multidisciplinary, team-based care environment. Health professions students need to learn the communication, collaboration, and coordination skills that will enable their success in future practice. The foundation for this kind of effective IPE has been laid; but there is more work to be done. In this section, we describe the foundation for IPE at AHCs, and the way forward.

The Foundation for Interprofessional Education is Laid

Today, most health professions faculty have moved beyond the point of debating the value and the imperative of IPE in their programs and are now actively seeking ways to build such competencies into their programs. Core competencies for IPE have been developed and published along with a guiding conceptual model to underpin them [10]. National meetings and forums have been convened to advance the IPE agenda and prestigious funding agencies, such as the Josiah Macy Foundation and the Robert Wood Johnson Foundation, have focused many of their programs in the area of building capacity for increasing the activities and study of IPE efforts [11,12]. In addition, agencies that accredit health profession schools, such as the Liaison Committee on Medical Education (LCME), have clearly incorporated expectations for IPE into their standards.

The design, implementation, and evolution of IPE programs face predictable structural challenges [13]. Scheduling difficulties created by the mismatched calendars of nonaligned health professions schools typically top the list of frustrations for IPE planners. Following closely, and often occurring in concert, are issues such as inadequate space to conduct large, combined group sessions; bringing in adequate numbers of engaged faculty from participating schools; and true integration of IPE material into curricula of the programs.

To address these structural challenges to IPE, many schools have convened key faculty leaders to comprise IPE steering committees. Many schools are showcasing new curricula that fully integrate IPE elements throughout all years of the participating programs. For instance, at Geisel an IPE Steering Committee composed of faculty and administrative leaders from nearby physician assistant and nursing schools collaborated to design an embedded IPE curriculum into each of their respective programs. In the early program years, sessions focus on role appreciation and communication. In the middle years, a variety of electives are offered in interprofessional formats. And, in the last year of the programs, IPE sessions which focus on ethics and system challenges are offered. The IPE Steering Committee works to evaluate new IPE proposals from the schools' faculty to assure curricular integrity and quality.

Next Steps: Toward Sustainable IPE

With much of the early IPE implementation work now done, it is critical that faculty in health professions programs turn their attention to the next and more difficult stage of IPE implementation: improving pedagogical quality. To do this, faculty members who are engaged in IPE require more focused development, incentives to remain engaged in IPE, and the ability to identify the impact of their work via well-defined educational outcomes.

Faculty Development

Faculty development efforts need to focus on the art of delivering high-quality IPE: the use of inclusive language, staging, and faculty role modeling for both students and other faculty.

The use of inclusive language is critical to the success of IPE efforts. It can be challenging for faculty to unlearn deeply enculturated, discipline-specific language and make intentional efforts to incorporate new ways of discipline and gender-neutral communication while teaching; but this helps IPE efforts in appealing to students from all fields. Words that today come naturally when teaching in disciplinary-specific settings such as "medical care," "physician," "nurse," "training," "he," and "she" should be consciously replaced with "health care," "clinician," "clinical education," and gender-neutral pronouns. Clinical situations used in IPE training must be inclusive of a variety of roles so each student can "find themselves" in the scenarios.

Faculty development in IPE can be complicated by AHC faculty skepticism about the need for IPE-specific training. To combat this skepticism, simple techniques such as videotape analysis of actual faculty teaching can point out areas in which language, for instance, can unintentionally undermine IPE. When followed up with coaching and mentoring from expert IPE educators regarding how to improve, faculty are better able to mindfully incorporate more appropriate language and staging techniques into future sessions.

IPE sessions should be staged in an inclusive fashion as well. Decisions as to which faculty members lead off a session and which close/wrap-up a session convey powerful messages to the student audiences and can send powerful nonverbal messages to participants from various fields. Care must be taken to assure a balance in faculty roles. The power of faculty role modeling these values of interprofessional collaboration, courtesy, and respect in session delivery cannot be understated.

Another aspect of staging that influences the success of IPE sessions is the introduction of IPE as either a "foreground" or "background" issue. When IPE is in the "foreground," knowledge, skills, and attitudes of teamwork, collaboration, communication, and disciplinary appreciation are the substance of the session. The benefits of such sessions are obvious and, thus, observable and measurable. When IPE is designed to be a "background" element of a session, the focus of the session is often one in which the gathered students all benefit from the topic being presented (e.g., ethics, QI, health policy) and, in most cases, work together in small interprofessional groups for discussion and problem solving during the session. The benefits of such IPE sessions are derived from the richness of perspectives from the various student disciplines that are garnered in the discussions.

Faculty Incentives for IPE Work

Most faculty involved in IPE describe their initial foray into the work as a "special project" that was either self-initiated or assigned by a leader within the institution. Few faculty members received workload releases or administrative support to conduct the early work of IPE, which often included extensive time investments in coalition building, program planning, and faculty recruitment.

Acknowledgment that IPE is a significant element of the teaching workload in health profession programs is rising. In leading AHCs (such as the University of Minnesota's National Center for Interprofessional Practice and Education and the University of Washington's Center for Health Sciences Interprofessional Education, Research and Practice), there have been appointments of leaders, committees, and staff to support such work. Curriculum committees now evaluate proposals that come forward as IPE initiatives to assure they are of high quality and are meaningfully integrated into the overall curriculum of the program.

Beyond the provision of an infrastructure to support IPE, AHC faculty members involved in IPE require complementary systems for tenure, promotion, and evaluation. The hard work of executing high-quality IPE within health professions schools must be acknowledged and rewarded in the same way as clinical and research productivity.

Evaluating IPE Outcomes

As with many educational interventions, it is daunting to measure IPE outcomes. While effective measures to evaluate short-term knowledge, skill, and attitude changes exist, long-term outcomes such as influence on postgraduation student behavior and practice patterns are still being assessed. Critical research questions for the years ahead include the following: Are efforts to introduce IPE in educational programs having the intended effect in practice settings? Are the cultures of practice settings transforming as the wave of newly minted pharmacists, nurses, physicians, and other professionals who have benefited from newer IPE training enter practice and do things "differently"?

The answers to such questions will provide the evidence to ensure that IPE efforts in the years ahead have the intended beneficial effects and the support from AHCs that they deserve.

Future Work

For the IPE work currently underway at AHCs to have sustained impact, IPE must occur in the context of interprofessional institutional leadership and day-to-day function. One great example of this model is the "Attending Nurse" model described by Fulmer and colleagues in which a doctorally prepared nurse is paired with an attending physician for patient-care rounds. The two attendings provide education to the learners on rounds from the different disciplinary views [14].

Senior leadership teams within AHCs, too, must absolutely be intentionally interprofessional themselves so that the value that the different disciplinary perspectives provide in the complex care delivery environment is realized throughout the institution. John Kotter writes, "In the final analysis change sticks when it becomes 'the way we do things around here,' when it seeps into the bloodstream of the corporate body. Until new behaviors are rooted in social norms and shared values, they are subject to degradation as soon as the pressure for change is removed [15]."

INTENSIFIED INVESTMENT IN EDUCATIONAL TECHNOLOGY TO SUPPORT EFFECTIVE LEARNING

As health professions education has evolved to incorporate not only biomedical training but training in communications, HCDS, and other fields in interprofessional and geographically distributed sessions, the need for sophisticated approaches to health professions education has intensified. Patient care benefits from broad adoption of electronic health records, increasing use of telemedicine, and the emergence of mobile technology platforms; and so too should the education of health professions students benefit from similar newer technologies. Using thoughtful development and deployment, health professions educators can leverage innovative technologies to meaningfully train students for successful future practice.

Technology Enables Distributed Learning in Authentic Practice Environments

Health professions education should be as technologically sophisticated as clinical medicine and health professions students. One major way technology-enhanced health professions education will improve the student experience is by providing an electronic "home" to an otherwise geographically distributed learning experience.

For the past two decades, medical educators have grappled with the educational implications of shifting patient care from the AHC to a network of ambulatory and community network providers. The use of learning technology now enables schools to provide quality instruction to students whenever and wherever they are training. Learning management systems, online learning, and televideo platforms maintain consistency and quality of core instruction and also allow broad and rapid distribution of expert teaching in emerging fields of medicine, such as population health, high-value care, and treatment of substance use disorders.

Student-use logs, assignment completion data, and learning dashboards provide students, course directors, and administrators at the home institution with the necessary information to monitor student learning and to meet accreditation requirements. Virtual learning environments can also facilitate interprofessional learning, even when schedules and geography make it difficult to arrange in-person meetings.

These same distance-learning methods may also be employed for AHC faculty development. Online modules and teleeducation reduce costly time away for practitioners and enable their participation as trained community preceptors. Distributed models provide equal cost savings for AHCs through more efficient use of faculty development professionals, and easy distribution of course materials.

Technology Enables Improved Teaching and Learning of Clinical Skills

Health professions students must learn huge amounts of information, often in geographically dispersed clinical contexts with potentially shifting AHC faculty member oversight. In this discontinuous and potentially fragmented teaching context, students must master increasingly complex, uncertain, and ambiguous clinical problems that stretch their reasoning skills to the edge of their competence.

Technology-enabled education can help tie these discontinuous learning experiences together into a continuous experience that is tailored to the individual learner's level of proficiency. The ubiquitous presence and ease of use of mobile phones and tablets is creating a renaissance in clinical skills education and authentic workplace-based assessment. Clinical skills checklists and mini-clinical evaluation exercises are being transformed into apps that can be downloaded and completed by faculty at the point-of-care with a learner on the fly, and then uploaded into an electronic portfolio. Thus technology-enhanced medical education can free AHC faculty members from wasting precious teaching time on repetitive content delivery, and instead promote a return to individualized evaluations at the bedside. Mobile video capabilities allow endless opportunities for faculty to directly observe clinical skills, with cloud storage for immediate or later review and feedback and advanced portfolio development. Imagine interviewing a recent graduate for a position, having first been granted access to their online video portfolio. Digital badging is another technology that can be harnessed to develop universal clinical skills badges to be awarded by certified faculty to qualified learners. A digital badge is an online symbol or indicator of a practitioner's accomplishment or skill and is awarded using standards accepted by the relevant community. Learning-outcome aligned badges have the potential to be shared publicly by a learner to indicate their accomplishments to other teachers, schools, or employers. This technology aligns well with the modern movement toward

entrusted professional activities in health professions education.

Virtual patients provide learners across the mastery continuum with an intentionally constructed yet realistic learning context, including lifelike interactivity informed by gaming interfaces in some programs that allow learners to safely practice their evolving skills via adaptive complexity and fidelity. Virtual patients and online simulations allow faculty to "immortalize" particularly instructive cases, model expert precepts, and create a safe practice environment for responding to communication, professionalism, and ethical challenges with targeted feedback. Furthermore, emerging innovations such as teleeducation—including online standardized patients and observed structured clinical exams—expose students to patient populations that may not be available in their practice locations and begin to provide learners with the comfort and skills needed to deliver effective telehealth in the future.

Technology Supports Application of Knowledge and Manipulation of Data

As the volume of clinical information mobilized by clinicians and learners alike grows yearly, both rely increasingly on online access to medical information, particularly via mobile platforms such as smartphones and tablets. Rapid mobile access to large amounts of up-to-date information is forcing educators to question what constitutes a "core knowledge base" for learners and what and how students should learn in the classroom. Online and mobile resources now provide rapid access to content that was once reserved for textbooks and reinforced through lectures. The new accrediting requirements for active and blended learning paradigms are shifting the burden of content learning to students through prerecorded lectures and online assignments, with the expectation that faculty will focus their teaching time on higher order applications of knowledge through authentic real-world problem solving. Basic science faculty members at AHCs, once the virtuosos of the lecture hall, are now finding new enthusiasm for the creation of animations to explain complex topics. For example, they are designing virtual laboratory experiments that can easily be altered and taking students to the simulation center to apply classroom concepts to simulated scenarios of cardiorespiratory shock.

Advances in information sciences and the availability of large data sets as research tools can lead to new classroom activities designed to teach students the data literacy skills they will need to provide effective population-based care. The development of mock data sets, computer models, and online data-based simulations affords another active learning tool to enhance students' facility with problem-solving through data sets, and creates opportunities for teamwork and leadership development through virtual team competitions.

Toward Universal Online Health Professions Education?

Do over 300 different biochemistry faculty at AHCs across the US really need to teach the Krebs cycle independent of each other each year? As online educational activities become increasingly utilized in health professions education, it becomes progressively easier to envision a future of basic science education as a blended learning model composed of multiinstitution online core courses delivered through video lectures by nationally recognized science educators. Thus, a core portion of health professions education could be delivered via massive open online courses (MOOC). These online learning platforms make best use of current cognitive sciences to design effective learning, and many can adapt to student learning needs and provide individualized cognitive coaching. These efficiencies are particularly appealing as reimbursements for teaching decline concurrent with rising expectations of clinical and research productivity, and they would allow local educational resources to be dedicated to high-yield teaching such as small group learning, labs, and simulations.

Challenges and Opportunities to Innovate

Faculty at AHCs, as well as affiliated community-based preceptors, must be given the time needed to explore the educational utility of emerging technologies and to participate and share in the development of online resources. Although initially expensive to develop, deployed technology-enabled resources can become cost-effective to deliver and maintain over time. National collaborative efforts have proved particularly successful and have provided important academic scholarly avenues for busy generalist educators. The MedU virtual patient cases,[1] developed by hundreds of clinician educators at medical schools throughout North America, are a leading example. Additionally, AHCs must invest in providing even a small dedicated team of education technologists and learning designers to support busy faculty educators in their efforts to design and employ technology in their teaching. High learner satisfaction, effective community placements, and innovative teaching and assessment programs provide a clear return on investment, although it is likely that even stronger evidence such as a reduction in other educational costs will make the case for educational technology even more effectively. Health professions schools within AHCs must resist the urge to limit legitimate uses of technology in education. For example, access to routinely available online resources should be built in to learner assessment (rather than banned from the classroom), and institutions

must grant learners access to authentic activities for them within the electronic health record. Finally, we must not forget that our learners have much to teach us about the innovative use of technology for both learning and improved patient care. All it takes is to see one student effectively teach a child and their family how a surgery will be performed using a tablet and a commonly available app.

ENHANCED FOCUS ON THE DEVELOPMENT OF LEADERSHIP SKILLS

For graduating health professions students to thrive in and improve the rapidly evolving context of health care delivery, health professions students need leadership training.

The traditional top–down, authority-based approach to leadership in academic medicine is increasingly ineffectual. This leadership approach is not wrong; in fact, there are times when authoritative decisions are important. But the complexity of challenges that confront medicine is too great for any one person or any single approach to address singlehandedly, which means today's leaders must function effectively in teams and have a broad range of possible responses to leadership challenges. As a result, our leadership paradigm must change along several dimensions.

We Must Continuously Expand Our Understanding of Leadership

In his autobiography, former Egyptian President Anwar El-Sadat wrote [16] "He who cannot change the very fabric of his thought will never be able to change reality, and will never, therefore, make any progress.... [No] change in the real world [can] be effected until a genuine change in ideas occur[s]." Unless and until we change the lens through which we understand and exercise leadership, our responses to the challenges we confront will reflect our old ways of thinking and are, as a result, unlikely to lead to a future that is substantially changed. If we define leadership as being about power and status, our ways of being and our actions will reflect that frame of reference.

The most powerful way to shift our worldviews, and thus to open ourselves to new ways of being, is to change the language we use to understand the challenges we face. The language we use includes spoken and written language, as well as posture, facial expressions, and tone of voice. Our memories of past experiences also shape our interpretations of new experiences, and these two are molded by language. Master leaders use language to recontextualize their leadership challenges in ways that allow them a wide range of responses that are nonetheless authentic. This gives leaders new and natural ways

---

[1] www.med-u.org.

of leading that are appropriate to the new challenges they face. This natural self-expression, duly informed by leadership training, gives leaders the power and freedom to lead effectively.

We Need to Revise and Expand the Way We Teach Leadership

Since health professions students need leadership training, and because true leadership requires leaders to recognize how language influences how they lead, health professions education at AHCs should prepare students to lead themselves and others effectively via the use of effective leadership language.

Teaching health professionals about leadership is distinct from *creating* health professionals who are leaders. Rather than approaching leadership as a knowledge problem that is solved by imparting a set of behaviors and traits (a third-person point of view), we must introduce pedagogies that allow the learner to master leadership as an "as lived" first-person experience. Just as health professions students learn clinical skills most effectively by seeing patients, health professions students must receive leadership training in the setting of leadership challenges.

By confronting real leadership challenges firsthand, students have the opportunity to identify and, if necessary, to change the ways of being and acting that limit their leadership effectiveness. That requires students to be given opportunities to lead in clinical contexts (with appropriate supervision), and to engage other kinds of leadership, such as through student government, service learning, and collaborative research projects.

To ensure students approach these first-person leadership experiences productively, they must learn to create a new "frame" and language that will facilitate effective leadership [17]. This new language grants leaders—including health professions students—a perspective from which they can orient their thinking, behaviors, and actions such that they are consistent with good leadership.

For example, it is well known that self-awareness is critical to tackling complex leadership challenges. Awareness involves bringing your full attention to the situation at hand and paying attention purposefully, nonjudgmentally, and with curiosity. A key aspect of awareness about which health professions students can be trained is mindfulness of the distortions existing in what has been called "your *already-always-listening*,"—that pervasive running commentary that's thinking for you and biasing you [18]. Awareness of this inner critic can help health professions students reduce its tendency to "control" their reactions to situations and thus behavior, opening up possibilities for new ways of being, thinking, and acting. This can give rise to a wider range of leadership approaches.

To support this kind of leadership learning among its medical students, the Geisel School of Medicine at Dartmouth has created a leadership elective for interested fourth-year students to discuss contextual and language influences on their personal leadership style, and how they can expand the range of leadership options available to them by learning a more effective language of leadership. This pilot program is being expanded into a core, required course in the new Geisel curriculum. The training, delivered by Geisel faculty as well as collaborating leaders from the Tuck School of Business, will be aligned with QI, service learning, and other practical collaborative projects to enable students to experience firsthand the potency of these leadership lessons through lived experience. We believe this unique opportunity to improve leadership skills during medical school, in conjunction with training in HCDS, will enable our graduates to be leaders and agents of change in the rapidly changing world of health care of the twenty-first century.

IMPROVED TRAINING IN PROFESSIONALISM AND PERSONAL RESILIENCE

The best college students enter the health professions often out of intense idealism and a desire to help others. Yet many health professions students experience a loss of idealism and a high incidence of burnout as their training progresses. This loss of idealism and surge in burnout rates can undermine future professionalism and plans to work with the underserved [19–23]. As a result, we believe a critical part of health professions education is the development of specific skills in professionalism and personal resilience.

Patient care is challenging work: clinicians make life-and-death decisions under time pressure and involving a vast variety of people and diseases. The difficult decisions and challenging conversations that arise in this context require a high degree of professionalism, including sharp ethical discernment and a firm commitment to compassionate care. When these age-old clinical stresses are joined by the newer challenges of working in interprofessional teams in a changing health care environment, the professionalism and personal resilience of health professions students can be severely tested. These challenges can lead to burnout, and from there a loss of ethicality and compassion [20,24]. Thus, health professions students require training to maintain professionalism and personal resilience.

Beyond the huge challenges of clinical training, another major contributor to the loss of idealism and to professional burnout is moral distress. Moral distress is common among health professions students [25]. In the classroom, health professions students come into contact and competition with diverse colleagues with a wide variety of attitudes about health care, and the resulting acculturation can challenge the original idealism that brought the students to their career in the first place. As they accrue clinical experience, health professions students may feel obliged to provide all care requested by a specific intensive care unit patient at the same time they know the expense of end-of-life care contributes to runaway health care costs. This too can create cognitive dissonance. Alternatively, students may recognize an ethical infraction on the part of the faculty member but worry that reporting it may adversely impact their career. In fact, students can perceive that their roles as learners along a continuum of professional power can create ethical dilemmas for them that may not be addressed by abstract discussions of philosophical principles [26]. Clinicians and learners confront these and myriad other ethical challenges and would benefit from training that sensitizes them to the priorities involved and how to balance them properly.

These challenges are not new in health professions education. Yet, the need of learners and clinicians for skills in ethical discernment and delivery of compassionate care is growing amid rapid health care change in which productivity pressures and clinician role uncertainty are intensifying. Such stressors can intensify the pace and complexity of health care decision-making, and the increasing stress can undermine provider ethical discernment [23,27–34]. These challenges may be amplified in AHCs where providers face multiple competing demands on their time, and clinical productivity can be influenced by the presence of learners.

To ensure students retain their ethical discernment and continue to deliver compassionate care to their patients despite these challenges, AHCs should support health professions students' development of skills in ethics and professionalism as well as personal resiliency.

Training in Ethics and Professionalism for Health Professions Students

There are many intriguing approaches to health professions education in professionalism and practice resilience. Examples of emerging approaches to teaching about ethics, professionalism, and compassion include the use of a professional formation model, integrated case-based teaching of "everyday ethics," and the use of techniques in narrative medicine. In the professional formation model of teaching about ethics and professionalism, skills and values needed for practice and growth in professional authenticity are aligned to match progressively greater stages of learner competency and autonomy, as described by MacIntyre, Dreyfus, and Palmer [35]. Increasingly, ethics and professionalism are taught not as abstract principles in separate courses but instead in integrated case-based sessions centering on so-called everyday ethics, that is, regarding practical skills needed to support ethical responses to specific ethical issues by students during health professions education [36].

There is no centralized guidance for the teaching of ethics and professionalism in health professions education, although there are innumerable institutionally-specific examples of how it might be done. David Doukas of the University of Louisville and other leaders of the recently established Academy for Professionalism in Health Care[2] are working via annual meetings and publications to redress this problem by identifying consensus topics, developing effective teaching approaches, and building a community of scholars who can make teaching in this field more cohesive [37,38].

Empathy for patients can support ethical behavior among students. Narrative medicine approaches, such as keeping a "parallel chart" on clinical rotations and close reading of clinically relevant works of literature as pioneered by Rita Charon at Columbia, can support empathy for patients by prompting students to deepen routinized consideration of patient experiences of illness, and thus should be used in health professions education to support student professionalism [39]. Given the high likelihood that complex, clinically relevant ethical and professional issues will arise during interdisciplinary team-based care, sessions in ethics, professionalism, and compassionate care are prime opportunities for IPE.

Importantly, despite the wide variety of potential approaches to teaching professionalism and ethics, none have been shown to predict future ethical behavior. Beyond the huge difficulty of tracking future ethical behavior among health professions students, the wide variety and multifaceted character of these teaching approaches make measurement challenging. Since these topics can be marginalized in the health professions curriculum, we must not shy away from this challenge and support teaching professionalism to health professions students that can be supported by hard data. It is worth mentioning that this standard is often not applied to other more foundational topics such as anatomy and biochemistry.

Recognizing that such teaching in the explicit curriculum can be undermined by the hidden curriculum [40], it is imperative for AHCs to support the development of robust professional identity among health professions students via interventions in the hidden curriculum, as well. Such interventions include the formation of centers in ethics, humanities, and compassionate care that can create community and dialog around the importance of ethics, humanities, and compassionate care in mainstream medicine. We also believe that collaborating with high-functioning teams engaged in the QI culture will help students retain their sense of personal efficacy [41,42].

At the Geisel School of Medicine at Dartmouth, we have developed a curriculum in ethics and social justice with strong roots in service learning to support medical student learning in this critical context [43,44]. Students will

---

[2] http://academy-professionalism.org/index_/Home.html.

learn the ethical foundations of professionalism and practical skills in personal resilience in the classroom, and then apply these concepts in real-life circumstances in clinics and service-learning environments, as well as in QI projects undertaken as part of their HCDS curriculum. We use narrative medicine throughout the curriculum to reorient students to the ideals that brought them to the health professions and ask them to engage with how experiences faced during their training influence those ideals.

Skills-Building in Personal Resilience for Health Professions Students

Stress and burnout can undermine ethicality among health care providers and students [45–48]. Similarly, struggling with ethical issues can lead to health care provider burnout [49]. Consequently, health professions education should include training in practice resilience skills that will allow students to enact behaviors they know are ethical during their training and eventually during clinical practice [50,51]. Teachable and measurable skills relevant to all health professions students that correlate with resilience include cultivation of a social support network, stress-coping skills, a sense of humor, an incrementalist mindset, high self-esteem and self-efficacy, setting values-based goals, optimism, and faith [52–54]. Mindfulness training, too, including meditation exercises, can improve empathy among health professions students [55]. Furthermore, there is early data suggesting that shifting to an efficient practice format utilizing interprofessional team-based care can improve health professional resilience [32]. These data suggest that firm training in practical skills in personal resilience can help health professions students retain and even deepen their sense of mission in their careers in a fashion that will support continued engagement with compassionate care. Thus, resilience training is baked into the core courses of the redesigned Geisel curriculum.

Today's health professions students will graduate into fast-paced and stressful careers in a changing health care environment. There is a significant risk that their professionalism will be endangered by burnout from these stresses. To counter these influences, to support their retention of the ideals that brought them into the helping professions and thus enhance the likelihood that they will continue to contribute to healthy change in medicine, it is critical to include training in practical skills in ethics, professionalism, and personal resilience.

THE CHALLENGE OF CHANGE IN ACADEMIC HEALTH CENTERS

In this chapter we have described six key transformative strategies that will help students who graduate from AHCs thrive and lead in the new health care environment. AHCs may be specially challenged to implement these changes. While change can be difficult for any organization, the culture of academia poses significant challenges for real transformational change. In the typical university setting, decision-making is usually one of "shared governance, academic freedom, extensive consultation, and widespread agreement before action—with limited tolerance for risks and mistakes [56]." The sense of urgency for change is often lacking, with faculty supporting the status quo and a return to the past. In most university settings, faculty members are said to care passionately about three things: money, space, and the curriculum. Any threat to these may ignite a reaction that opposes any significant change.

US medical centers are large and complex with the locus of power often diffused across several areas, such that no one person has the power to effect change. On the other hand, AHCs have been forced to adapt to the transformation of our health care system over the past 30 years with an increasing emphasis on efficiency, cost, quality, patient safety, patient satisfaction, and outcomes. Thus, it may be argued that AHC faculties are better poised to facilitate and implement change compared with academic faculties in other fields.

AHCs are united around three major missions: patient care, research, and education. In the past, AHCs often relied on revenues from the clinical mission to subsidize the education and research missions; however, the current realities of declining reimbursement of hospital and physician services, resulting in lower profit margins, provide less opportunity for this cross-subsidization. Such financial pressures on the AHC may impede the willingness of faculty or senior leadership to invest additional time or financial resources to the necessary curriculum redesign effort.

Despite these challenges, we believe that leaders of AHCs must be willing to do their part in improving the health care system through meaningful education reform that will be more patient—and learner—focused. New models of health professions education should align with new models of health care, such as the patient-centered medical home. Improvements in health professions education do not necessarily imply greater costs. For example, the new curriculum at the Mayo Medical School reduced the number of contact hours for faculty with students in several blocks by as much as 30%, thus lowering costs [57]. Leaders of AHCs must build a sense of urgency for local health professions education reform, create and communicate a vision for their new curriculum model, and empower broad-based action from both faculty and students. Certainly, there will be bumps and hurdles along the way, so finding creative solutions that work at the local level will be necessary. Strong leadership and effective organizational structures that promote high standards, monitor performance, and align resources are keys to a successful transformation.

CONCLUSIONS

AHCs must adapt in response to recent rapid changes in the health care environment. Health professions students need strong training in HCDS in the geographically distributed and interprofessional context of clinical care. Technology-enhanced health professions education can help create a learning "home" for these experiences and the training in professionalism, resilience, and leadership skills necessary to thrive and lead health care into the twenty-first century.

REFERENCES

[1] Fisher ES, Shortell SM. Accountable care organizations: accountable for what, to whom, and how. JAMA 2010;304(15):1715–6.

[2] Gerstner, L, IBM's leadership and competitiveness. BrainMass. https://brainmass.com/business/403142. [accessed 21.10.14].

[3] Pronovost P, Needham D, Berenholtz S, Sinopoli D, Chu H, Cosgrove S, et al. An intervention to decrease catheter-related bloodstream infections in the ICU. N Engl J Med 2006;355(26):2725–32.

[4] Srinivasan A, Wolfenden LL, Song X, Mackie K, Hartsell TL, Jones HD, et al. An outbreak of *Pseudomonas aeruginosa* infections associated with flexible bronchoscopes. N Engl J Med 2003;348(3):221–7.

[5] Kirschke DL, Jones TF, Craig AS, Chu PS, Mayernick GG, Patel JA, et al. *Pseudomonas aeruginosa* and *Serratia marcescens* contamination associated with a manufacturing defect in bronchoscopes. N Engl J Med 2003;348(3):214–20.

[6] Ludmerer KM. Time to heal : American medical education from the turn of the century to the Era of Managed care. New York: Oxford University Press; 1999.

[7] Schlesinger LS, Helms CM. Cost-conscious care, housestaff training, and the academic health center. Acad Med 1995;70(7):561–2.

[8] Morgenstern BZ, editor. Guidebook for clerkship directors, 4th ed. J Med Libr Assoc 2013;101(2):161–2.

[9] Asch DA, Nicholson S, Srinivas S, Herrin J, Epstein AJ. Evaluating obstetrical residency programs using patient outcomes. JAMA 2009;302(12):1277–83.

[10] AAMC. Core competencies for interprofessional collaborative practice: report of an expert panel. In: Colleges AoAM. 2011.

[11] Josiah Macy Jr. Foundation. Conference on interprofessional education. 2012, http://macyfoundation.org/publications/publication/ipe-conference (accessed 2.1.15).

[12] Robert Wood Johnson Foundation. Advancing interprofessional education. 2011, http://www.rwjf.org/en/about-rwjf/newsroom/newsroom-content/2011/05/advancing-interprofessional-education.html (accessed 2.1.15).

[13] Oandasan I, Reeves S. Key elements for interprofessional education. Part 1: the learner, the educator and the learning context. J Interprof Care 2005;19(Suppl. 1):21–38.

[14] Fulmer T, Cathcart E, Glassman K, Budin W, Naegle M, Devanter NV. The attending nurse: an evolving model for integrating nursing education and practice. Open Nurs J 2011;5:9–13.

[15] Kotter JP. Leading change: why transformation efforts fail. Harv Bus Rev March–April 1995;73(2):59–67.

[16] Sadat A. In search of identity : an autobiography. 1st ed. New York: Harper & Row; 1978.

[17] Souba W. Perspective: a new model of leadership performance in health care. Acad Med 2011;86(10):1241–52.

[18] Souba W. The science of leading yourself: a missing piece in the healthcare reform puzzle. Open J Leadersh 2013;2(3):45–55.

[19] Hojat M, Vergare MJ, Maxwell K, Brainard G, Herrine SK, Isenberg GA, et al. The devil is in the third year: a longitudinal study of erosion of empathy in medical school. Acad Med 2009;84(9):1182–91.

[20] Morley CP, Roseamelia C, Smith JA, Villarreal AL. Decline of medical student idealism in the first and second year of medical school: a survey of pre-clinical medical students at one institution. Med Educ Online 2013;18:21194.

[21] Mader EM, Roseamelia C, Morley CP. The temporal decline of idealism in two cohorts of medical students at one institution. BMC Med Educ 2014;14:58.

[22] Meltzer LS, Huckabay LM. Critical care nurses' perceptions of futile care and its effect on burnout. Am J Crit Care 2004;13(3):202–8.

[23] Dyrbye LN, Massie Jr FS, Eacker A, Harper W, Power D, Durning SJ, et al. Relationship between burnout and professional conduct and attitudes among us medical students. JAMA 2010;304(11):1173–80.

[24] Riess H. Empathy in medicine–a neurobiological perspective. JAMA 2010;304(14):1604–5.

[25] Berger JT. Moral distress in medical education and training. J Gen Intern Med 2014;29(2):395–8.

[26] Feudtner C, Christakis DA. Making the rounds. The ethical development of medical students in the context of clinical rotations. Hastings Cent Rep 1994;24(1):6–12.

[27] Linzer M, Manwell LB, Williams ES, Bobula JA, Brown RL, Varkey AB, et al. Working conditions in primary care: physician reactions and care quality. Ann Intern Med 2009;151(1):28–36. W26-29.

[28] Stone PW, Harrison MI, Feldman P, Linzer M, Peng T, Roblin D, et al. Organizational climate of staff working conditions and safety-an integrative model. In: Henriksen K, Battles JB, Marks ES, Lewin DI, editors. Advances in patient safety: from research to implementation. Concepts and methodology, vol. 2. 2005, Agency for Healthcare Research and Quality (US), Rockville (MD).

[29] Hyman SA, Michaels DR, Berry JM, Schildcrout JS, Mercaldo ND, Weinger MB. Risk of burnout in perioperative clinicians: a survey study and literature review. Anesthesiology 2011;114(1):194–204.

[30] Putnik K, Houkes I. Work related characteristics, work-home and home-work interference and burnout among primary healthcare physicians: a gender perspective in a Serbian context. BMC Public Health 2011;11:716.

[31] Iliceto P, Pompili M, Spencer-Thomas S, Ferracuti S, Erbuto D, Lester D, et al. Occupational stress and psychopathology in health professionals: an explorative study with the multiple indicators multiple causes (MIMIC) model approach. Stress 2013;16(2):143–52.

[32] Sinsky CA, Willard-Grace R, Schutzbank AM, Sinsky TA, Margolius D, Bodenheimer T. In search of Joy in practice: a report of 23 high-functioning primary care practices. Ann Fam Med 2013;11(3):272–8.

[33] Fiabane E, Giorgi I, Sguazzin C, Argentero P. Work engagement and occupational stress in nurses and other healthcare workers: the role of organisational and personal factors. J Clin Nurs 2013;22(17–18):2614–24.

[34] Zwack J, Schweitzer J. If every fifth physician is affected by burnout, what about the other four? resilience strategies of experienced physicians. Acad Med 2013;88(3):382–9.

[35] Daaleman TP, Kinghorn WA, Newton WP, Meador KG. Rethinking professionalism in medical education through formation. Fam Med 2011;43(5):325–9.

[36] Fox E, Arnold RM, Brody B. Medical ethics education: past, present, and future. Acad Med 1995;70(9):761–9.

[37] Doukas DJ, McCullough LB, Wear S, Lehmann LS, Nixon LL, Carrese JA, et al. The challenge of promoting professionalism through medical ethics and humanities education. Acad Med 2013;88(11):1624–9.

[38] Doukas DJ, McCullough LB, Wear S. Project to R, Integrate Medical Education I. Perspective: medical education in medical ethics and humanities as the foundation for developing medical professionalism. Acad Med 2012;87(3):334–41.

[39] Charon R. Narrative medicine : honoring the stories of illness. New York: Oxford University Press; 2006.

[40] Hafferty FW. Beyond curriculum reform: confronting medicine's hidden curriculum. Acad Med 1998;73(4):403–7.

[41] Liao JM, Thomas EJ, Bell SK. Speaking up about the dangers of the hidden curriculum. Health Aff 2014;33(1):168–71.

[42] Lahey T. The hidden curriculum: changing the water in which we swim. Health Affairs Blog; 2014. http://healthaffairs.org/blog/2014/01/31/the-hidden-curriculum-changing-the-water-in-which-we-swim/. [accessed 25.04.14].

[43] Coria A, McKelvey TG, Charlton P, Woodworth M, Lahey T. The design of a medical school social justice curriculum. Acad Med 2013;88(10):1442–9.

[44] Lahey T. Perspective: a proposed medical school curriculum to help students recognize and resolve ethical issues of global health outreach work. Acad Med 2012;87(2):210–5.

[45] Nelson WA. Ethical uncertainty and staff stress. Moral distress has negative consequences for healthcare organizations. Healthc Exec 2009;24(4):38–9.

[46] O'Donnell P, Farrar A, BrintzenhofeSzoc K, Conrad AP, Danis M, Grady C, et al. Predictors of ethical stress, moral action and job satisfaction in health care social workers. Soc Work Health Care 2008;46(3):29–51.

[47] Branch Jr WT. Supporting the moral development of medical students. J Gen Intern Med 2000;15(7):503–8.

[48] Dyrbye LN, Thomas MR, Shanafelt TD. Medical student distress: causes, consequences, and proposed solutions. Mayo Clin Proc 2005;80(12):1613–22.

[49] Wlodarczyk D, Lazarewicz M. Frequency and burden with ethical conflicts and burnout in nurses. Nurs Ethics 2011;18(6):847–61.

[50] Howe A, Smajdor A, Stockl A. Towards an understanding of resilience and its relevance to medical training. Med Educ 2012;46(4):349–56.

[51] Dunn LB, Iglewicz A, Moutier C. A conceptual model of medical student well-being: promoting resilience and preventing burnout. Acad Psychiatry 2008;32(1):44–53.

[52] Kjeldstadli K, Tyssen R, Finset A, Hem E, Gude T, Gronvold NT, et al. Life satisfaction and resilience in medical school–A six-year longitudinal, nationwide and comparative study. BMC Med Educ 2006;6:48.

[53] Connor KM, Davidson JR. Development of a new resilience scale: the Connor-Davidson Resilience Scale (CD-RISC). Depression Anxiety 2003;18(2):76–82.

[54] Dweck CS. Self-theories : their role in motivation, personality, and development. Philadelphia: Psychology Press; 1999.

[55] Barbosa P, Raymond G, Zlotnick C, Wilk J, Toomey R, 3rd, Mitchell J, 3rd. Mindfulness-based stress reduction training is associated with greater empathy and reduced anxiety for graduate healthcare students. Educ Health (Abingdon) 2013;26(1):9–14.

[56] Azziz R. What is the value and role of academic medicine in the life of its university? Acad Med 2014;89(2):208–11.

[57] Lindor KD, Pawlina W, Porter BL, Viggiano TR, Grande JP, Barrier PA, et al. Commentary: improving medical education during financially challenging times. Acad Med 2010;85(8):1266–8.

ABOUT THE AUTHORS

Leslie Fall, MD serves as the Associate Dean for Faculty Development at the Geisel School of Medicine at Dartmouth and is an inaugural member of Geisel's Academy of Master Educators. She is a Professor of Pediatrics (Hospital Medicine) and also serves as the Vice Chair for Education in the Department of Pediatrics.

Tim Lahey, MD, MMSc is an associate professor of medicine at the Geisel School of Medicine at Dartmouth, where he sees patients with HIV and infectious diseases. Dr Lahey is also chair of the Clinical Ethics Committee of Dartmouth-Hitchcock Medical Center.

Greg Ogrinc, MD, MS is the Director of the Office of Health Systems and Clinical Improvement (OHSCI), and an Associate Professor of Community and Family Medicine of The Dartmouth Institute at the Geisel School of Medicine, Dartmouth. He is also Director of the Quality Literature Program at Dartmouth, which developed the Standards for Quality Improvement Reporting Excellence (SQUIRE) guidelines.

Susan Reeves, EdD, RN is Associate Academic Dean for the Dartmouth-Hitchcock Partnership Program, The Gladys A. Burrows Distinguished Professor of Nursing, and Chair, Nursing & Public Health at Colby-Sawyer College.

Rich Simons, MD, MACP is the Senior Associate Dean for MD Programs at The George Washington University School of Medicine and Health Sciences. Prior to this, Dr Simons served as the Senior Associate Dean for Medical Education and Associate Vice President for Health Affairs at the Geisel School of Medicine at Dartmouth.

Wiley "Chip" Souba, MD, ScD, MBA is the former Dean and Vice President for Health Affairs at the Geisel School of Medicine at Dartmouth, where he also holds a faculty appointment as Professor in the Department of Surgery at the Tuck School of Business. Before moving to Dartmouth, Dr Souba held similar positions at the Ohio State University.

The Challenge of Discovery

Chapter 14

The Changing Spectrum of Biomedical and Clinical Research

Robert Clarke, Elliott Crooke and Howard J. Federoff

FUNDING RESEARCH AND THE VALUE PROPOSITION

Biomedical and clinical research at US institutions continues to have a major beneficial impact on human health and well-being. The ability to achieve this outcome has been largely the result of a consistent level of publicly (federal, state, and local government) and commercially (mostly pharmaceutical, biotechnology, and medical device industries) funded research. Private (philanthropic) support has also contributed to advances in research and treatment. In 2012, 41% of public and commercial biomedical research and development (R&D) expenditures were supported by public funding [4]. The value of this investment to human health is readily evident—between 1988 and 2005, research supported by public funding led to almost half of all drugs approved by the Federal Drug Administration [25]. However, the value of research universities in general (not only those involved in biomedical and clinical research) is not restricted to the products of their research. On behalf of the Association of American Universities, Berdahl [2] provided a brief but compelling overview of their impact on society, public policy, humanities and the arts, civic and economic life, and as agents of a free society. The critical importance of research universities, and the current threats to their continued dominance, was acknowledged in the 2012 report, *Research Universities and the Future of America: Ten Breakthrough Actions Vital to Our Nation's Prosperity and Security* ([5]; see also [6] for an overview of this report).

Most publicly supported research has been investigator-initiated (the ideas having come from individual scientists or research collaboratives) and passed rigorous, independent peer review. Publicly funded biomedical research has been performed mostly at academic health centers (AHCs) and research institutes associated with universities. Consequently, tightly integrated with this research enterprise is an education mission. A critical objective of the education mission is to train the next generation of researchers and practitioners—those who will learn, advance, and implement biomedical and clinical research discoveries. However, the research experience is often a critical component of both undergraduate and postgraduate education, independent of whether the degree is expected to lead to a career in biomedical or clinical research.

Despite the consistent upward trend over several prior decades, the significant contribution to local and national economies, and the international reputation of the United States as the definitive world leader in biomedical and clinical research and education, funding for US biomedical and clinical research has recently begun to falter. Overall, the US share of expenditures in global biomedical and clinical research fell from approximately 50% to just over 40% between 2007 and 2012 [4]. A significant component of this decline in total dollars reflects a reduction in commercially funded R&D [4]. For AHCs, the primary focus of this chapter, changes in the trend of public funding have the greatest impact on their ability to sustain a research mission and their contribution to human health and well-being.

In response to faltering public funding in support of research, the contribution by AHCs and universities from their institutional resources has increased substantially. A recent study by the Association of Academic Health Centers (AAHC) estimates that 35% of total research expenditures at AHCs are supported by internal funds [1]. With the added pressures on tuition costs and student debt for full-time university students, the challenges continue to increase with respect to how US research universities can and should balance their respective investments in research and education. Most institutions have primary income streams from tuition, philanthropy (current use and income from endowment and other financial investments), and services such as rental of facilities, residences, and access to other resources.

For research universities, the indirect cost recovery from research grants and contracts provides financial support for the research infrastructure. However, this income is not intended to cover all research-associated costs. While the margin on clinical care and services at university hospitals is generally modest, it is common practice for some component

The Transformation of Academic Health Centers. http://dx.doi.org/10.1016/B978-0-12-800762-4.00014-1

to be reinvested in biomedical and clinical research. In this context, the ability of research to drive reputation for clinical care and offer patient access to clinical trials can affect the ability to compete effectively for market share. Return from royalties and intellectual property rights that are licensed for financial return also support the research enterprise, but this is generally a relatively small income source.

Overall, universities and AHCs face the challenge of a limited number of significant revenue streams from which to support research. In any given fiscal year, some key revenue streams can be uncertain and/or affected by external financial forces that can be beyond an institution's immediate control. For example, the Affordable Care Act (ACA) heralds a decline in the generation of clinical margin among AHCs owing to shifting economics [27]. The need for financial diversification for greater sustainability of research universities has been widely acknowledged [8].

Given these challenges, it might seem counterintuitive that an institution would choose to maintain or increase its support for an activity that requires over one-third of its expenditures while experiencing increasingly constrained internal sources. However, universities benefit from the rewards of successful research enterprises in many of the same ways as do local and national governments. Research drives reputation, with several key university ranking systems incorporating research metrics in their assessments. AHCs and schools of engineering are the most common drivers of the funding metrics that are incorporated into these rankings. In 2013, 44 (86%) of the top 51 colleges and universities, as ranked by *US News and World Report*, were classified by the Carnegie Foundation for the Advancement of Teaching as RU/VH (very high research activity). Of these, 36 have AHCs; 12 of the remaining 15 institutions have schools of engineering. Attracting the best students into undergraduate programs in the natural and physical sciences is frequently dependent upon the quality and reputation of the research programs in the professional schools of medicine and engineering and the opportunity to participate directly in the research enterprise. The process of research, from exploring the literature and generating a hypothesis through constructing an approach to hypothesis testing, teaches general skills and principles that extend well beyond the sciences.

Universities exist within communities where they are often a major employer. Supporting and executing the biomedical and clinical research enterprise provides significant advantages to both the local community and its government. In 2010, NIH investments resulted in the creation of approximately 485,000 quality jobs and produced $69.190 billion in new economic activity across the country [9]. Given many of these diverse advantages, once internal financial issues are set aside, the value proposition to academic institutions of biomedical research becomes unquestionable.

During this time of competing pressures on the costs and value of higher education at research universities, it is becoming more challenging for the very institutions that are poised to have the greatest impact on human health. Moreover, these challenges have become constraining at a time when the pace of technological advances has increased the potential for major medical advances to a level unprecedented in recent decades. Paradoxically, it is the past investment in research that has brought us to where we can imagine advances that previously were unimaginable; yet, the ability to realize these advances and translate them into new standards of care for patients and into new preventive interventions is now compromised.

HISTORICAL TRENDS IN PUBLIC RESEARCH FUNDING (NIH)

As the major public source of funding for US biomedical and clinical research, NIH is a primary focus for this chapter. The general trends in NIH research funding over time are often also reflected in other public funding sources, e.g., the National Science Foundation. The data presented are obtained from public sources and can readily be accessed online; we have generally used data that exclude the one-time budget increase NIH received from the American Recovery and Reinvestment Act of 2009.

NIH funds support research on the various NIH campuses (intramural research) and the research performed in the United States at universities and research institutions (extramural research). In 2003, the intramural program represented 9.6% of the total funds available, increasing to 11.2% by 2013. Most funding of extramural biomedical research occurs at academic entities, with the greatest proportion at medical schools (Figure 1).

When examining the total NIH budget from 1982 through 2013, with and without adjustment for inflation, the clear and continuing national support for this component of the US R&D enterprise shows the more recent slowing in the rate of increase, and even a decline when adjusted for inflation (Figure 2). Figure 3(a) shows the impact on overall success rates from 1995 to 2013. From a recent peak in 2000, when the NIH reported success rates of 32% for R01 equivalent awards, there has been a steep fall to 17.5%, with the 2013 success rate at a historic low.

R01 Funding Success Rates

R01 and equivalent grants have been the primary individual investigator-initiated awards throughout the history of NIH funding. Most research laboratories at AHCs are supported by R01 and equivalent awards. For many faculty members, their employment, tenure, and promotion prospects are largely determined by these awards and by their ability to be competitively renewed or replaced with new awards. Figure 3(b) shows the inverse relationship between the increasing

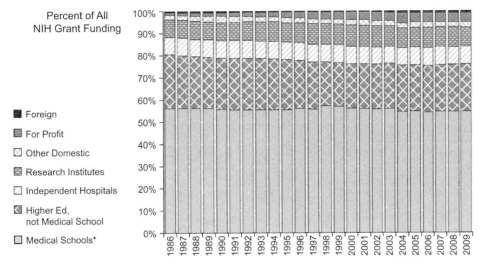

FIGURE 1   **Percent of all NIH grant funding by institution type.** *From Ref. [22].*

number of applications and the success rate; the increase in applications comes despite a change in policy in January 2009 that restricted investigators to only two submissions of the same application. Policing of the policy ensured that substantially similar work could not be resubmitted as a new application after it had failed to receive funding after two rounds of review. This policy was very recently replaced (2014) to effectively allow unlimited submissions because unsuccessful grants could now be resubmitted as new applications without NIH screening for adequate changes from the second unsuccessful submission. The impact of this change is likely to substantially increase the number of submissions for the foreseeable future, driving success rates even lower. A further pressure on funding is the ever-increasing cost of wet laboratory-based research, particularly for those using many of the newest research technology platforms. The Biomedical R&D Price Index,

which measures such costs, generally runs slightly ahead of the Consumers Price Index. Perhaps not surprisingly, the average amount awarded for R01 equivalent applications almost doubled between 1995 and 2013.

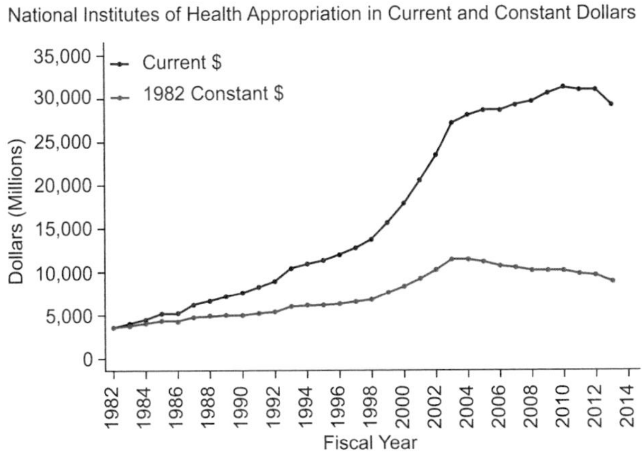

Funding the Investigator Pipeline

A major concern has been the loss of almost an entire generation of young scientists during this contraction in funding. NIH introduced a policy that identified early stage and new investigators and dictated that review of their applications be separate from the review of more established investigators. The goal was to level the playing field for less-experienced researchers attempting to obtain their first R01 equivalent award. Two categories of new researchers are defined with respect to obtaining their first R01 and elapsed time since obtaining their terminal degree. Early Stage Investigators are within 10 years of completing their terminal research degree or within 10 years of completing their medical residency. New Investigators have not previously competed successfully for an NIH-supported research project other than small research or training/career awards. Success rates for new versus established investigators have trended downward since 2009 when they were equivalent at approximately 18% (see Figure 4). This trajectory is causing leadership at AHCs to closely examine the faculty composition and the associated success of sustaining the biomedical research mission.

NIH has a general target of 50% of new investigators also being Early Stage Investigators. This policy has been generally successful in narrowing the success rate gap between early and new investigators and established investigators (see Figure 4). While the most recent policy changes were instigated in 2009, the success rates are largely indistinguishable between these two groups of investigators from 2007 through 2011. In 2010, almost 38% of all competing R01s were by

National Institutes of Health Appropriation in Current and Constant Dollars

FIGURE 2   **NIH appropriations in current and constant dollars from 1982 through 2013.** These data do not include American Recovery and Reinvestment Act funds. *Republished with permission of Ref. [10] Figure 1; permission conveyed through Copyright Clearance Center, Inc.*

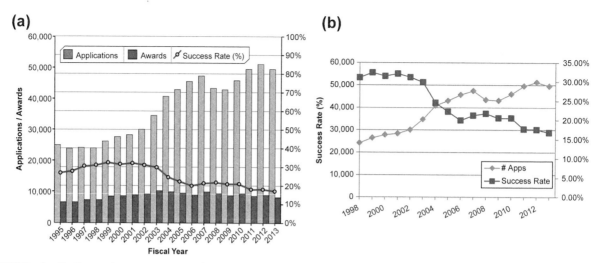

FIGURE 3 Applications and success rates. (a) The number of applications, awards, and success rates from 1995 to 2013 *(from Ref. [14]).* (b) The number of applications and success from 1998 to 2014 demonstrating an inverse relationship.

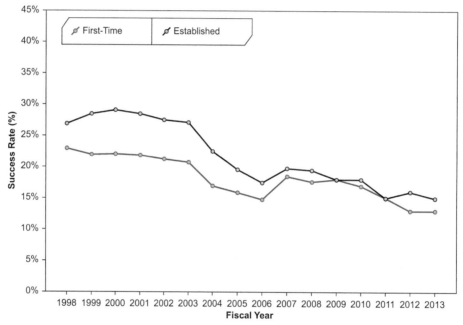

FIGURE 4 The Success rates for R01 equivalents for first-time and established investigators. The data show the percent success rate for applications submitted by new and established investigators from 1998 to 2013. *From Ref. [15].*

new investigators, an increase from approximately 28% in 2006. This favorable change has not been sustained recently; the percentage of first-time R01 investigators fell to 33% in 2013. Nonetheless, with failure rates approaching 85% for all investigators, the ability to recruit and sustain any researcher and enable her or him to experience a reasonably predictable and productive career is becoming increasingly untenable.

Investigator Life Trends

The final trend of note is the age at which investigators obtain their first R01 equivalent award—the mark that they have placed their foot firmly on the first step of

their research career ladder. For investigators with a PhD degree, the age for first R01 has remained at approximately 42 for over a decade; it is slightly older and increasing for those with an MD or MD/PhD degrees. For most other professions, this is an age that would mark a mid-career, not an early stage, achievement. At a time when most are attempting to raise young families, buy homes, and invest in retirement, significant uncertainty in career security and advancement are major disincentives to remain in science and the pursuit of new knowledge. Those obtaining their first R01 will likely only remain in research if they can subsequently renew or replace this with another comparable award. However, they then face the challenge of further attempts to obtain funding when their likelihood of success remains equally limited; and now they must also compete with fully established investigators.

What remains unclear is whether trends over the last decade will ultimately prove to be noise (over a prolonged period of analysis), an inflection point (where future trends may either recover or further degrade), or a stabilization (where the level of funding adjusted for inflation remains constant). The generally conservative and short-term nature of university financial planning, relative to the longer-term trends shown above, will likely have the effect of planning for the worst but hoping for the best. The outcome of which, should it continue for perhaps five more years, is likely to have adverse long-term consequences for the ability of the United States to remain competitive and to drive and maintain its economic recovery through R&D.

THE PHYSICIAN-SCIENTIST: DIMINISHING CONTRIBUTIONS?

Physician-scientists typically pose clinically relevant questions in research settings and often utilize scientific inquiry in patient care. Accordingly, they are key contributors to the biomedical research workforce. In the past 10 years, NIH grants to investigators with an MD degree—compared to those scientists with a PhD degree—were more likely to involve research with human subjects [24]. A recent analysis [10] of physician-scientists indicates that the numbers of physicians have nearly doubled since 1980, accompanied by a greater than 50% reduction (from 3.6% to 1.6%) reporting research as their primary activity (from 1982 to 2011). The data reveal declining number of medical students who express strong interest in research. This is paralleled by declining numbers of medical doctors supported by NIH T32 and F32 awards during their training. In addition, the trend for clinical doctorate holders for K08 and K23 awards has also declined, particularly after the period of NIH doubling (1998–2003).

As an extension to the earlier discussion regarding overall NIH funding of R01s, applicants with an MD degree declined from approximately 20% in the late 1980s to approximately 13% in 2011. In the aggregate, the trajectory is not surprising given declining medical student interest, fewer applications for postdoctoral training, and falling success rates for R01 applications. Among those successfully competing for an R01, the age at receipt has increased substantially—as noted above. Some of the drivers contributing to this trend include medical student indebtedness, job security and salary dependence on funded grants, perceptions that the scientist career is too competitive, and lifestyle choices; these variably compel consideration of a purely clinical career. The growth of loan forgiveness by the NIH has not as yet produced a measurable impact on the physician-scientist pipeline.

This pattern will continue as the pipeline is depleted. The implications for the performance of the biomedical research enterprise are substantial, perhaps most notably in the area of translational and clinical research.

Institutional Factors: Investigator Faculty

Research universities grapple with an array of issues to sustain faculty success as investigators. In addition to underwriting a larger "research gap," owing to declining HHS funding and NIH pay lines (see above), research universities also struggle with several other issues: graying of faculties, augmented regulatory burden, increased downward pressures on clinical and/or educational revenues, and inadequate maintenance and updating of vital infrastructure needed to remain competitive. The graying of faculties roughly parallels the aging demographic in the United States From 1990 to 2011 the percentage of faculty over the age of 60 has grown steadily, as shown in Figure 5, reflecting a sample of public and private research-intensive universities [12]. This change in age distribution can impact the research and educational missions.

Grant funding has also mirrored the demographics of faculty in this same period (Figure 6). Analysis of NIH R01s awarded between 1990 and 2010 reveal that the proportion of grants to principal investigators (PIs) 66 years or older grew from 2% to 7%. In this same period, the proportion of grants awarded to PIs 36 years or younger shrank from ~11% to ~3%.

Given level or contracting resources, AHCs continue to support research, education, and clinical care with faculties that are unlikely to grow. The presenting issue is how to balance the faculty composition to meet all missions. Declining coverage of salaries from extramural grants is forcing AHCs and research universities to confront a substantial financial challenge that threatens the future of the biomedical research enterprise. Absent intervention, further graying of its faculties will eventually result in reductions in grant successes. To address this issue, research universities and AHCs are offering faculty incentives to induce long-standing tenured faculty to consider retirement [20]. These incentives take the form of buyouts or phased retirements for tenured faculty. The recovery of salary dollars from senior faculty transitions can allow AHCs and universities to invest in the recruitment of junior faculty, gradually changing the age distribution of faculties, and consequently promoting the remodeling of the research engine. Alternatively, some universities have sought to disproportionately hire nontenure-track faculty, including those devoted to the research mission. Recent data indicate that the predominance of faculty hiring by research universities is of nontenure line faculty, although many are oriented to education [13]. Whether this approach will ensure the recruitment and retention of the brightest and best in support of the research mission remains unclear. These data suggest a strategic redistribution of faculties to meet the mission areas and concomitantly mitigate financial exposure associated with

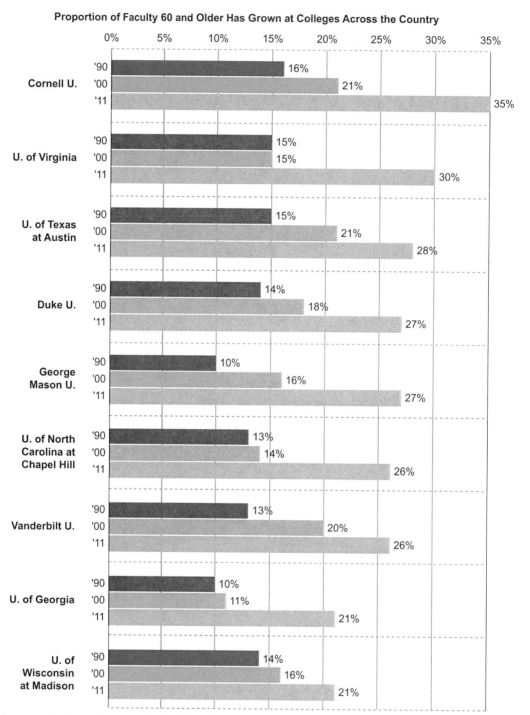

FIGURE 5 Change in faculty age in a sample of public and private institutions between 1990 and 2011. The data show the marked increase in the percentage of faculty greater than the age of 60 over the reported period. *("Aging Professors create a Faculty Bottleneck," The Chronicle of Higher Education, March 18, 2012 (chronicle.com/article/Professors-Are-Graying-and/131226)). Used with permission of The Chronicle of Higher Education Copyright© 2014. All rights reserved [12].*

lagging performance. We anticipate a gradual shift in faculty distribution by type and rank over the next decade with the arrival of a new steady-state funding environment. Going forward, AHCs will need to refine processes to finely tune a leaner and more efficient research enterprise.

Related is the ongoing discussion on what is the value and role of tenure at AHCs. Many medical schools are working to revise metrics and guidelines so that tenure review committees will be able to better recognize and reward collaborative accomplishments made by team-science participants, roles

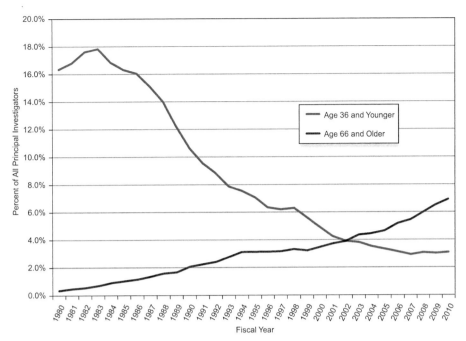

FIGURE 6   Percentage of NIH R01 principal investigators (PIs) age 36 and younger and age 66 and older (FY 1980–2010). The percentage of R01s awarded to PIs as a function of two age ranges, 36 and younger and 66 and older. *From Ref. [23].*

often played by junior faculty [11]. The "job-for-life" aspect of tenure is also under question. Ten years after implementation of posttenure review at one institution resulted in over 96% of the faculty receiving ratings of satisfactory, while also leading to a threefold increase in extramurally supported research and scholarly productivity. Much of the success of this program may lie with the nonpunitive nature of the review process, with the institution offering faculty development resources and mentoring to faculty identified as underperforming [26].

Funding and Support for Biomedical Trainees

To offset risk to AHCs' research and education missions brought on by the aging and eventual transitions of faculty and federally funded PIs, the next generation of biomedical scholars needs to be trained and poised to assume similar roles. Yet, the annual number of first-time, full-time students enrolling into biomedical sciences graduate programs has increased only modestly over the last few decades (Figure 7), with the total enrollment in the programs slightly outpacing initial enrollment largely due to a longer time-to-degree average over the same period.

In 2012, of the approximately 60,000 students pursuing graduate training in the biomedical sciences, only 15% were supported primarily by teaching assistantships and another 15% were either self-supported or supported by other sources. The remaining 70% were supported by research assistantships, fellowships, or traineeships [17]. Of these, about 20,000 were full-time students who were federally

supported, with approximately 15,000 supported by NIH and the remainder split between NSF, HHS (excluding NIH), and other sources (Figure 8).

Research assistantships, positions for graduate students that are most often funded from a Principal Investigator's R01 or equivalent grant, are the predominant NIH-based mechanism of graduate support in the biomedical sciences. However, over the last few years during the decline in R01 application success rates, the number of research assistantships has in itself declined, as can be expected. Unfortunately, the decline in research assistantships has not been offset by an increase in NIH-traineeships, which have remained quite constant over the last 30 years, and thus their percentage of the graduate students supported by NIH has significantly declined. While NIH-based fellowships have increased, they make up such a small percentage of the NIH-funded graduate student positions that their increase has failed to compensate for the loss of research assistantships (Figure 9).

Moreover, while the number of fellowship awards has gradually doubled over the last 15 years, the number of applications has more than tripled, leading to a decline in fellowship success rates from 50% to 60% at the turn of the decade to 30% in 2013.

Also of note is the limited number of NIH-funded trainees and fellows, especially given the need to transition new scholars into positions currently held by aging faculty and PIs. Students who qualify for and complete the traineeships and fellowships have a remarkably stronger record at becoming successful independent investigators as compared to other PhD recipients from the same fields and institutions,

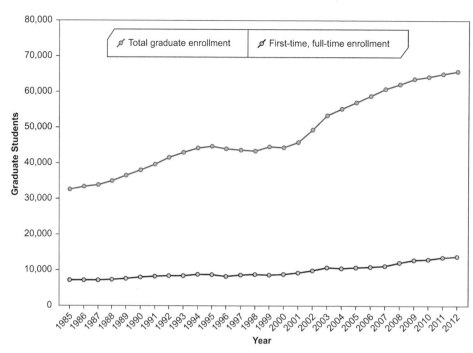

FIGURE 7 **Total graduate enrollment and first-time, full-time enrollment in the biomedical sciences.** Total and new, first-time enrollments, in biomedical sciences are shown. *From Ref. [16].*

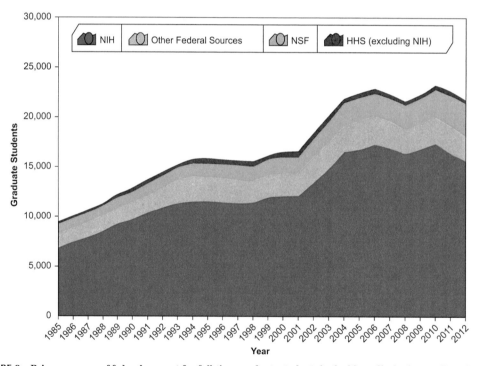

FIGURE 8 **Primary source of federal support for full-time graduate students in the biomedical sciences.** *From Ref. [18].*

and even more so when compared to recipients at institutions that do not have NIH training grants. Specifically, 37% of NIH-funded trainees and fellows apply for a significant NIH grant within 15 years of getting their degree, and over 23% will obtain such an award within that same 15-year period. This is in contrast to 13% and 7%, respectively, for

other PhD recipients at the same institution, and only 6% and 2.6% for students at institutions without training grants (NIH Trainee and Fellow File, IMPAC II).

There is little doubt that, over the next 5–10 years, many AHC faculty and major contributors to knowledge in the basic, translational, and clinical sciences will no longer be

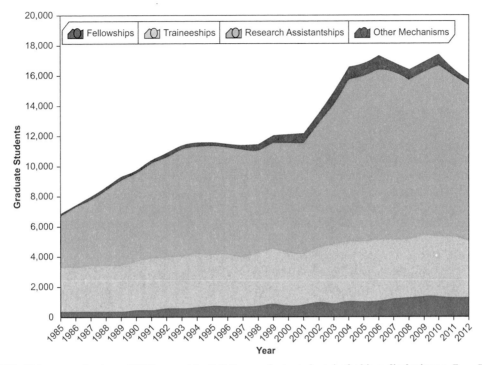

FIGURE 9 **Primary mechanisms of NIH support for full-time graduate students in the biomedical sciences.** *From Ref. [19].*

available to advance this mission. This leaves to question, what is the role of AHCs in training, recruiting, and supporting the next generation of biomedical scholars? Given the variety of financial pressures and uncertainties facing AHCs, will the training of students with a strong foundation in biomedical and clinical research remain a priority and, if so, can it still be achieved? While some institutions have made this a high priority in their strategic plan [21], many others are still struggling to address and answer the question.

While there is a growing need to engage the next generation of biomedical scholars at AHCs, what career options are there for doctoral and postdoctoral trainees who do not secure AHC appointments? The proportion of PhDs who obtain tenured or tenure-track faculty positions has dropped from ~34% 20 years ago to ~26% in 2012; the percentage of nontenure line faculty appointments has remained relatively constant. Likewise, opportunities to perform doctoral-level research in industry and government have also remained proportionally constant. One area of prospective job growth has been in science-related occupations that do not directly involve research [3]. In 2009, for US-trained biomedical PhDs who remained in the United States, 18% were in science-related, non research jobs—such as program and review officers for NIH and other government agencies (13%), managers in biotechnology companies, or other activities (5%). Of the remaining US-trained biomedical PhDs remaining in the United States, 6% were in government research, 43% in academic research or teaching, and 18% in private sector research. A small proportion (13%) was in nonscience-related occupations, and 2% were unemployed [3].

Despite these trends, biomedical graduate and postdoctoral training still focuses heavily on preparing individuals for research positions in academia. The 2012 NIH Biomedical Research Workforce Working Group recommended that biomedical doctoral students and postdoctoral fellows should be exposed to a broader range of experiences in preparation for expanded career options. Such options include project management, bioscience entrepreneurship, biomedical education at liberal arts institutions, and bioscience and regulatory policy development [3]. However, given that doctoral students and postdoctoral fellows play such a critical role in the performance of research at AHCs, the challenge remains as to how best to divert some of their effort from research toward training in these other activities, and who should be responsible for providing the necessary financial resources to do so.

The Workforce Working Group also examined whether more doctoral-trained staff scientist positions should be created. While such individuals would not be expected to become independent investigators, they would play a more important role in biomedical research by supporting research teams with specialized and sophisticated technologies and bringing long-term continuity and stability to research institutions. An example where this has worked well is with staff scientists who constitute an essential part of the NIH intramural research program. Still, how such positions could be supported at AHCs in a sustainable fashion is problematic. If these positions are to depend extensively on extramural funding, they may be too transitory to provide the above benefits; if the positions are to be funded by the academic institutions, at the expense of what other activities or initiatives?

INTERDISCIPLINARITY AND CONVERGENCE: FUTURE AHC RESEARCH MODEL

Interdisciplinary research has been recognized as an effective means to address questions that require multiple different disciplines to provide solutions. Much modern translational research falls into this category as investigators, most often with basic and clinical expertise, collaborate to bring observations from the laboratory into clinical application. The Clinical Translational Science Awards, a robust new NIH-based funding mechanism to address this emerging research area, promises to deliver more drugs, diagnostics, and devices to improve patient care and clinical outcomes. The interdisciplinary approach provides a foundation for further elaboration of collaboratives.

Convergence is problem-solving that cuts across disciplines, i.e., it is transdisciplinary. Convergence integrates knowledge, instruments, and concepts across the sciences, mathematics, and engineering to form a framework for scaling intrinsically complex problems. Implicit in convergence is that each discipline brings a different, perhaps unique, perspective to the problem—but alone cannot address the wide range of issues that impinge on multiple domains: health and wellness, security, economics, political stability, climate, and others. The goal is to merge these diverse areas into partnerships that effectively catalyze innovation from fundamental discovery to application. Convergence is a platform for support of collaborations among academia, national laboratories, industry, clinical enterprises, regulatory bodies, and funding entities.

Case Example: Georgetown University Medical Center

Georgetown University Medical Center (GUMC) has developed several convergence projects. One, termed the Argus Project, assembled a team with expertise in social anthropology, medicine, global infectious disease, epidemiology, computer science, and analysts fluent in more than 30 languages. The mission of Argus was to digest publicly available web and other content focused on evidence of social disruptions that herald emerging infections, possible pandemics. Operating over approximately five years with funding from the intelligence community, the nonclassified work of the Argus Project was responsible for early detection of SARS and H5N1. As anticipated for convergence projects, they have the potential to migrate. Before being decommissioned, at GUMC the Argus Project transitioned to monitoring political instability. Some of the faculty from this period have nucleated or joined other efforts.

The transient nature of convergent projects is a function of a time-limited problem. In larger organizations, the potential for convergence faculty to reorganize is

greater, with the institution being able to sustain their salaries as they migrate from problem to problem. In smaller organizations, the convergence faculty may need to seek additional outside experts from other research entities in order to build the transdisciplinary ensemble to be highly effective and competitive. This latter model suggests that "the faculty" may be more fluid and, in some instances, only transiently associated with the AHC or research university. This approach has some appeal for smaller research entities as it seeks to strategically leverage its assets to engage other experts that can commit to participate in a time-limited collaborative.

In the steady state, this strategy intends to iteratively redeploy the diversity of faculty expertise and bring into apposition those other experts who complement the employed faculty. Whether the scale of problem-solving will grow proportionate to the size of the convergence group is an attractive and testable hypothesis. Early support for this concept will be revealed by sponsors willing to underwrite key projects at scale. Demonstration of success, using both traditional and nontraditional metrics, will be required for faculty engaged in this model to be incented and rewarded with the expectation that convergence can be sustained. Faculty review committees and their criteria for promotion and tenure will need to evolve.

Adoption of convergence requires a cultural change for faculty, staff, administration, and leadership at AHCs and research universities—institutions that harbor traditionally organized, discipline-based departments. The nearly exclusive focus on individual contributions to the academy currently dominates the work undertaken by promotion and tenure committees. It is far easier to have external referees with disciplinary expertise provide rigorous input to the candidate's contribution *within that discipline*. In practice, inter- or transdisciplinary faculty members are most often recognized as those who uniquely "glue together" the collaborative. These faculty may not receive the proper attribution for their key roles and may fail to progress in the academy. This issue will become more vexing for promotion and tenure committees when the convergent group contains not only faculty representing different domains of science and engineering, but also the inclusion of others from the social sciences and possibly management. In those institutions where convergence is endorsed, it will be necessary to reframe the practices of promotion and tenure committees.

SUMMARY AND CONCLUSIONS

We find US AHCs buffeted by a confluence of factors that impinge on the research mission and which require consideration of alternative approaches and models in order to sustain the mission. Historically, AHCs provided subsidies from clinical revenues to support the research mission. Today, with few exceptions, this approach is not financially sustainable. Reductions in HHS allocation to NIH, beginning after the period of NIH doubling in 2003, have resulted in many changes that have adversely affected AHCs. These include diminished interests in MD and MD/PhD holders entering research, overall reductions in the pipeline of biomedical scientists, and greater regulatory and fiscal burdens on AHCs as they attempt to maintain faculty salary coverage and programmatic bridge-funding for PIs who are between grants.

These factors are also occurring in the context of ACA implementation, with potential implications for acute care facility revenues, financing of charity care, and, most recently, the possibility of changes in underwriting for graduate medical education [7]. In addition, the graying of AHC faculties is a demographic that challenges management to revitalize a workforce capable to compete in the currently constrained environment.

Research across the continuum from basic to translation to clinical remains the purview of research universities and AHCs. Many of the most critical and medically relevant discoveries emerged from fundamental basic science. Unfortunately, the decline in NIH, and particularly NSF, funding will slow the pace of US-sponsored basic investigation. It will be important for research universities to continue engaging in fundamental research; they are the training ground for next-generation scientists. This too requires a recalibrated economic model to sustain their efforts, suggesting that AHCs should focus on translational and clinical research. Inter- and transdisciplinary approaches offer potential promise to problems confronting the academic medical community. Some of these span disciplines of clinical medicine, ethics, economics, and policy. Whether convergence will be embraced by AHCs as a research strategy requires some institutions to pioneer the approach and demonstrate greater success than the more traditional, siloed approach.

AHC leadership experiencing a changing competitive context must concurrently manage the legacy of the research enterprise and execute new strategies in order to remain vital in addressing the tripartite mission that our society requires.

REFERENCES

[1] Association of Academic Health Centers. AAHC research data. 2014. Washington (DC). http://www.aahcdc.org/Resources/BenchmarksandMetrics/TheCostofResearch.aspx.

[2] Berdhal R. Thoughts of the current status of American research universities: a presentation to the national academy's board on higher education and work force. Association of American Universities (AAU); 2009. https://www.aau.edu/WorkArea/DownloadAsset.aspx?id=970.

[3] National Institutes of Health. Biomedical research workforce working group report: a working group of the advisory committee to the director. June 14, 2012. http://acd.od.nih.gov/biomedical_research_wgreport.pdf.

[4] Chakma J, Sun GH, Steinberg JD, Sammut SM, Jagsi R. Asia's ascent—global trends in biomedical R&D expenditures. N Engl J Med 2014;370:3–6.

[5] Committee on Research Universities: National Research Council of the National Academies. Research universities and the future of America: ten breakthrough actions vital to our Nation's prosperity and security. Washington (DC): National Academies Press; 2012. http://www.nap.edu/openbook.php?record_id=13396.

[6] Duderstadt J. Research universities and the future of America. Trusteeship 2012;5. http://agb.org/trusteeship/2012/9.

[7] Eden J, Berwick D, Wilensky G, editors. Graduate medical education that meets the nation's health needs. Washington (DC): The National Academies Press; 2014. http://books.nap.edu/openbook.php?record_id=18754.

[8] Estermann T, Pruvot E. Financially sustainable universities II: European universities diversifying income streams. In: Curaj A, Scott P, Vlasceanu L, Wilson L, editors. European university association in European higher education at the crossroads: between the bologna process and national reforms. Springer; 2011.

[9] Ehrlich E. An economic engine: NIH research, employment, and the future of the medical innovation sector. United for Medical Research; 2011. http://www.unitedformedicalresearch.com/wp-content/uploads/2012/07/UMR_Economic-Engine.pdf.

[10] Garrison HH, Deschamps AM. NIH research funding and early career physician scientists: continuing challenges in the 21st century. FASEB J 2014;28:1049–58.

[11] Glicksman E. Is tenure good for science: re-evaluating and academic tradition. AAMC Report 2014;23:6–7.

[12] June AW. Aging professors create a faculty bottleneck. Chron High Ed 2012;58:A1–8.

[13] Kezar A, Gehrke S. Why are we hiring so many non-tenure-track faculty? Lib Ed 2014;100(1). http://www.aacu.org/liberaleducation/le-wi14/kezar_gehrke.cfm.

[14] NIH Data Book. Success rates and funding rates: research project grants- competing applications, awards, and success rates. 2014. http://report.nih.gov/NIHDatabook/Charts/Default.aspx?showm=Y&chartId=20&catId=2.

[15] NIH Data Book. Success rates and funding rates: R01-equivalent grants, new (type 1) - success rates, by career stage of investigator. 2014. http://report.nih.gov/NIHDatabook/Charts/Default.aspx?showm=Y&chartId=136&catId=13.

[16] NIH Data Book. National statistics on graduate students in the biomedical, behavioral. social, and clinical sciences: total graduate enrollment and first-time, full-time enrollment in the biomedical sciences. 2014. http://report.nih.gov/NIHDatabook/Charts/Default.aspx?showm=Y&chartId=228&catId=19.

[17] NIH Data Book. National statistics on graduate students in the biomedical, behavioral. social, and clinical sciences: primary mechanisms of graduate support in the biomedical sciences. 2014. http://report.nih.gov/NIHDatabook/Charts/Default.aspx?showm=Y&chartId=236&catId=19.

[18] NIH Data Book. National statistics on graduate students in the biomedical, behavioral. social, and clinical sciences: primary source of federal support for full-time graduate students. 2014. http://report.nih.gov/NIHDatabook/Charts/Default.aspx?showm=Y&chartId=239&catId=19.

[19] NIH Data Book. National statistics on graduate students in the biomedical, behavioral. social, and clinical sciences: primary source of national institutes health (NIH) support for full-time graduate students in the biomedical sciences. 2014. http://report.nih.gov/NIHDatabook/Charts/Default.aspx?showm=Y&chartId=244&catId=19.

[20] Pencavel J. Faculty retirement incentives by colleges and universities. New York: TIAA-CREF; 2004. http://www1.tiaa-cref.org/ucm/groups/content/@ap_ucm_p_tcp_docs/documents/document/tiaa02029363.pdf.

[21] Pizzo P. Case study: the Stanford university school of medicine and its teaching hospitals. Acad Med 2008;83:867–72.

[22] Rockey S. Rock talk: looking to the future of the biomedical workforce. NIH; 2011 (online only). http://nexus.od.nih.gov/all/2011/01/31/future-biomedical-workforce/.

[23] Rockey S. Rock talk: age distribution of NIH principal investigators and medical school faculty. NIH; 2012 (online only). http://nexus.od.nih.gov/all/2012/02/13/age-distribution-of-nih-principal-investigators-and-medical-school-faculty/.

[24] Rockey S. Rock talk: does your academic training destine your choice of research subject? 2013. http://nexus.od.nih.gov/all/2013/02/01/does-your-academic-training-destine-your-choice-of-research-subject/.

[25] Sampat B, Lichtenberg F. What are the respective roles of the public and private sectors in pharmaceutical innovation? Health Aff 2011;30:332–9.

[26] Schweitzer L, Eells TD. Post-tenure review at the university of Louisville school of medicine: a faculty development and revitalization tool. Acad Med 2007;82:713–7.

[27] Stimpson J, Li T, Shiyanbola OO, Jacobson JJ. Financial sustainability of academic health centers: identifying challenges and strategic responses. Acad Med 2014;89:853–7.

ABOUT THE AUTHORS

Robert Clarke, PhD, DSc is the Dean for Research at Georgetown University Medical Center and Co-Director of the Breast Cancer Program at the Georgetown Lombardi Comprehensive Cancer Center. Dr Clarke has served as chair of several NIH peer-review study sections, and currently serves as a Senior Editor for the journal *Cancer Research*.

Elliott Crooke, PhD is Professor and Chair of Biochemistry and Molecular & Cellular Biology at Georgetown University Medical Center. In 2012, he assumed the role of Senior Associate Dean of Faculty and Academic Affairs for the School of Medicine and the School of Nursing and Health Studies at the Medical Center.

Howard J. Federoff, MD, PhD is Executive Vice President for Health Sciences at Georgetown University and Executive Dean of the School of Medicine. Prior to Georgetown, he held appointments as Senior Associate Dean; Professor of Neurology, Medicine, Microbiology, and Immunology; and Professor of Oncology and Genetics at the University of Rochester School of Medicine.

Chapter 15

Managing, Funding, and Supporting Research

Craig Johnson, Rex L. Chisholm and Eric G. Neilson

FINANCIAL MANAGEMENT OF MEDICAL SCHOOLS

University medical schools and their affiliated hospitals have spent decades experimenting with various models for financing the academic health center. A variety of mechanisms exist today. None are perfect.

Many public universities fund medical schools through an appropriation model supplemented by clinical practice income. In this approach, state legislatures from time to time issue bonds to support capital projects that keep interest payments and depreciation off the general ledger of the school. Nevertheless, many legislatures in the last decade have downwardly adjusted annual support, which in turn caused state tuition to rise and required clinical revenue to fill the resulting gaps. This zero-sum game leaves many public schools little wherewithal to stimulate new programmatic growth.

States with a legacy of prosperity tend to adequately support their medical institutions, and in some localities even new public schools have sprung up to increase the number of physicians per capita. Those state schools that have migrated from state-supported to state-assisted appropriations find it increasingly more difficult to attract an accomplished labor force to replace retirees.

Private university medical schools, without state appropriations as a source of revenue, have a long history of operating as tubs on their own bottom. They have managed finances through responsibility-centered budgets. Responsibility center management refers to a decentralized budgeting system designed to provide incentives and empowerment to individual deans and other managers of schools and institutes within a university with the goal of achieving mission in a more efficient manner (e.g., annual tuition; endowment; and clinical, hospital, and research revenues and expenses are locally distributed and applied toward strategic goals). These schools finance their capital projects with philanthropy and by borrowing on the debt market—or occasionally using clinical revenue when available. The goals of these medical schools vary with the expectations of their university or health system and, in particular, to the degree with which medical education can be intertwined with the research interests of the faculty as a matter of scope and reach of the academic enterprise. Endowment is often mentioned as a useful opportunity for private schools to invest strategically. But, in reality, most endowment is restricted by donor intent and new programs tend to follow giving (what are often termed "gifts that eat"), sometimes skewing the strategic plan.

The challenge for medical schools today is growing bioscience for the benefit of patients in an environment where 65% of medical or graduate training is experiential. These clinical or laboratory environments desperately need more cost-effective models to remain solvent in the face of rising expenses and diminished funding opportunities. In recent years, medical schools using either appropriated or responsibility-centered models have found more and more clinical income going to support nonsalary operating expenses. And, as insurance payers tinker with various clinical reimbursement models, the amount of funds available for faculty salary, education, and research has become more precarious.

Ironically, medical schools and their affiliated hospitals are the only places where improving health beyond the individual patient is possible—the medical profession, with the support of government and industry, in effect creates its own future. Out of 141 US medical schools extant today, about 45 hold most of the externally sponsored funds available for research. This group of 45 schools, a mix of public and private, is considered research-intensive and has substantially more complex budgets and programs than those schools that do little research. Research-intensive schools use their annual results from operations to grow educational programs or their scientific enterprise. Many operate as responsibility centers and appropriate revenues to their departments, divisions, and research units. These appropriations are rationalized around unit goals that support a common strategy for growth and attainment. Such an approach tends to even out the vagaries of lopsided accumulations from clinical reimbursement at the unit level.

The Transformation of Academic Health Centers. http://dx.doi.org/10.1016/B978-0-12-800762-4.00015-3

As a frame of reference for the discussion that follows, the Feinberg School of Medicine at Northwestern University is a research-intensive medical school affiliated historically with three hospitals—the Northwestern Memorial Hospital, the Ann and Robert H. Lurie Children's Hospital, and the Rehabilitation Hospital of Chicago. The school comprises 18 clinical departments, 7 basic science departments, 8 institutes, and 27 centers; all faculty have a primary appointment in a department, but work in a variety of clinical and research environments across the Chicago campus. The school has ~710,000 sq ft of research space, of which 75% is wet and 25% dry. Faculty in the school of medicine and various affiliated hospitals provide ~$420 million dollars of sponsored research annually in an environment that rolls up to a zero-based budget.

Feinberg has experimented with a variety of financial models for allocating resources. The school is a responsibility center and now appropriates its funds to its various units to meet a long-term financial plan for growth and accomplishment. The goals of a responsibility center can only be met by laser focus on its long-range financial plan. One of our principal aspirations is to double the size of the research enterprise as a way of increasing intellectual diversity that benefits education and patient care. What follows is our approach to managing the research enterprise using data and dashboards to meet this objective. Nothing can be managed without data.

CASE STUDY OF MODELS AFFECTING GROWTH OF RESEARCH

At Feinberg, like many other schools of medicine, our approach to effective research management began with the restructuring of the financial model between the medical school and the central administration of the university. Prior to 2001, the school was on an appropriated financial model, where the university center provided a net expense budget or appropriation to the medical school and retained nearly all revenues. The main income and expense components of research—indirect cost recovery, faculty and staff salary expenses, and research facilities space costs—were managed and funded by the university. Several of the school's clinical departments were allowed to supplement this support with their excess clinical revenues. This, however, did not occur that often, and doing so invited the university administration to reduce departmental appropriations in subsequent years.

Efficient management of the medical school's research portfolio was the responsibility of central administration as they provided the financial means to support it. However, the feeling among medical school faculty and leadership was that university administration was somewhat dislocated from the medical school, both physically and knowledgeably, to properly evaluate performance and objectives, and

the university did not create incentives for clinical departments to apply clinical surpluses to further the academic mission.

Thus, in 2001, the university agreed to embark on an initiative to operate the medical school as a tub on its own bottom—as a fully decentralized financial entity. Under this version of responsibility-centered budgeting, the medical school and its units would retain all of its revenues and be responsible for all of its expenses, along with paying a shared services fee for certain centralized functions (such as information technology, human resources, and research administration). The idea was to unleash the school to manage the productivity of its research portfolio and reap the economic benefits of successfully doing so.

By all measures, decentralization drove a good deal of growth. From 2001 to 2011, awards at Feinberg grew to $317 million at a compound annual growth rate of 5.9%. The medical school's share of the total university research portfolio grew from 51% to 62%.

While successful in terms of absolute research growth, school leadership felt that the model still needed improvement. The 2001 construct extended the tub on its own bottom approach down to departments, institutes, and centers. Each operating unit received 60% of its indirect cost recovery revenue, a modest dean's allocation for the basic science departments, and a dollar per square foot charge for the space it occupied. In addition, all clinical surpluses in the departments were under the discretion of the chairs and could be applied to the research mission as they saw fit. This approach attempted to mirror what the university had done for the medical school. However, when applied to the individual operating units of the medical school, it created what became known to us as "irresponsibility-centered budgeting."

There were several fundamental limitations arising from decentralizing the operating units of the school. The model left the central administration of the school largely insolvent; it created an environment favorable for hoarding and facilitated suboptimal long-range planning that perpetuated a system of haves and have-nots. The absence of central planning stymied the collective vision for academic research. An example of suboptimal decision-making came in the form of revenue walls that disincentivize scientific collaboration. Departments were often reluctant to create multi-investigator grants without first negotiating the terms around sharing of indirect cost recovery. Occasionally, meritorious collaborations would be scuttled over disagreements on one or more terms applying business rules set around the unit advantages of the haves, which perpetuated hoarding.

The school's units would also occasionally leave portions of their space allocations fallow, not wanting to give up research space—even if there were potential collaborations or other active researchers needing more space. Space removed from a portfolio was considered a diminution in assets.

Perhaps the most detrimental aspect of our "irresponsibility-centered budgeting" was the perpetuation of unequal wealth. Under the tub model, certain clinical departments, due to the vagaries of medical reimbursement, were able to generate significant revenues above the costs of their clinical operations. Often these funds were sent to support their research program, but occasionally—where the department did not have strong research opportunities—funds were used to support pet projects, high administrative staffing ratios, robust faculty incentive compensation, or other clinical mission priorities not shared by the whole. Other less well-off clinical departments and the basic science departments did not have access to this stream of surplus revenue. To balance their operations, they often had to reduce their administrative staffing levels, rely more heavily on limited dean's office support, or curtail their research portfolio.

The school also had no ability to use resources to take advantage of opportunistic hiring targets or exciting funding opportunities. The school had to hope for a happy coincidence that exciting opportunities arose in the well-resourced departments. Though the tub model had allowed the school to create tremendous strides in growing its research portfolio, the suboptimal decision-making and unequal expansion of wealth convinced school administration of needed change.

In 2011, Feinberg—while still a responsibility-centered budget with the university—began migrating its internal financial units toward an appropriated model. After the faculty practice was integrated into the affiliated hospital system in 2014, the transition was complete. All indirect cost recovery and clinical margin revenue flowed to the dean's office and departments, and other units received an appropriation to support their operating expenses. The dean's office assumed all space management responsibility—including assignment authority and support for operating and capital expense.

Given that our prior model had ushered in a period of growth in research activity, we were careful to construct new incentives to perpetuate that growth. Our performance expectations and incentives were crafted to target both the chair leadership and the individual faculty. This initiative established a detailed financial planning process and created performance expectations and incentives crafted to target both chair leadership and individual faculty. The compound annual growth rate for extramural research awards from 2012 to 2014 grew at a stronger rate of 12.1% with expectation of higher rates of growth going forward.

To adequately manage these revenue and expense streams, the school established a robust long-range financial planning process (LRFP). The crux of any planning process is the means by which an institution manages two types of financial variables—those that can be controlled by administration and those that cannot. For controllable variables, the LRFP provides a guide for future decisions.

Annual inflation adjustments—such as annual salary increases, tuition, and financial aid increases—are conditioned to the LRFP guidelines. In addition, the plan must contain assumptions underpinning the major growth initiatives at the school.

These assumptions then become the hurdles that must be cleared during the execution phase. For example, our plan at Feinberg assumes a new faculty recruit at the professor level will: (1) have 70% of his or her salary covered on grants up to the NIH salary cap; (2) require variable millions in start-up packages depending on accomplishment; and (3) have grant dollars supporting space allocation of appropriate dollars per square foot. If one of the units in the school proposes metrics outside of these guidelines, the recruitment will go on hold for reconciliation.

For variables outside the direct control of administration—for example, endowment growth, indirect cost recovery rates, and clinical funds flow—the school builds detailed financial models and prepares sensitivity analyses. A large component of Feinberg's revenue is clinical funds flow from the health system. Each year, the hospital, clinical practice, and the medical school update their LRFPs and reconcile common variables. This allows the entities to incorporate intelligence from whatever business unit is closer to front-line revenue and expense activity. In addition, for each of these variables, sensitivity analyses are prepared showing the bottom line impact of different assumptions. For instance, one such analysis might show the impact of each percent change in net patient revenue growth.

At the chair level, a metric-driven model was instituted for incentive compensation. 30% of salary was available according to how they performed against a group of 10–15 performance metrics. Included in this list were: (1) year-over-year growth in research awards; (2) year-over-year growth in research proposals; (3) dollar per square foot research productivity in space in their department; and (4) percent of investigators salary covered by external grants. With this methodology, the chairs' personal compensation, rather than the financial returns to the department, would motivate attention to growing the school's research mission.

At the faculty level, incentives were implemented for sustaining an active research portfolio. Across the school, we implemented a standardized incentive compensation plan for investigator faculty. Up to 10% of the NIH cap salary from nongrant revenue was available as incentive based on the type of award, how much funding investigators had, and how much of their salary they supported with the grants. In addition, annual merit increases were available only to those faculty who either had a significant portion of their salary covered on grants or who were actively seeking grants through the submission of new proposals in the previous year. Beyond salary considerations for faculty, we also looked at laboratory size for research faculty. Our target efficiency metrics were around $1000 per square foot for dry laboratory investigators and $625 per square foot for wet bench research. As new space was assigned, old space renovated, or new grants obtained, we reconfigured space within these guidelines.

Although these changes have only been in place for 3 years, we are seeing continued growth in our research portfolio, with new awards in 2014 up 12.2% and the pipeline of new proposals up another 15.9% above the previous year. More importantly, "have" or "have not" departments no longer exist; all receive appropriations consistent with their focus on our long-range financial plan.

ESTABLISHING AND USING RESEARCH METRICS

Like many research-intensive medical schools, we adopted a management approach that depends on using appropriate metrics to evaluate our research enterprise, followed up with executing a quality improvement process that maximizes efficient use of our resources, maintains compliance with all regulations, and minimizes the administrative burden on investigators. In our view, the fundamental principles driving the use of metrics are transparency and adherence to best practices. Key metrics we have found useful include proposal and award data, space productivity, and clinical trials activity. What follows below describes these metrics and how we have used them.

Proposal and Award Metrics

Extramural proposals and awards are key metrics tracked at most academic health centers. These measure the activities of the faculty who enable research. We provide our department chairs a monthly report of year to date proposals submitted, both dollar amounts and numbers of independent submissions and of year to date awards, again showing both dollar amounts and numbers of submissions. These are compared with the similar period in the previous fiscal year. The report shows all departmental results and is distributed to all chairs, providing broad transparency.

The award amounts represent the funds that enable discovery across our research enterprise. We think of this as a research capacity. Like many academic health centers, we also track and report expenditures. Expenditures typically lag behind awards due to factors such as delays in staff recruitment following an award as well as inherent fiscal conservatism of many investigators.

To assist departments in managing their portfolios, we provide them with a quarterly report of expired or expiring awards. This report shows, at an individual award level, each award that has expired in the past quarter, or which will expire in the next 12 months. Such a report allows department chairs to follow up with their faculty to better understand and aid them in managing their individual research portfolios. It also gives chairs an easy tool to monitor trends in the productivity of individual faculty, providing an early warning for faculty who may be having difficulties in maintaining their research portfolio. For such faculty, chairs can offer additional support—such as presubmission grant review, studios to improve competitiveness, or advice on altering research directions—as needed.

We also track funding profiles to monitor the percentages of our overall portfolio from federal sources; voluntary health organizations; industry, including pharmaceutical and biotech companies; and foundations and philanthropy. The trends in these metrics provide another valuable indicator to help assess if our research portfolio is appropriately balanced to maximize indirect cost recovery. While our portfolio has continued to show compound annual growth rates of 7.5% since 2011, we have also seen significant growth in the percentage that is funded by industry. This is useful in an environment where emphasis on clinical trials is important to our translational mission.

Within the NIH family of award mechanisms, we believe there is value to also tracking the distribution of awards in categories such as training, research, multi-investigator, multiproject, and clinical trials. We explicitly encourage more team-based awards—such as P01 and P30/P50 Center grants and U01 and U54 cooperative agreements. These multi-investigator awards reflect an interest in team science, but also diversify mechanisms to maximize faculty salary support. Most importantly, they also represent projects that are likely to have large scientific and societal impact.

Finally, we track the balance of wet, dry, and damp research activities. Wet research represents traditional laboratory research—including cell and molecular biology, neuroscience, pharmacology, microbiology, immunology, and biochemistry. Dry research includes activities such as epidemiology, health care studies, outcomes, and community health research. Damp research represents a mix of these activities within a single project. Examples of this include those at our Alzheimer's Center that perform anatomical studies on human material along with clinical cognitive assessments. To better track this, we require our principal investigators to indicate in sponsored project applications which of these categories best describes the work being proposed in their applications. This is captured by our electronic proposal submission workflow such that, should a proposal be funded, we are able to categorize the studies that result.

Space Productivity Metrics

Beyond human talent, space is probably the next most important resource in an academic health center. Space is expensive to construct, renovate, and maintain; thus, it is essential that the school assure efficient and productive use of this resource. We track several space metrics, which we believe give us different insights into how space is employed at the school, departmental, and investigator level.

A key element of tracking space productivity is to have an effective database for space assignments. Most academic health centers will have some system that tracks space. This is typically driven by the need to designate research space for the purposes of negotiating the Federal Facilities and Administration Rate that reimburses the institution for overhead costs. At Northwestern, we use a system called SIMS (Space Information Management System) and we have developed methods to integrate data from SIMS with our research awards data.

Perhaps the most traditional measure of space productivity is expenditure dollars per square foot of space assigned to research. This metric measures actual activity that occurs in the space. While this space metric is generally useful, it is imperfect and we have found it helpful to evaluate several other productivity measures.

The first space productivity metric we calculate is total research awards divided by the square footage of space that is designated as research space. This is important because it provides a measure of research capacity—how much research can we accomplish with the space we have and the research funds that have been awarded to the medical school.

The traditional expenditure dollar amount per square foot is typically somewhat less than the award amounts. For Feinberg, this number often runs at 80% or less of the award amount per square foot. There are a number of reasons for this, including the fiscal conservatism of many faculty who seek to make their grant awards go as far as possible. There are also some practical reasons that expenditures lag behind awards dollars. Delays in staffing-up following an award and the concomitant effects on expenditures for supplies and other research expenses may be contributory factors.

We also track awards assigned to individual principal investigators as a function of their individually assigned research space. This includes a proportional share of shared-equipment space based on their proportion of assigned space to the total for that floor or departmentally assigned space. This metric is an indicator of when faculty may be likely to need additional space to accomplish their funded studies, or when faculty needs for assigned space might require reduction. At Feinberg, all space is assigned by the dean's office and is assigned—in consultation with department chairs—to individual investigators. As individual research portfolios wax and wane, so too should space. Those with growing portfolios are candidates for more assigned space, when requested. In an effort to buffer the inevitable fluctuations due to meritorious grants that miss a pay-line, we use a 3-year average to identify space productivity trends for our faculty. This also assures that faculty have the space they need to stay as productive as they can, to do additional work needed to address reviewer concerns, or to pivot to new research directions should their previous directions evolve.

The next space metric we employ is a roll-up of space individually assigned to investigators plus space assigned to the department chair. This typically includes space designated as administrative space or space used for laboratory meetings and seminars, such as conference rooms. This roll-up metric helps us understand how departments utilize space assigned to them. For a department whose space utilization metric based on award portfolio shows a significant reduction relative to those of the investigator whose space is included, a review might be triggered to better understand efficient utilization of the administrative space.

As noted above, we widely publicize the institutional targets for research productivity so faculty and department chairs are aware of expectations. The dean's office provides a performance evaluation in each of these metrics to the department chairs annually. This facilitates discussion between chairs and their faculty and the dean's office about how to most efficiently utilize space going forward. Following these discussions, space may be reassigned to improve efficiency. Faculty are transitioning from a past—where one measure of their standing was how much space they had—to a future where efficient use of space is a more important metric. This has the advantage of allowing small-focused laboratories that use their space efficiently to be viewed as equals of larger laboratories based on cost-effectiveness.

Clinical Trial Metrics

For most academic medical schools, clinical trial activity is a critical measure of translational success. Management of clinical trials activity at many academic health centers has received less attention than more traditional research activity. It is not uncommon for clinical trials awards to have dollar volumes 2–4 times the actual amount of billed and recovered revenue for clinical trials. To better track our clinical trials activity at Feinberg, our Center for Clinical Research (CCR)—a critical unit of our Clinical Translational Science Award housed in the Northwestern University Clinical and Translational Science Institute—has established a Clinical Trials Dashboard that tracks a variety of metrics in real time.

Two key tracking elements of our clinical trials metrics are an online eIRB database that captures all approved studies, authorized personnel, and enrollment targets, as well as a participant tracking system called eNOTIS. All studies at the Feinberg School of Medicine are required to record consented participants in the eNOTIS database within 48 hours of obtaining consent. The eNOTIS system allows us to monitor the number of participants in each IRB-approved study in real time.

We have also developed a clinical trials dashboard that integrates information from several enterprise systems—including eNOTIS, the eIRB system, our financial system, and our enterprise data warehouse (EDW). The EDW captures clinical information from electronic health records and other information systems at our primary affiliate—Northwestern Memorial Health care—which includes both inpatient and outpatient clinical practice. The dashboard presents a variety of data elements, including

- Newly opened studies
- Number of studies
- Number of participants enrolled
- Target enrollment
- Number of days study has been open
- Accrual rate
- Ratio of award to invoice dollar amount

This dashboard allows CCR staff to review progress of individual studies. By focusing attention on studies that are not meeting accrual targets, it is often possible to remove barriers, deploy additional recruitment resources, or modify recruitment strategies to increase participation. Another important benefit of the EDW is that the CCR can use it to help investigators prospectively determine the feasibility of any particular project by assessing how many patients receiving health care through our system meet enrollment criteria. This data—when combined with financials for the units serving as clinical research offices for the projects—allows school administration to monitor the costs of recruitment and clinical trials administration. This data brings administrative attention to trials teams that are outliers in terms of recruitment costs per participant. It also provides improved data for future clinical trials budgeting.

IMPROVING EFFICIENCY OF THE RESEARCH ENTERPRISE

Efficient and effective administration is a critical component to improving the research enterprise. In addition, effective research administration improves compliance, thereby minimizing risk to the institution. We have developed six metrics that we track for each department, center, and institute that manages research grants and projects. This project has been rolled out as a quality improvement process. The goal is to identify potential audit triggers and to improve efficiency by reducing activities that create additional administrative work across the institution. Success at reducing these costs frees up administrative time and reduces the administrative burden on investigators.

Effort Reports

Achieving 100% effort report certification by the institutional deadline assures that the school is meeting its reporting obligations. By tracking this measure, we have been able to improve compliance and reduce the follow-up effort needed to track down incomplete effort reports.

Research Administration Compliance

To make grant deadlines more predictable, reduce stress on our Office for Sponsored Research, and minimize grant submission failures, a second metric measures research administration compliance among groups with institutional internal deadlines for grant submission. This is especially important for system-to-system transfers—such as to grants.gov. At one time, we saw as much as 15–20% of our system-to-system transfers fail for a variety of reasons, ranging from improperly formatted PDF documents to presence of nonstandard characters in the submission. Each of these is easily addressed—provided there is sufficient time before sponsor deadlines. Tracking this metric and encouraging faculty and research administrators to adhere to institutional deadlines maximizes our ability to address potential submissions issues. Consequently, we have seen significant improvement in timeliness of application submission to the sponsored projects office.

Financial Management

The four remaining metrics are focused on financial management. Two of these metrics measure salary cost transfers onto grants. We are measuring both the number of post 90-day salary transfers and the dollar volume of salary transfers onto awards. The post 90-day transfers are potential audit flags, but even more importantly, each requires significant additional administrative effort—not only from the originating department but also from multiple departments across the university. Reducing their number multiplies the administrative savings far beyond those achieved in the originating department. The dollar amount of salaries transferred onto grants, calculated as a percentage of the total dollar amount of salaries on original grant budgets, reflects the amount of salary that is not being directly charged to grants. Our goal is to maximize to the extent allowable the direct charging of grants. Again, the theme is reducing the amount of administrative activity related to managing salary on grants by encouraging departments to undertake robust salary planning and direct charging to grants initially.

As for the salary transfers, we seek to encourage departments to appropriately direct charge these expenses to grants in the budgeting system. This reduces the administrative churn associated with posting such expenses more than once. Every unnecessary transfer results in wasted administrative effort, the elimination of which offers the possibility of cost-effective administrative support that can be redirected toward more productive activities.

The second pair of financial management metrics parallels the salary transfers, but focuses on nonsalary expenses. We measure the number of post 90-day nonsalary transfers and the dollar amount of nonsalary transfers as a percentage of total nonsalary expenses on grants.

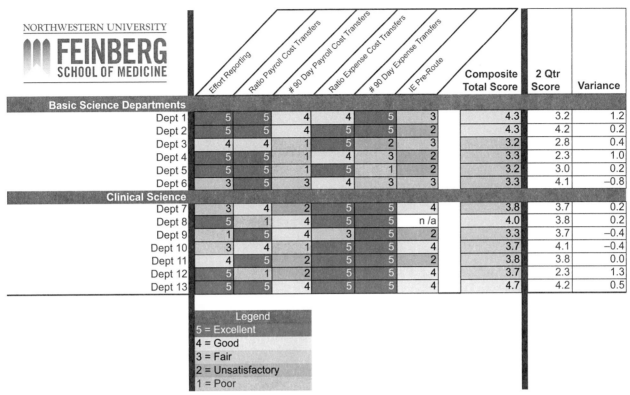

	Effort Reporting	Ratio Payroll Cost Transfers	# 90 Day Payroll Cost Transfers	Ratio Expense Cost Transfers	# 90 Day Expense Transfers	IE Pre-Route		Composite Total Score	2 Qtr Score	Variance
Basic Science Departments										
Dept 1	5	5	4	4	5	3		4.3	3.2	1.2
Dept 2	5	5	4	5	5	2		4.3	4.2	0.2
Dept 3	4	4	1	5	2	3		3.2	2.8	0.4
Dept 4	5	5	1	4	3	2		3.3	2.3	1.0
Dept 5	5	5	1	5	1	2		3.2	3.0	0.2
Dept 6	3	5	3	4	3	3		3.3	4.1	−0.8
Clinical Science										
Dept 7	3	4	2	5	5	4		3.8	3.7	0.2
Dept 8	5	1	4	5	5	n /a		4.0	3.8	0.2
Dept 9	1	5	4	3	5	2		3.3	3.7	−0.4
Dept 10	3	4	1	5	5	4		3.7	4.1	−0.4
Dept 11	4	5	2	5	5	2		3.8	3.8	0.0
Dept 12	5	1	2	5	5	4		3.7	2.3	1.3
Dept 13	5	5	4	5	5	4		4.7	4.2	0.5

Legend
5 = Excellent
4 = Good
3 = Fair
2 = Unsatisfactory
1 = Poor

FIGURE 1   Research Administration Dashboard. This sample of the Feinberg Research Administration dashboard shows department performance in each of six categories (described in the text), a composite score, and change from the previous quarter. The goal is for all units to have scores of good to excellent.

For each department, center, or institute, each of these six metrics were initially scored in a dashboard with a 1–5 rating (Figure 1), based on the quintile of the distribution across units, with a rating of 5 being best.

In addition, an average across all six metrics was used to provide an overall rating for the unit. In subsequent fiscal quarters, the bins remained as initially established, allowing departments to improve their ranking with the goal of all departments receiving a rating of 5. As can be seen in Figure 2, overall performance by departments has improved significantly over time.

To increase the cost-effectiveness of research administration, we also consolidated this activity into several pods. This has the benefit of allowing more cross-training and consistent implementation of best practices, providing backup for times when research administrators take vacation or family leave and aligning teams so that research volume is sufficient to assure expertise—thus, minimizing times when individual administrators perform tasks on only an occasional basis. One example of this is the consolidation of the research and departmental administration for our wet laboratory-oriented basic science departments into a single unit named Basic Sciences Administration. In this structure, each principal investigator still has their go-to research administrator focused on providing high-quality and service-oriented preaward and postaward grants management. Similarly, we consolidated research administration for some of our clinical departments and Centers/Institutes that have smaller research portfolios into a pod called Clinical Research Administration. This group has grown over the past 4 years from 2 highly competent individuals to 14 today—far fewer than would be expected from a decentralized model. Implementation of these research administrative pods, which report to the dean's office, has also had the benefit of reducing the time department chairs need to spend worrying about administration, and thus allowing them to focus on faculty development, their clinical practice, and their education responsibilities.

Another means to improve efficiency in a research enterprise is by providing robust, service-oriented, and state-of-the-art research core facilities. These provide faculty, staff, and students with specialized resources or instrumentation that are used on an occasional basis, easily shared, prohibitively expensive for single laboratories to purchase and maintain, or require specialized expertise. At Feinberg, our research core facilities include the Center for Advanced Microscopy, which houses: multiple light, confocal, and electron microscopes with a range of technologies; a flow cytometry core that provides standard quantification and

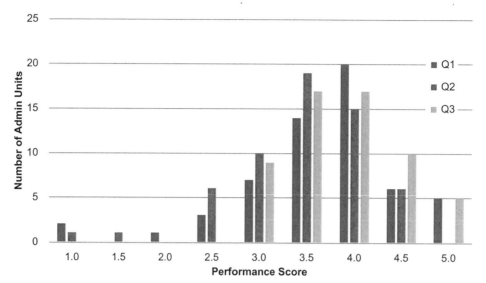

FIGURE 2   Trend over time of Department, Center and Institute Composite Research Administration scores. Providing these reports and working with individual departments led to a significant improvement with scores below 3 eliminated, and the mean shifting significantly toward better performance.

cell sorting; a transgenic and targeted mutagenesis core for producing genetically engineered mice; and a Next-Gen sequencing core. Each core operates on a first-come-first-served basis and is managed typically by a PhD level director who has specialized expertise with the instrumentation or service.

Several elements contribute to successfully and efficiently managing these facilities. Perhaps the most important is a robust annual review process. Each spring, the cores are required to provide a report on the services they provide and a 3-year budget that compares past, current, and proposed future-year budgets. In addition to the report, our centralized office for core facility planning uses a sophisticated adaptive survey tool to capture user satisfaction, user impressions of the quality of service, and user impressions of the quality and state-of-the-art nature of the instruments or services provided by the core. Over 1000 respondents from across our institution who used one or more of the core facilities complete this survey. Figure 3 represents a graphic of the survey results that are provided to the core directors.

We have also developed an online core facilities management system, called NUcore, which allows users to reserve instrument times, as well as request services. Thus, the system captures all activities of that core facility. Because of this, NUcore can provide an objective usage report—a key element of the research core review process. The NUcore system interfaces directly with the university financial system, significantly automating the billing process and allowing the core directors to focus on providing service without expending time on administrative trivia, such as tracking down expired account numbers.

Applying this data, a peer review committee of research-active faculty evaluates each core with regard to

services, user satisfaction, budget, and value to the institution. Underperforming cores are given recommendations to improve. Since many of the cores receive support from the dean to subsidize their budgets, they are typically motivated to recommend improvements. Those that consistently underperform may see leadership changes or even loss of funding—especially in the case where utilization no longer justifies subsidy. By retiring core facilities that no longer provide added value, we have been able to reinvest funds to create new cores and services.

Seed funds that allow groups to hold retreats to develop plans for multi-investigator grants are also an important mechanism to stimulate new research directions and team-based research projects. At Feinberg, we have a two-stage process for receiving seed funds. The first focuses on a mini-retreat. Up to $1000 can be provided to support refreshments, lunch, or even an outside speaker. These funds are accessed by a short application, with approval usually within 24 hours. Following a retreat, with an outline of an application, applicants can access up to an additional $14,000 to support gathering limited preliminary data or receiving administrative support for participant enrollment or project management. From this school investment of $264,000, investigators have received over $37.4 million in awards, equaling a return on investment of ~$142 for every dollar spent on the program.

Finally, even the most accomplished faculty can find themselves narrowly missing a pay-line. We have found it quite useful to provide bridge funding to faculty to enable them to gather data to address the concerns noted in their reviews, hold together a research team, or support students in the lab so as not to disrupt their training. Our goal for bridge funding is to support the group for

Panel A:

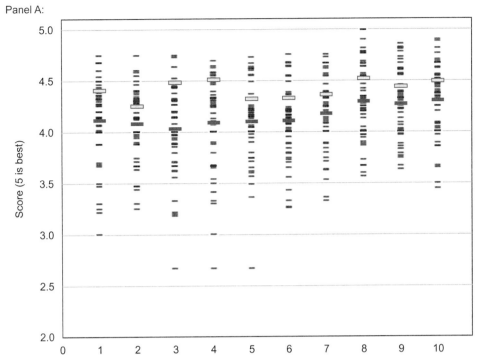

Panel B:

	Category	Question	Your Core	All Cores
1	Instrumentation	It is easy to schedule instruments in this core	4.41	4.11
2	Instrumentation	Instruments that I need are generally available	4.25	4.08
3	Instrumentation	Instruments are in good operating condition	4.48	4.03
4	Instrumentation	Instrument training is available	4.51	4.09
5	Services	The core has an easy / efficient process to request services	4.32	4.10
6	Services	Services are provided in a timely fashion	4.33	4.11
7	Services	Provided services are of high quality	4.36	4.18
8	Staff	The staff are accessible / responsive	4.52	4.30
9	Staff	The staff are knowledgeable in experimental design	4.44	4.27
10	Staff	The staff are capable of resolving technical issues	4.49	4.31

FIGURE 3   Research Core Survey Report. Sample report provided to each research core in the Feinberg School of Medicine. Over 1000 faculty, student, and staff members provide feedback on the research cores they use. The results of this survey are provided to the Research Core directors and to the Research Core review committee as part of their annual review of core facilities. Panel A shows responses graphically. Blue line (light gray in print version) represents the score of the particular core being presented, with the red line (dark gray in print version) showing the mean score for all cores. Panel B shows a subset of the individual questions asked in the survey and the numerical score for the core compared to the average score for all cores.

up to one grant review cycle. The dean provides up to $80,000, with each dollar from the bridge funding program to be matched by up to $20,000 from the recipient's department. To date, the program has seen an outstanding return on investment with approximately 60% of those supported eventually getting the grant for which

the bridge funds were provided. We have even seen faculty with bridge funding return unused funds to the pool when it was clear that they could establish a prespending account. Clearly, it is more efficient for the institution to provide this level of support than it is to risk losing a laboratory.

CONCLUSION

How medical schools maximize cost-effectiveness in managing their research enterprise is one of the most important insulators from insolvency. Every organization has their own way of accomplishing this task. What we outline here is a set of business rules or guideposts to preserve growth while remaining solvent that works for us. Change is not easy, but for the right reasons can become a shared imperative. Group discussion and transparency-guiding change management is the solvent of necessity.

ABOUT THE AUTHORS

Eric G. Neilson, MD is the Vice President for Medical Affairs and the Lewis Landsberg Dean at Northwestern University Feinberg School of Medicine. He previously served as the Hugh Jackson Morgan Professor and Chairman of the Department of Medicine at the Vanderbilt University School of Medicine.

Craig Johnson, MBA is the Vice Dean for Finance and Administration at Northwestern University Feinberg School of Medicine.

Rex L. Chisholm, PhD is the Vice Dean for Scientific Affairs and Graduate Education and Adam and Richard T. Lind Professor of Medical Genetics at Northwestern University Feinberg School of Medicine. Dr Chisholm previously served as Founding Director of the Center for Genetic Medicine at Feinberg.

Chapter 16

Transformative Changes to Embrace, Manage, and Exploit "Big Data"

Michelle L. Kienholz, Rebecca S. Crowley, Jeremy M. Berg and Arthur S. Levine

In many sectors of society, we are in the midst of a digital revolution, moving from analog and mechanical devices to digital technology. In the context of academic health centers (AHCs), tools in both the clinic (e.g., electronic health records) and the laboratory (e.g., next-generation sequencing methods) are generating large amounts of data that are, in principle, useful for research and patient care. The major challenge now is to extract from these large data sets insights that can guide future directions, whether in the realm of research, education, or clinical care.

As we increasingly move beyond small labs performing experiments to team science employing high-throughput data collection technologies, and from paper-based single practitioner offices to electronically integrated health care systems, our problem shifts from having not enough data to having too much data to be grasped by the human mind. On the research side, we have vast data sets at the molecular, protein, organelle, cell, tissue, organ, system, human, and population scales. Advances in computing, patient monitoring systems (both inpatient and outpatient), and imaging and "omics" technologies have enabled hospitals to electronically record an ever-increasing volume and variety of variables. These advances, coupled with the proliferation of electronic health records (EHRs) incentivized by the Health Information Technology for Economic and Clinical Health (HITECH) Act,[1] have resulted in the massive accumulation of data.

Rather than pare these data down to become manageable on the human scale, we can now embrace their size and complexity through the use of machine learning and other tools to computationally discover predictive and causal relationships that could not have otherwise been imagined or hypothesized. These analytic tools can have a similar transformative impact on the efficiency and efficacy of health care delivery—including the full implementation of personalized medicine. Being able to merge disparate data

to discern causality will advance our knowledge remarkably, more than any one single technological advance; but we must transform our environment and workflow—and our workforce—to take advantage of these expanding technical capabilities.

As an AHC that has already undergone considerable transformation [1], the University of Pittsburgh (Pitt) and the UPMC (University of Pittsburgh Medical Center) can look with confidence at the challenges and opportunities in harnessing the power of big data in academic medicine, and in this chapter we will share our experience with two major "big data" initiatives to date, one national and one local, as a roadmap for others.

TRANSFORMING THE ACADEMIC HEALTH CENTER ENVIRONMENT

The Institute of Medicine's (IOM) concept of a continuous learning health system is one that leverages the health data infrastructure to access and apply evidence in real time [2]. Indeed, in April 2014, the IOM Roundtable on Value & Science-Driven Health Care and the Patient-Centered Outcomes Research Institute (PCORI) engaged health system leaders in a discussion of how to accelerate the integration of clinical practice and research by harnessing advances in data infrastructure and analysis [3]. Grantees of the PCORI National Patient-Centered Clinical Research Network (PCORnet[2]) brought to this meeting their plans and experience in working to make such integration possible at an institutional, regional, and national scale.

Pitt and UPMC lead one of the 11 Clinical Data Research Networks (CDRN) in PCORnet, which also includes 18 Patient-Powered Research Networks [4]. Our CDRN integrates EHR data from Pitt/UPMC and the UPMC Health Plan, Pennsylvania State University/Hershey Medical Center, Temple University/Temple Health, and Johns Hopkins University/Johns Hopkins Health System and Johns

---

[1] Public Law 111-http://www.hhs.gov/ocr/privacy/hipaa/understanding/coveredentities/hitechact.pdf.

[2] http://pcornet.org.

The Transformation of Academic Health Centers. http://dx.doi.org/10.1016/B978-0-12-800762-4.00016-5

Hopkins Health Care, collectively known as PaTH, in a centralized data center housed on the secure servers of the Pitt Comparative Effectiveness Research Center [5]. Using i2b2, an open-source informatics framework that integrates clinical and research data, we de-identify and extract common data elements from EHRs—including patient-reported outcomes—using standard health terminologies (e.g., RxNorm, SNOMED CT, and LOINC) to create a unified PaTH dataset. In turn, our data will be merged at a national level through the PCORnet Coordinating Center in Boston led by Harvard Pilgrim Health Care Institute and Duke Clinical Research Institute (Figure 1).

Throughout PaTH and the entire PCORnet, patient, health care professional, and health care system representatives participate in all aspects of Network governance and data use. The overarching goal of PCORnet is to build patient-centered, rather than investigator-driven, clinical and comparative effectiveness research into the health care process, allowing the conduct of fast, focused studies that do not strain institutional resources. Indeed, the focus of

PCORnet studies, to be reviewed by the broadly representative Network Steering Committee, will align with AHC priorities at all levels and thus enhance sustainability. The ability to secure new knowledge that can be implemented at scale in the real-world practice settings from which it was generated will likewise enhance acceptance and incorporation of evidence-based practices. To get there, PCORnet will jointly work to harmonize technical and regulatory aspects of making a national patient-oriented research network a reality.

Even outside PCORnet, AHCs can follow the same path to harness the benefits of big data in ways that align with their research, clinical, and education missions. The process begins with resource mapping to assemble a taxonomy of available data sources with clear descriptors of each that are meaningful to all participants in the process [6]. Utility mapping can then be used to assess how questions in need of answers map to these existing resources and methods, both to focus data collection and management in ways that address institutional priorities and to identify unmet areas of need.

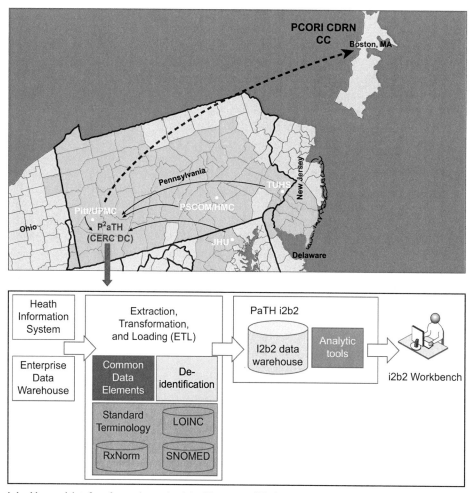

FIGURE 1 Electronic health record data from four major academic health centers will be integrated regionally as well as nationally through the Patient-Oriented Outcomes Research Network (PCORnet).

These findings in turn can be used in planning an institutional data center or distributed data system that provides [7]:

- open and accessible data for immediate and secondary use;
- sharable data, including well-defined consent (patient) and data use (institutional) procedures;
- queryable data (both with and without data release following the query);
- integration across biological scales, biomedical disciplines, and research and clinical care domains;
- metadata and annotation to ensure the provenance of all digital assets (datasets, papers, software, lab notes, clinical notes, and so on) by having them uniquely identified with access control; and
- user-friendly interface, including the presentation of data to maximize value.

A fully open and shared data space would allow an AHC to track, for example, that a cardiology and a cancer researcher are querying the same gene in the context of different disease processes with different patient outcomes (concept from Bourne [8]). Their efforts, unintentionally complementary, could potentially enhance each project and develop new knowledge that might otherwise be contingent on someone reading both published reports in discipline-specific journals and linking the gene data.

Finally, AHC leaders must recognize that biospecimens themselves are a form of big data storage that must be managed and potentially shared in the same way (i.e., accessible, queryable, derived from standardized collection and storage protocols). In addition to research biorepositories, archived clinical specimens and images represent a wealth of data accessible through informatics tools capable of mining unstructured text data in EHRs using methods from Natural Language Processing. Informatics scientists at Pitt developed the Tissue Information Extraction System (TIES) [9] to de-identify (in compliance with HIPAA) and process more than 25 million pathology and radiology documents in the UPMC system. TIES allows researchers to query, identify, and retrieve formalin-fixed, paraffin-embedded, and fresh-frozen specimens and/or radiological image sets. The TIES software further supports cross-institutional data sharing using Globus grid services, based on a strong privacy model, including integration with human honest broker systems. TIES has been deployed at multiple National Cancer Institute-designated cancer centers and Clinical and Translational Science Award (CTSA) Consortium sites and will also be used by PCORnet.

TRANSFORMING THE ACADEMIC HEALTH CENTER WORKFLOW

The National Institutes of Health (NIH) [10] and National Academy of Sciences [6] have issued reports on the big data priorities and challenges in biomedical research and

health care, respectively. The overwhelming size and scope of available data raises the need to determine which data are most important to capture and how these data should be recorded. The NIH Big Data to Knowledge (BD2K) initiative[3], in which Pitt will be one of the inaugural Big Data Centers of Excellence (Center for Causal Discovery), will address standardized methods and tools for data sharing and software development [11], and PCORnet will contribute to EHR data harmonization on a national scale. However, AHCs must also consider their own EHR capabilities and research program needs locally.

For those EHR data currently recognized as relevant, such as laboratory, imaging, and pathology reports, potential problems with how the data are recorded must be considered, including errors, inconsistent definitions or nomenclature, missing values (including type of "missingness"), and incomplete data [12]. Establishing structured report templates (whether developed de novo or to fit commercial EHR software already in use) can ensure complete data collection and maintain a format with standardized data fields that support rapid queries. Structured data collection must also use uniform definitions and classifications, particularly with regard to lifestyle measures (e.g., smoking, alcohol use, exercise, and dietary intake) and free-text entry [13]. As per the PCORnet model, a group discussion among biomedical informaticians, researchers, health care workers, and health system executives of what data are needed and how workflow might be affected by data collection policies is important for the smooth adoption of data standards that enhance the productive use of big data.

As a starting point, throughout every AHC both researchers and clinicians should be encouraged to query BioPortal[4], the most comprehensive repository of biomedical ontologies, at the National Center for Biomedical Ontology, which was funded through the NIH Roadmap as one of the National Centers for Biomedical Computing. Researchers should do so prior to developing data collection forms or annotating any data (in vitro, in vivo, or clinical) to ensure that their data can be meaningfully shared and merged with other datasets [14]. Drawing upon established nomenclature will also make valuable data and tools more readily discoverable and shared, such as through the Resource Discovery System[5], whose development by a consortium of CTSA and National Center for Biomedical Computing sites was led by Pitt [15].

Although efforts at harmonization may be perceived by faculty and staff as inconvenient, the long-term payoffs are enormous. The intersection of knowledge among diverse disciplines will be more readily discoverable and actionable, and duplication of effort will be reduced. Although

---

[3] http://bd2k.nih.gov.
[4] http://bioportal.bioontology.org.
[5] http://biositemaps.org/rds/index.html.

further off, application of big data in real time as part of the clinical workflow could include comparison of individual patients with aggregated data for risk assessment (including risk of utilizing more health care resources), clinical decision support, and personalized point-of-care information and treatment plans [16].

TRANSFORMING THE ACADEMIC HEALTH CENTER WORKFORCE

Even with the environment and workflow optimized, AHCs will not realize the full potential of their big data unless personnel at each stage understand and are rewarded for good data practices, such as using ontologies. Up front, the need for accurate and complete data entry by everyone who interacts with the EHR must be emphasized. Data entry fields can be structured to enforce valid data entry (e.g., rejecting out of bounds values, refusing to advance without data entered, and providing pull-down choices rather than free text entry), but more important is the training as to why data entry is so important—and valued. This goes beyond simple data entry to the process of phenotyping itself as well. Just as clinical staff are evaluated, sometimes with financial consequences, based on patient satisfaction surveys, a data-entry grade automatically generated at regular intervals could alert personnel to practices that need improvement—perhaps also with a financial carrot or stick attached. Trust will be gained by both clinicians and researchers if they know that the data they access are valid and usable and by patients if they know that their records will be accurate and their confidentiality will be maintained [17].

Another workforce challenge will be clinician acceptance and implementation of data-driven recommendations that might not make immediate empirical sense to them, although such hesitation may be alleviated as clinical informaticians become integrated throughout AHCs. The creation of a clinical informatics subspecialty began in 2007, with the first cohort of clinical informaticians sitting for the board examination in 2013 (90% passed, despite having had only informal training in many cases, although eligibility to sit for the board after 2018 will require completion of an accredited fellowship in clinical informatics) [18]. These clinical informaticians span a wide range of primary medical specialties, offering the potential to integrate this level of expertise across all departments and to provide leadership in implementing good data practices and use. Having such experts embedded within the clinical workflow will also enhance utilization of computational tools capable of predicting risk or providing treatment guidance in real-time as these come online.

Among researchers, concerns about data sharing predominate, particularly in advance of publication. In addition, the process for making data publicly available when the time is right is often limited if the data are not appropriate

for a specific repository. In the latter case, venues such as *Nature Scientific Data*[6] will likely expand in the coming years, and some data may be appropriate for merging into the PCORnet data stores.

In addition, the NIH is considering developing the capability to allow researchers to upload data appendices and their metadata into a resource linked to PubMed, such that literature searches would also return links to the associated data [10]. The NIH is likewise considering expanding the current data sharing policy to encompass data dissemination with tracking statistics that enable grant applicants to cite the number of times their data have been accessed and/or downloaded for use as a metric of research productivity (also applicable for promotion and tenure decisions, particularly when the data are not published). In the meantime, individual Institutes are expanding their own data sharing policies, such as the National Institute for Mental Health[7], which built the National Database for Clinical Trials in Mental Health[8] as a platform for the sharing and querying of both positive and negative clinical trial data.

Toward this end, AHC leadership can and should take measures to promote a data sharing culture at their institutions [19]. In addition to providing the computing infrastructure and expertise needed to support data sharing, they can recognize and reward data sharing when making hiring and promotion decisions and encourage data sharing in curricula and publication policies. Data shared outside one's AHC increases opportunities for collaborative research, grant funding, and publications and can increase citation rate [20]. Data shared within the AHC, per the Bourne model of the digital enterprise [8], likewise enables new and unforeseen research collaborations, new knowledge from different perspectives, and new linkages that align the interests of research, education, health care, and industry.

A TRANSFORMATIONAL CASE STUDY

Prior to our participation in PCORnet, Pitt–UPMC sought to develop a local resource through which multiple data types could be accessed, merged, and analyzed. Recognizing how the growing speed and capacity of genomic sequencing technologies would support the use of individual genetic and other omics profiles for anticipating risk and personalizing treatment, we set about creating the Pittsburgh Genome Resource Repository (PGRR). Our goals were to support the integration and analysis of omics, clinical, laboratory, treatment, and outcomes data for:

- prediction of disease states and outcomes;
- classification of phenotypes;
- development of genomic and clinical models of disease;

---

[6] http://www.nature.com/sdata/.
[7] http://grants.nih.gov/grants/guide/notice-files/NOT-MH-14-015.html.
[8] http://ndct.nimh.nih.gov.

- association of genes and variants of interest with treatment responses; and
- development of new methods for personalizing care.

Our initial driver was a more effective and efficient use of an existing public resource of great value to Pitt researchers, The Cancer Genome Atlas (TCGA)[9], which contains about 1 petabyte of exome (variant analysis), SNP, methylation, mRNA, miRNA, and de-identified clinical data from about 10,000 cases across 34 tumor types. Through leadership discussions, we were aware that individual investigators and groups faced considerable challenges both in gaining access to TCGA, including significant regulatory hurdles, and in managing and analyzing the large data files. We envisioned developing the PGRR as a framework to address these barriers (e.g., data use agreements, data security, data transmission and storage, and data updates) and to allow the central storage and analysis of large national datasets, including customized tools and end-user training in the actual use of the data. Further, because Pitt–UPMC is the single largest contributing institution (~10% of all cases), we also had the opportunity to re-link the rich tumor sequencing and microarray data from TCGA with the treatment and long-term outcomes in the UPMC EHR for those patients whose tumor samples were contributed for analysis and who provided informed consent to have their clinical data re-associated with the TCGA data generated from their tissues.

Thus, we saw and seized the opportunity to merge and exploit diverse, rich, valuable data that could support our research, education, and clinical missions. In the course of 18 months of planning and negotiating legal and regulatory hurdles, our approach to making the PGRR possible took us through the issues discussed earlier in the chapter: access, trust, integration, annotation, and analysis.

On the individual investigator level, access to and use of the TCGA is constrained by three important barriers:

1. Investigators must complete a Data Access Request[10] through the database of Genotypes and Phenotypes (dbGaP) from the National Institutes of Health, which takes time to be approved and is often limited to a small number of investigators. In contrast, our PGRR supports wide-scale collaboration among Pitt investigators under an institutional data use certificate.
2. Downloads of TCGA are limited by the slow transfer rates from distribution sources as well as the slow movement of files between Pitt and UPMC. PGRR reduces transfer burden by centralizing TCGA using a unique distributed file system shared between the Pitt Center for Simulation and Modeling (SaM) and the Pittsburgh Supercomputing Center, minimizing the time needed to transfer files for analysis.
3. TCGA data require hundreds of terabyte- to petabyte-level storage, which is expensive to establish and maintain. PGRR limits the duplication of files and thus also simplifies control over provenance.

Figure 2 provides a conceptual overview of the PGRR data management and computing infrastructure.

PGRR functions as a "data trust" based on a regulatory foundation. All Pitt users of the TCGA data are co-investigators and collaborators on a single dbGAP Data Use Certificate. In addition, the creation and use of the PGRR itself has been approved by the University of Pittsburgh Institutional Review Board (IRB; PRO12090374). The use of these data at UPMC is covered by a Data Use Agreement between Pitt and UPMC, while the use of these data at Pittsburgh Supercomputing Center is covered by a Data Use Agreement between Pitt and the Center. Requests by new researchers to collaborate as part of the PGRR dbGAP Data User Certificate are considered on a quarterly basis. For those AHCs seeking to establish their own institutional data use agreements, the text of the Research Use Statement for our dbGAP Data Use Certificate can be found in the Appendix at the end of this chapter.

Data available to researchers are completely de-identified. Per our IRB-approved PGRR protocol, UPMC patients who gave consent both for their tissue to be submitted to TCGA and for their clinical data to be re-associated with the sequencing data can be re-identified by UPMC Honest Brokers, who in turn extract and then again de-identify the relevant clinical data so it can be merged with the patient's TCGA data in the UPMC Enterprise Analytics Data Warehouse. Archived and real-time patient EHR data are derived from more than 200 software components serving UPMC's 22 hospitals, two surgery centers, 35 cancer centers, 17 long-term care facilities, more than 70 rehabilitation facilities, and more than 400 outpatient offices. Located behind the UPMC firewall, the Data Warehouse is subject to UPMC security scans and meets all dbGaP and HIPAA security requirements.

We have already linked clinical and TCGA data for breast cancer patients from UPMC and will continue to merge de-identified patient datasets. UPMC had proactively established a Data Governance Program, whose mission is to collect, change, store, move, consume, and release UPMC data assets efficiently, accurately, and legally. Thus, UPMC was in a position to readily participate in the PGRR. AHCs that have not undertaken a similar data governance initiative would be well advised to do so.

To simplify access to and analysis of these data, our bioinformaticians modified and significantly extended the TCGA Roadmap software [21] to manage the download of all TCGA data and metadata, both open and protected access, from National Cancer Institute repositories. A summary of the goals and functionality of software specially

---

[9] http://cancergenome.nih.gov.

[10] https://dbgap.ncbi.nlm.nih.gov/aa/wga.cgi?login=&page=login.

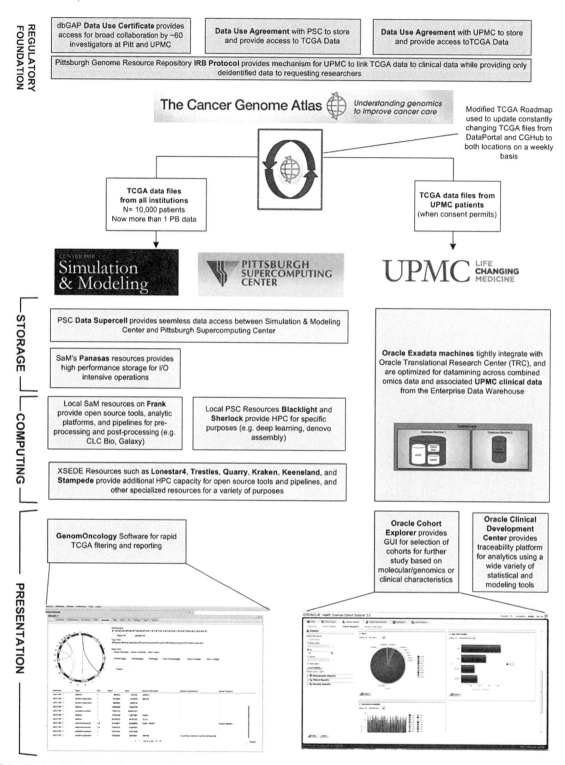

FIGURE 2 The Pittsburgh Genomic Research Repository provides the computing and oversight framework for the central storage, integration, and analysis of big data from national and local datasets.

developed by our Department of Biomedical Informatics to facilitate use of the TCGA data can be found in this chapter's Appendix. To support Pitt investigators in their use of these data, PGRR personnel offer group training sessions as well as on-demand help to individual researchers. The computing infrastructure and expertise to support an initiative such as PGRR is not trivial. We are fortunate to have access to the Pitt SaM and the Pittsburgh Supercomputing Center, both of whose computing capabilities are summarized in this chapter's Appendix.

AHCs not currently partnered with a supercomputing facility can access high-performance computing resources through the Extreme Science and Engineering Discovery Environment (XSEDE)[11], which currently supports 15 supercomputers or computational clusters and well over 100 petabytes of storage located at 17 partner institutions [22].

NEXT STEPS TOWARD A TRANSFORMED ACADEMIC HEALTH CENTER

As big as the PGRR is, it is a baby step. Having successfully developed and negotiated the PGRR framework to support TCGA omics data merged with UPMC patient phenotype and clinical data, we look forward to developing additional big data resources focused on other large public data-sharing repositories[12] as well as large national studies (e.g., Framingham Heart Study and the Women's Health Initiative) and global datasets (e.g., International Cancer Genome Consortium). The payoff for these efforts will be rapid and significant: within months of merging TCGA and UPMC breast cancer data, investigators at Pitt discovered molecular differences in the pattern of pre- versus post-menopausal cancer, which could guide the development of treatment targeting these distinct cohorts.

A long-term goal includes merging big data from throughout our institution. Dozens of separate genomic and other omic studies are scattered throughout the six schools of the health sciences, each limited by the ability of available department infrastructure to house data tailored to the immediate project and by sustainability only as long as grant funding continues. Harmonizing data and biorepository practices among so many individual projects will be difficult but rewarding, particularly if we can achieve the vision of the first NIH Associate Director of Data Science, Philip Bourne, of having all data generated by University researchers (e.g., electronic lab notebooks, protocols, raw data, and analyzed data) maintained in an open-access data commons so that links between interests and discovery can be identified automatically via sweeps of the total data set [8].

We look forward to national data sharing and analyses through PCORnet and possibly to a future in which unique patient identifiers allow us to accurately track data across the lifespan—making for some very big data indeed.

APPENDIX

Technical details related to the establishment of the PGRR are provided below.

Data Use

As an illustration of the level of details needed, the Research Use Statement for our dbGAP Data Use Certificate reads as follows:

University of Pittsburgh Institute for Personalized Medicine (IPM), University of Pittsburgh Cancer Institute (UPCI), and UPMC are collaborating to develop a central repository and enterprise data warehouse (EDW) to support future personalized medicine initiatives at our institution. As part of this larger initiative, the current TCGA data request will support the formation of a large collaborative effort at our institution to develop novel research methods and algorithms that leverage both clinical and genomic data to personalize cancer care. As described in our University of Pittsburgh IRB protocol, for patients originating from UPMC, we seek to link all levels of the TCGA data on samples originating from University of Pittsburgh to the richer clinical data available for these patients through UPMC clinical systems, as well as additional research data available. Additional clinical data will include tumor recurrence, tumor metastasis, co-morbid conditions, treatment, treatment response, and outcomes such as disease-free survival. Additional research data will include manually abstracted clinical data, results of multigene testing, and novel biomarkers. Collaborators named on this DUC will use the resulting "enhanced" TCGA dataset for: (1) prediction of disease states and outcomes; (2) classification of phenotypes; (3) development of genomic and clinical models of disease; (4) association of genes and variants of interest with treatment response; and (5) development of new methods for individualizing cancer treatment.

We will control access as follows: (1) data will be maintained in two secure locations and analytic tools will restrict download of protected levels of data; (2) data will be available only to the named collaborators and their staff and graduate students; (3) for data originating from the University of Pittsburgh, we will combine TCGA data with de-identified clinical, laboratory, and outcomes data derived from medical records of these patients, using an "honest broker" (neutral third party) to maintain the linkage; and (4) data accessed by investigators will be de-identified.

Software

In creating the PGRR, the Pitt Department of Biomedical Informatics developed software to provide the following functionality:

- Extract essential metadata from TCGA Code Table Reports and update the Resource Description Framework (RDF) graphs for Disease Study, Tissue Source Site, Center, Sample Type, and Portion Analyte

---

[11] https://www.xsede.org/.
[12] http://www.nlm.nih.gov/NIHbmic/nih_data_sharing_repositories.html.

- Use the open access and controlled directories to obtain all clinical and omics data
- Use scripts to download BAM files (binary format for storing sequence data) on demand; BAM files are stored as a separate analysis type for a corresponding sample, and metadata about BAM files are available
- Manage generation of the PGRR version by taking appropriate action after comparing the existing file version in PGRR storage (if any)
- Create PGRR data stores and archives using standard directory structure, naming, and version conventions

Within this central storage location, PGRR manages the use of TCGA files:

- Creates an RDF store of all TCGA metadata
- Annotates the file's metadata based on the TCGA Roadmap code (url, label, type, first-Seen date, last-Seen date, and last-Modified date plus directory level: DiseaseStudy/ CenterType/CenterDomain/Platform/DataType/Archive/ File)
- Creates PGRR RDF Store Schema with the following data types:
  - DiseaseStudy
  - Tss [tissue source site]
  - Patient
  - Sample
  - AnalysisType [i.e., "Clinical," "Protected Mutations"]
  - CenterName [short name in Code Table Report]
  - Platform
  - File

With the following properties of entities of any data type:

- label
- type
- path [in Pitt storage]
- dateCreated
- dateModified

In addition, each data file itself has the following properties:

- barcode
- dataSource [i.e., TCGA or UPMC]
- diseaseStudy
- tss [tissue source site]
- patient
- sample
- sampleType
- analyte
- centerName
- platform
- level
- public [false for controlled data]
- version [Pitt version]
- refGenome

- regGenomeSource
- pittURL
- tcgaURL [pass to the original archive]
- dateArchived
- reasonArchived ["Modified" or "Deleted"]
- sizeInBytes
- checksum

Users can request automated email updates when data files of interest have been updated to new versions in PGRR.

We also maintain a user interface through which investigators can access TCGA datasets using the GenomOncology (GO) analysis and visualization software. Intended for translational research end-users, the GO software provides an easy-to-use analytics portal that we can leverage as a platform to visualize results of any analytic process. GO software allows investigators to explore annotation-based filtering of next generation sequencing-based datasets and to quickly compare results among several different individuals within a project and against public (e.g., dbSNP or the "1000 genomes" data) and custom databases. Users can interactively view and analyze all types of next generation sequencing data—including variation data from SNPs, structural variants, copy number variants, loss of heterozygosity regions, relative expression (from RNA-Seq), and/or epigenetic data—and analyze their combined impact on genes and pathways. They can also dynamically filter variations to focus on features of biological interest. Specific to the use of TCGA, investigators can analyze protected TCGA data and partition access to these data based on the Data Use Certificate and user roles, and they can annotate all Pitt data with information from analogous variation data from the full TCGA or other reference databases.

Hardware

The PGRR is made possible through the availability of two exceptional high-performance computing resources: the Pitt SaM and the Pittsburgh Supercomputing Center.

SaM, the premier shared high-performance computing facility in the University community, represents investments of hardware and human capital from several University Schools and Departments. SaM operations are overseen by three directors (senior faculty members from the Schools of Engineering, Arts and Sciences, and Health Sciences) and are facilitated through the activities of five consultants (faculty members from various departments). The Center also serves as a collaboration portal, having assembled a group of more than 50 collaborators from across the University who are engaged in computational research in Chemistry, Biology, Physics, Astronomy, Mathematics, Computer Science, Economics, and several of the departments in the Swanson School of Engineering, as well as faculty from the School of Medicine, the Graduate School of Public Health, and the Graduate School of

Public and International Affairs. SaM provides user support, training, and project management services on a continual basis through Web 2.0-based platforms[13,14] and offers workshops and training sessions on cluster usage, parallel programming, and various topics in high-performance computing-based research. The Center also acts as liaison for national computational resources through partnerships with the Pittsburgh Supercomputing Center and the NSF/XSEDE Campus Champions program.

Computational resources at SaM include a heterogeneous grid/cluster comprised of 200 8-core Intel Nehalem and Harpertown, 45 12-core Intel Westmere, 23 48-core AMD Magny-Cours, 82 16-core Intel Sandy Bridge, and 110 32-core AMD Interlagos compute nodes, for a total of 8076 computation-only CPU cores, with a maximum of 128 GB per node shared memory (1.5–8 Gb per core). Several Nehalem nodes also have general purpose NVIDIA GPU accelerator cards, for a total of 16 GPU cards comprising 5504 GPU cores. Most nodes are connected via fast Infiniband low-latency network fabrics. Local (temporary) storage on the computer nodes is typically 1 to 3 TB. Users' home directories are maintained on an 80-TB RAID5 NAS unit, with a redundant array providing online snapshots/backup, and a high-performance 120-TB Panasas storage array is available for processes requiring fast distributed disk access.

All cluster nodes at SaM are secured behind the University firewall, which permits only encrypted communication from the external network. All extraneous services are controlled at the firewall (e.g., email, file sharing, and printing), so that the cluster is as secure as possible. All user accounts are password-restricted with strong password policies dictating the content of passwords and requiring password rotation. All data on the cluster are restricted using access-control lists so that only appropriate project members have access to data from a particular project.

As a full partner in the Pittsburgh Supercomputing Center, we have access to three specialized computing resources for managing and analyzing TCGA data: Blacklight, the world's largest shared memory system (enabled the largest metagenomic assembly to date); Sherlock, a graph analytics system with unique architecture for discovering patterns in large data (built by YarcData, subsidiary of Cray); and the Data Supercell, a scalable (currently ~4 petabytes), low-cost, reliable disk-based data management and storage solution that replaces slower commercial tape-based storage and is ideal for data analytics with very large datasets.[15]

For applications requiring very large shared memory, Blacklight provides a total of 4096 cores and 32 TB across the whole system.[16] Each core has a clock rate of 2.27 GHz,

supports two hardware threads, and can perform 9 Gflop/s, for a total system floating point capability of 37 Tflop/s. Blacklight is part of the National Science Foundation XSEDE integrated national system of cyber-infrastructure. Sherlock enables large-scale, rapid graph analytics through massive multithreading, a shared address space, sophisticated memory optimizations, a productive user environment, and support for heterogeneous applications.[17]

REFERENCES

[1] Levine AS, Detre TP, McDonald MC, Roth LH, Huber GA, Brignano MG, et al. The relationship between the university of Pittsburgh school of medicine and the university of Pittsburgh medical center – a profile in synergy. Acad Med 2008;83:816–26.

[2] Green SM, Reid RJ, Larson EB. Implementing the learning health system: from concept to action. Ann Intern Med 2012;157:207–10.

[3] Health system leaders working toward high-value care through integration of care and research – workshop in brief. June 2014. Accessed at: http://www.iom.edu/Activities/Quality/VSRT/~/media/Files/Activity%20Files/Quality/VSRT/Health%20System%20Leaders/VSRT-HealthSystemLeadersWIB.pdf.

[4] Fleurence RL, Curtis LH, Califf RM, Platt R, Selby JV, Brown JS. Launching pcornet, a national patient-centered clinical research network. J Am Med Inf Assoc 2014;21:578–82.

[5] Amin W, Tsui F, Borromeo C, Chuang CH, Espino JU, Ford D, et al. Path: towards a learning health system in the Mid-Atlantic region. J Am Med Inf Assoc 2014;2014(21):633–6.

[6] Institute of Medicine. Digital data improvement priorities for continuous learning in health and health care: Workshop summary. Washington, DC: The National Academies Press; 2013.

[7] National Research Council. Frontiers in massive data analysis. Washington, DC: The National Academies Press; 2013.

[8] Bourne PE. What *big data* Means to Me. J Am Med Inf Assoc 2014;21:194.

[9] Crowley RS, Castine M, Mitchell K, Chavan G, McSherry T, Feldman M. caTIES: a grid based system for coding and retrieval of surgical pathology reports and tissue specimens in support of translational research. J Am Med Inf Assoc 2010;17:253–64.

[10] Data and Informatics Working Group. Draft report to the advisory committee to the director. June 15, 2014. Accessed at: http://acd.od.nih.gov/Data%20and%20Informatics%20Working%20Group%20Report.pdf.

[11] Ohno-Machado L. NIH's *big Data to knowledge* initiative and the advancement of biomedical informatics. J Am Med Inf Assoc 2014;21:193.

[12] Hoffman S, Podgurski A. The use and misuse of biomedical data: is bigger really better? Am J Law Med 2013;39:497–538.

[13] Schneeweiss S. Learning from big health care data. N Engl J Med 2014;370:2161–3.

[14] Musen MA, Noy NF, Shah NH, Whetzel PL, Chute CG, Story MA, et al. The national center for biomedical ontology. J Am Med Inf Assoc 2012;19:190–5.

[15] Tenenbaum JD, Whetzel PL, Anderson K, Borromeo CD, Dinov ID, Gabriel D, et al. The biomedical resource ontology (BRO) to enable resource discovery in clinical and translational research. J Biomed Inf 2011;44:137–45.

---

13 http://core.sam.pitt.edu.

14 http://collab.sam.pitt.edu.

15 http://www.psc.edu/index.php/computing-resources/data-supercell.

16 http://www.psc.edu/index.php/computing-resources/blacklight.

---

17 http://www.psc.edu/index.php/computing-resources/sherlock.

[16] Simpao AF, Ahumada LM, Galvez JA, Rehman MA. A review of analytics and clinical informatics in health care. J Med Syst 2014;38. Article 45. Accessed at: http://link.springer.com/article/10.1007/s10916-014-0045-x.

[17] Costa FF. Big data in Biomedicine. Drug Discov Today 2014;19: 433–40.

[18] Detmer DE, Shortliffe EH. Clinical informatics: prospects for a new medical subspecialty. JAMA 2014;311:2067–8.

[19] Piwowar HA, Becich MJ, Bilofsky H, Crowley RS. Towards a data sharing culture: recommendations for leadership from academic health centers. PLoS Med 2008;5:e183.

[20] Piwowar HA, Day RS, Fridsma DB. Sharing detailed research data is associated with increased citation rate. PLoS ONE 2007;2:3308.

[21] Robbins DE, Gruneberg A, Deus HF, Tanik MM, Almeida JS. A self-updating road map of the cancer genome atlas. Bioinformatics 2013;29:1333–40.

[22] LeDuc R, Vaughn M, Fonner JM, Sullivan M, Williams JG, Blood PD, et al. Leveraging the national cyberinfrastructure for biomedical research. J Am Med Inf Assoc 2014;21:195–9.

ABOUT THE AUTHORS

Jeremy M. Berg, PhD is Pittsburgh Foundation Chair and Director of the Institute of Personalized Medicine, Associate Vice Chancellor for Science Strategy and Planning in the Health Sciences, Professor of Computational and Systems Biology in the School of Medicine, and Professor of Chemistry in the Dietrich School of Arts and Sciences at the University of Pittsburgh. Prior to coming to Pittsburgh in August 2011, Dr Berg served as the director of the National Institute of General Medical Sciences (NIGMS) at the NIH.

Rebecca S. Crowley, MD, MS is a Professor of Biomedical Informatics at the University of Pittsburgh School of Medicine, with secondary faculty appointments in the Intelligent Systems Program, Clinical and Translational Science Institute, University of Pittsburgh Cancer Institute, and Department of Pathology. Dr Crowley is the Chief Information Officer for the Institute for Personalized Medicine and also Director of the Graduate Training Program in Biomedical Informatics.

Michelle L. Kienholz is a science writer-editor with 30 years of experience in academic biomedical research. Ms Kienholz assists investigators at the University of Pittsburgh with grant applications and manuscripts and works closely with faculty, postdocs, and graduate students in developing and communicating their research.

Arthur S. Levine, MD is the Senior Vice-Chancellor for the Health Sciences and the John and Gertrude Petersen Dean of the School of Medicine at the University of Pittsburgh. He is also Professor of Medicine and Molecular Genetics. Prior to his leadership appointment at the University of Pittsburgh in 1998, Dr Levine served as Scientific Director of the National Institute of Child Health and Human Development at the NIH.

Chapter 17

Bridging Science and Practice—A Case Study: The Military Translation of Innovative Responses to Urgent Military Medical Needs into Widespread Clinical Practice

Kent E. Kester and Charles L. Rice

INTRODUCTION

A major challenge in the advancement of medical care and its practice is the timely dissemination of new and relevant information and its hoped-for adoption by practitioners. The translational gap between the identification of important clinical findings and translation into everyday practice is a continuing problem in medicine [1–4]. The reasons for this gap are multiple, ranging from cost and complexity to ineffective communication and publicity, as well as individual, institutional, and practitioner biases. Indeed, the continued existence of this gap has led, with some urgency, to a major emphasis in translational research by senior leaders in academic medicine and industry, with the desired end result being a better translation of scientific discovery into products and practice [4,5]. As in many areas of medicine, there are certain situations of clinical urgency whereby the field can actually move forward—quickly and with widespread acceptance.

While the emphasis in this volume addresses the essential role of academic health centers (AHCs) in the development and dissemination of medical innovation, it is relevant to consider another parallel system that, in many ways, mirrors key aspects of the traditional academic health center—the US Department of Defense's Military Health care System, or MHS. While the major mission for this system is to deliver care to military health care beneficiaries (e.g., active duty members of the military, their family members, and military retirees, among others), there are also robust and diverse medical education and medical research components that directly complement the health care delivery system. These mission-areas train the next generation of military physicians and allied health professionals, as well as conduct militarily-relevant medical research in order to provide solutions (e.g., medical products, actionable medical knowledge, new policies, and practice guidelines) needed by deployed and garrisoned US military forces and those whom they support. In effect, they are in many ways a massive AHC (or more properly, a network of AHCs supporting a larger base of primary and secondary care centers), albeit with some obvious differences from the traditional civilian AHC.

It is in this context that we discuss a number of important operationally-driven innovations in medical care championed by the US military to care for ill and injured patients deployed to the wars in Iraq and Afghanistan. The emphasis of these innovations has always been to provide better care to the US and allied military personnel—whether in key areas of combat casualty care; in the identification and treatment of those with behavioral health problems, such as posttraumatic stress disorder (PTSD); or in the identification and control of contributors to nosocomial infections. However, in a variety of ways, these same innovations—whether new or optimized practices, policies, or "things"—have rapidly permeated the nonmilitary medical practice community in the US and elsewhere, suggesting that in spite of ongoing problems in translating science into practice, these same elements of operational urgency have served as a catalyst for change in medicine.

While there are a large number of important research areas relevant to military medicine, the following sections will focus on key areas for which the military has always played a leading role (pre-hospital care, emergency trauma resuscitation and surgery, mental health, and infectious diseases), in addition to serving as points of reference for further dissemination into the larger civilian health care world.

The Transformation of Academic Health Centers. http://dx.doi.org/10.1016/B978-0-12-800762-4.00017-7

MEDICAL ADVANCES FROM MILITARY TO CIVILIAN PRACTICES

Emergency Trauma Resuscitation and Surgery: Optimized Physiologic Replacement of Blood Products during Trauma Resuscitation

A key area of operationally-driven innovation in the area of medical practice is that of transfusion support in the resuscitation of multiple-traumatized patients—one of the core competencies of military medicine. In the civilian sector, at least 10% of deaths after traumatic injury are likely preventable, and approximately 15% of these are related to hemorrhage [6,7]. In fact, hemorrhage ranks second to central nervous system injuries in the overall cause of death and remains the leading preventable cause of death [8]. Over the past 10 years, discovery, or rediscovery, of key aspects related to transfusion and resuscitation from traumatic injury in military operations has led the way for a significantly improved process throughout the larger nonmilitary, trauma-care world.

Although most deaths (~80%) that occur during combat are not preventable, the large majority of those deaths that are preventable are a result of hemorrhagic shock [9]. A recent study estimated that nearly 30,000 wounded personnel were transfused with nearly 300,000 units of blood products during the conflicts in Iraq and Afghanistan [10], suggesting a sizable number of hemorrhagic incidents requiring resuscitation. Historically, severely injured patients with hemorrhage were typically resuscitated with whole blood and normal saline [11]. Other additives, such as sodium bicarbonate, were provided to deal with the acidosis that often resulted. It was thought, correctly so, that the use of whole blood—ideally fresh whole blood—would assure provision of proper amounts of clotting factors, often lacking in stored blood.

Guidelines developed by the American Blood Banking Association (ABBA) over time, and mostly based on practices associated with elective surgery, moved to favor the use of selective component therapy, the idea being that this would allow for very focused correction of anemia and clotting factor deficiencies, rather than using the more blunt tool of whole blood. Thus, a variety of components became available to supplement packed red blood cells (PRBCs) that included clotting factors and fresh frozen plasma (FFP). Unclear at that time, however, was what ratio of FFP to PRBCs was the most efficacious in terms of stabilizing bleeding patients, as well as preventing re-bleeding in the post-resuscitation time period. While not rigorously tested, the practice of providing FFP to any patient who required a massive transfusion (MT), that is 10 or more units of PRBCs, became the de facto standard of care. Later adjustments led to FFP being recommended for those who had received 6 units of PRBCs with continued bleeding, with other options proposed depending on laboratory measurements of circulating clotting factors, as well as the physical presence of active bleeding, oozing from wounds, or clotting [11].

Over time, emergency care of trauma patients became codified in the practices of Advanced Trauma Life Support (ATLS). An increasing emphasis was placed on the use of crystalloid therapy early on in the resuscitative process in order to maintain perfusion with blood losses being replaced with a 3:1 ratio of isotonic crystalloid fluid—even as the use of FFP continued to increase—albeit not in an evidence-based manner [8]. Owing to ongoing uncertainty in the field as to the best approach, the NIH sponsored a Consensus Development Conference in 1984 where it was concluded that the use of FFP should be curtailed and provided only to those patients with demonstrable coagulopathy, as defined by abnormal tests of coagulation function [12]. Still, the concept of the lethal triad of coagulopathy, hypothermia, and acidosis have long been recognized as markers of increased mortality after traumatic hemorrhage requiring MT [13,14], an issue incompletely addressed by the NIH Consensus Development Conference.

It is against this background that US and allied military surgeons began caring for combat trauma patients wounded in Iraq and Afghanistan [11]. Early in the Iraq conflict, military surgeons in Baghdad began adopting different methods of managing hemorrhage and its associated physiologic impacts in their trauma patients. An aggressive approach was adopted to deal with hemorrhage in concert with a surgical concept (now termed damage control resuscitation), characterized as a combination of hypotensive resuscitation (ideally, to prevent the dislodgement of newly-formed clots) and a more unified approach to hemostatic resuscitation [6,8]. The deployed surgeons, a number of whom were researchers assigned to the US Army Institute of Surgical Research, began to explore optimizing aspects of the transfusion and fluid support for these critically-injured patients. In particular, it became quickly apparent that the standard ATLS approach to fluid resuscitation was associated with an increase in dilutional coagulopathy and acidosis.

Results from a number of retrospective and observational studies in trauma patients demonstrated improved survival and decreased death rates from hemorrhage when transfusions using increased ratios (>1:2) of plasma to PRBCs and platelets to PRBCs were administered [9,15,16]. This and other related work, coupled with a directed focus on learning more about combat injuries and optimal ways to address them, led to the establishment of a Joint Theater Trauma Registry (JTTR) along with a larger Joint Trauma System or JTS [17,18]. This system and its associated registry have served to disseminate best practices and lessons learned to deploying military medical personnel. Further, through the ongoing development, review, and dissemination of focused clinical practice guidelines (CPGs), the

JTS has served as a model for the advancement of surgical knowledge throughout the military and beyond [19].

In large part owing to the work of engaged military surgeons, the current JTS guidelines for damage control resuscitation, based both on expert opinion and outcomes data for MT, focus on the use of a 1:1:1 ratio of plasma, PRBCs, and platelets [20]. With the establishment of this as the standard of care, subsequent analysis of military trauma outcomes in Iraq and Afghanistan demonstrated 30-day survival data ranging as high as 90% in patients requiring MT [21]. This compares favorably against companion data of 59% survival in a recent multicenter US civilian trauma study [16]. Because of the unique logistical aspects associated with combat trauma care and blood supply, the use of whole blood transfusion has also been championed, although its availability and utilization has been much less [10,22].

Moving forward, and mindful of the need for full evidence-based recommendations, the military and civilian trauma communities are now conducting a number of prospective, multicenter studies intended to evaluate more fully these important aspects of trauma-related transfusion practice, with early results suggesting that earlier attainment of high transfusion ratios may ultimately reduce the requirement for MT [23]. In fact, it appears that in patients who receive more aggressive physiologic component transfusion, which includes FFP and platelets, the overall requirement for blood products is less, suggesting better hemostasis in the face of a new resuscitation paradigm—which may now also include elements of damage control surgery [23].

More recently, major military/civilian collaborative studies, such as the Prospective Observational Multicenter Major Trauma Transfusion (PROMMTT) study and others, have demonstrated in mixed trauma populations that plasma transfusion early in resuscitation had a protective association with mortality and that transfusion ratios approaching 1:1:1 were associated with decreased short-term mortality [16,24,25]. As the best approach to trauma-related blood product support remains to be defined, further collaborative, multicenter studies—such as the recently completed and soon to be reported Pragmatic Randomized Optimal Platelet and Plasma Ratios (PROPPR) study (ClinicalTrials.gov number NCT01545232)—will likely contribute more to advancing the field beyond expert opinion as the basis for clinical practice [26,27].

Pre-hospital Care: Use of Optimized Tourniquets to Control Severe Extremity Bleeding

Not all combat casualty care innovations have involved new or improved medical/surgical practices in the emergency department. A companion area of major interest to US military leaders has been the continued enhancement of the pre-hospital component of trauma care. This ranged from point of injury interventions to innovative training programs for deploying general medical officers, physician assistants, and combat medics—as well as new approaches for rapid aeromedical evacuation of injured patients. An important example of a point of injury treatment whose use has been revived by the military is that of the field tourniquet to stop major hemorrhage.

While tourniquets have been effectively used for centuries as an emergent approach to control battlefield or other trauma-related hemorrhage, over the past 100 years their use in the US has fallen into disfavor for a variety of reasons. In both World Wars, there were many limbs amputated due to improper application of tourniquets in emergency situations [28]. Unsubstantiated comments by senior experts and thought leaders in orthopedics and casualty care relating tourniquet use to both the unnecessary loss of limbs and the loss of life did little to enhance the standing of this intervention in Western medicine [29,30]. Additionally, the standard issue military tourniquets then available were fairly rudimentary and rather ineffective at controlling significant hemorrhage [28]. Similarly, first-aid classes intended for the larger civilian community typically focused on tourniquets as a risky intervention to be used only for patients *in extremis*, typically relying on field expedients (such as pieces of cloth) to encircle the affected limb and rigid objects (such as sticks) to provide the tightening windlass function. Predictably, such approaches, along with the mindset that valid considerations for tourniquet use were limited, made them of questionable value in the eyes of practitioners for the pre-hospital or hospital care of trauma patients suffering life-threatening hemorrhage from an extremity wound.

For years, both in the military and in the civilian medical community, the prevailing wisdom for stopping severe bleeding remained the application of direct pressure and early surgical intervention, with the latter often limited by access to definitive surgical care. This negative view of tourniquet use was bolstered by the poor design of commonly available tourniquets, along with multiple accounts of inappropriately applied tourniquets (whether applied properly in inappropriate situations or improperly applied in appropriate situations) leading to unnecessary limb loss or worse [29,31]. Presciently, and nearly coincident with the onset of combat operations in Iraq and Afghanistan, the US Army Medical Department conducted a review and re-evaluation of a number of tactical combat casualty care practices and interventions—among them the combat tourniquet.

After a comprehensive review of the available published military and civilian data, as well as an evaluation of new combat tourniquet prototypes, the US Army Office of the Surgeon General approved the procurement of a new type of combat tourniquet, suitable for self-application, which could be issued to each soldier [32]. With its issue to deploying US military personnel bound for the conflicts in Iraq and Afghanistan, coupled with an aggressive pre-deployment training program that emphasized the timely application of tourniquets for severe limb bleeding (self-applied by the injured patient, or applied by his/her combat buddies, or by field medical personnel), pre-hospital control of trauma-associated hemorrhage was noted to be greatly enhanced—with data demonstrating better survival rates when tourniquets were appropriately applied. Essentially, this intervention became an important adjunct in the larger approach of damage control resuscitation [33,34].

The combination of better tourniquets, consistent and effective training for all personnel (medical and nonmedical) that emphasizes control of major hemorrhage over airway and breathing assessments [35], and rapid aeromedical evacuation of injured patients has significantly impacted the long-term morbidity and mortality of combat-injured military personnel. With this type of data being generated in military combat operations, along with the more common (and frequent) deployment of civilian reservist medical and paramedical personnel, it was only natural that adjuncts to care—such as the combat tourniquet—would be evaluated in civilian pre-hospital care [36]. In fact, the use of tourniquets is now viewed much more positively in the civilian trauma care community [37–39]. Thus, the effectiveness and availability of well-designed and well-tested pre-hospital tourniquet devices, along with the fairly rapid dissemination of their proper use in the civilian emergency medical care community, has likely contributed to an increasing number of lives being saved in the US civilian EMS sector—again demonstrating the value and power of operationally-driven military medical research as a timely and important catalyst for change in the larger medical practice community.

Mental Health: Identification of PTSD and the Development of Treatment Methodologies

Unlike most other combat or deployment-related medical or surgical problems, the broad area of behavioral health and all its ramifications is one that is very difficult to fully detect and treat. As combat and stabilization operations continued in both Afghanistan and Iraq, there were predictably a number of mental health problems identified with some regularity. These diagnoses—including PTSD, chronic anxiety, and short-term and chronic depression—were not necessarily different from those seen in earlier conflicts. However, for the first time, military researchers were able to frame the risk for these various diagnoses in the context of ongoing and multiple deployments and combat activities, rather than the more typical analyses of prior conflicts that were conducted mostly after the cessation of military operations [40].

In a process analogous to that adopted by military surgeons in their approach to the resuscitation of patients with combat injuries, military psychiatrists and psychologists at the Walter Reed Army Institute of Research (WRAIR) and the Uniformed Services University (USU) proactively developed plans to regularly evaluate the state of mental health for deployed forces. Through the use of deployed mental health researchers and rigorous data capture and analysis of surveys and focus group feedback, these Mental Health Advisory Teams (MHAT) were able to regularly define key aspects related to the impact of deployed military operations (with or without combat) on population mental health [41,42], providing actionable feedback to the commanders in the field as well as to senior military policymakers. The data generated by Hoge et al. [40] regarding the prevalence of PTSD and its variable presence in personnel who were deployed to either Iraq or Afghanistan (18% and 12%, respectively) publicized the very real importance of this and related diagnoses to the US military. Because of the large number of military reservists who were deployed to these war zones, an increasing number of these combat-related mental health diagnoses became much more common in non-military communities as those Reservists and Guardsmen returned home following completion of their deployments [43].

Media stories focusing on mental health aspects of military operations have led to a high degree of public and professional (psychiatric) visibility of deployment and combat-related PTSD [44]. Along with continued prospective real-time assessments of deployed and garrisoned military personnel and their families, this has led to significant new interventions—implemented on a large scale and intended to reduce the incidence of these disorders, as well as to enhance the diagnostic and treatment options available [45]. Importantly, the efforts of the Department of Defense and the Department of Veterans Affairs have been augmented by major work supported by the National Institute of Mental Health (NIMH), further broadening the impact on the larger civilian medical care community.

In all of these programs, meaningful data have been captured that can be applied in the civilian health care world. As a result of the dissemination of this information, civilian providers have a much better understanding of the prevalence of certain mental health-related diagnoses in their patients, regardless of their current or prior affiliation with the military. With a much better recognition of the development of PTSD resulting from different types of stressful situations—such as natural disasters, criminal or domestic violence, or accidents—providers are using tools derived from military medicine to screen their patients [46,47]. In addition, a variety of interventions for mental health diagnoses resulting from military medical research are now in common use in the military, veteran, and civilian health care settings, whether to address PTSD, suicidal ideation, depression, or other associated diagnoses and conditions [48–52]. The inclusion of PTSD and related disorders in the new Diagnostic and Statistical Manual-5 is in large part related to work done by military researchers [53]. Importantly, from a public health standpoint, key insights and approaches related to psychological trauma pioneered by the military have been embraced by those tasked with addressing population-related impacts of, and response to, natural disasters [54–56]. The USU-based Center for the Study of Traumatic Stress routinely provides consultative support focused on mental health evaluation and response related to a variety of situationally-based traumatic events (see http://www.cstsonline.org/our-work/disaster-mental-health/), military and civilian.

In alignment with the real-time assessment of deployment-related mental health disorders by military researchers, follow-on longitudinal Framingham-type studies—such as the Millennium Cohort Study and the Army STARRS program (the latter being the largest study of suicide involving the military ever undertaken)—are now in place and will significantly impact how we diagnose and treat certain mental disorders, military and civilian [57–59].

Response to Complex Antibiotic-Resistant Nosocomial Infections

Early in the conduct of the wars in Afghanistan and Iraq, and continuing for a number of years, it was noted that a variety of severely-wounded patients who transited the aeromedical evacuation chain through Germany and ended up for definitive care at large military medical referral centers in the continental US, such as the Walter Reed Army Medical Center and the National Naval Medical Center (now combined as the new Walter Reed National Military Medical Center in Bethesda, MD), were either colonized or frankly infected with a variety of antibiotic-resistant bacterial species [60–62]. Importantly, as in the case of other aspects of combat casualty care described previously, the JTTR was able to help providers in a near-prospective manner to learn more about the incidence and types of wounds and other infections (often related to Gram-negative organisms) associated with many of these patients who were evacuated from the war zone [63,64]. While infections associated with the highly antibiotic-resistant bacterium *Acinetobacter baumanii*—especially when associated with transmission to health care workers [65]—received much publicity in the public media, the reality was that these patients were often infected with multiple-resistant bacteria species, leading to many complications and delays in their definitive care, as well as contributing to increased morbidity and mortality [60–62]. As became apparent over time, the isolation of *A. baumanii* often served as a marker for the presence of other resistant bacteria as well, with isolates sometimes sharing resistance mechanisms [66,67]. Eventually, it became apparent that the presence of these difficult-to-treat bacteria in the US military hospitals led to further spread into other nonmilitary facilities, including those belonging to the Department of Veterans Affairs as well as multiple civilian hospitals. This led to increased scrutiny by the US Congress [68,69].

In efforts analogous to those recounted previously, research-focused US military Infectious Disease physicians developed plans to: (1) identify the source of the multiple-resistant bacteria; (2) optimize infection control practices in the deployed environment to reduce or limit colonization and infection with these strains; and (3) establish a prospective effort to fully characterize the bacteria associated with these infections. The goal was to develop this data into a platform suitable for identifying clustered outbreaks, as well as to serve as a resource to evaluate novel diagnostic and therapeutic measures—especially important in this era of limited antibiotic options for resistant bacterial infections.

While a variety of hypotheses were advanced regarding the source of antibiotic-resistant bacterial colonizations and infections, diligent work by military infectious disease physicians and microbiologists determined that the key factor associated with nearly all of these situations had to do with the presence, often a long-term presence, of ill or injured host nation patients in the same facilities, especially in ICUs, where military casualties were being treated. This, coupled with often inadequate adherence to infection control practices, led to cross-contamination of patients. Further work showed that Gram-negative bacteria recovered from host nation patients had high rates of antimicrobial resistance [70–72]. Subsequent studies documented changes in the antimicrobial susceptibility patterns in US-based facilities—all related to the importation of colonizing and infecting organisms from patients evacuated from the war zones [73]. Key to this was enhanced infection control practices, something not always easy in a deployed, wartime environment, and efforts were undertaken to limit the exposure of US and allied military casualties to resistant bacteria. In addition to cohorting and timely transfer of host nation patients to nonmilitary medical facilities, intentional efforts were established to optimize infection control practices in the field to include enhanced emphasis on basic infection control practices, establishment of electronic infectious disease and infection control consultation systems, generation of broadly-applicable infectious disease and infection control CPGs, and the development of a deployment-focused infection control course [74–76].

Although the deployment infection control course was focused on the urgent operational needs of the military, its value clearly transcended the military with obvious applications for disaster and humanitarian relief operations and civilian mass casualty events. As a supplement to these interventions, measures were also undertaken to limit the use of broad-spectrum antibiotics at the earliest levels of care. Codified in a number of regularly updated clinical practices, and supported by the larger military pharmacy supply chain, the guidelines led to reduced antibiotic pressure in the deployed medical setting [77,78]—a key element in any program intended to reduce the presence of antibiotic-resistant organisms and their impact on health care [79].

Further evaluations were considered as part of the novel Trauma Infectious Disease Outcome Study, an effort led by the USU-based military Infectious Disease Clinical Research Program (see http://www.idcrp.org/). Over the projected life of this innovative study, patients infected with resistant organisms will be evaluated for risk factors, short and long-term outcomes, and other factors that may contribute to the impact of bacterial colonizations and infections [80]. Finally, in an effort to prospectively identify problematic bacterial isolates and associated trends in colonization and infection, researchers at the WRAIR developed a surveillance and repository program for resistant organisms that is unique within the entire US Federal Government. Named the Multidrug-resistant Organism Repository and Surveillance Network (MRSN), this initiative coupled with the power of the Department of Defense's largest biomedical research laboratory and its research clinicians and microbiologists in order to fully characterize (phenotype and genotype) resistant organisms submitted by medical facilities. This basic and applied microbiologic analysis is also coupled with associated clinical data in order to serve as a key participant in the identification of outbreaks and relatedness of isolates [81,82].

Since its establishment, the MRSN has played a key role in a number of military and civilian outbreak investigations, in addition to alerting the military medical leadership as to the presence of significant resistant bacterial strains in the military health system—strains with the potential to enter the wider US health care system [83–85]. Now coupled with the broader mandate of antimicrobial-resistance surveillance of the Armed Forces Health Surveillance Center and its Global Emerging Infections Surveillance and Response System (GEIS), the MRSN has become the lynchpin for the larger DoD response to antimicrobial resistance through the development of the Antimicrobial Resistance Monitoring and Research Program (ARMoR) [86,87]. All of these efforts, resulting from the identification of resistant bacteria in returning ill and injured service members from the wars in Iraq and Afghanistan, have led to significant innovation—innovation that has natural spillover into the wider US health care system. The MRSN, in particular, has no parallel elsewhere. Its example and value clearly transcend the military.

PERSONNEL CONSIDERATIONS: INNOVATION IN EXPANDING THE POOL OF PROVIDERS

Through all of the advances described above, the military has also taken advantage of its most important asset—intelligent and motivated people—to broaden its medical capabilities and responses to urgent needs. While the traditional model of military combat health care has historically been physician- (surgeon) and nurse-centric, this approach

has continued to evolve over the past century to complex systems, relying in large part on escalating echelons of care—including basic "buddy care" (otherwise known as combat lifesaver care), combat medics, and physician assistants. Further, the application of these care extenders into flight medicine and special operations (often in remote and austere environments) has expanded the availability of multiple levels of medical care and effectively broadened the types of military operations possible.

In many ways, one can see parallels of this approach in the community-based emergency medical service organizations that provide urgent and emergency care throughout the US, with upfront care provided by nonmedical personnel, followed by the progressive involvement of emergency medical technicians, paramedics, and others throughout the emergency and trauma care system. In an era of increasing medical care complexity, coupled with an uneven distribution of health care providers, these "modular" substitutions for certain key aspects of emergency medical care—as pioneered by the military—continue to make a positive difference in our modern health care system.

SUMMARY AND CONCLUSIONS

There are many challenges associated with the effective translation of medical advances, innovation, and new knowledge into practice—this in spite of a good recognition of problems and proposals for improvement [1,4,5]. In the end, it is often the clinical urgency (or government edict or incentives) that wins the day in terms of effecting change in a timely manner throughout a large health care system. For example, under the provisions of the Affordable Care Act, the Centers for Medicare and Medicaid Services have a number of incentive programs intended to move practitioners from traditional paper medical records to electronic medical records, an initiative that is accelerating the adoption of this type of technology throughout the US health care system [88].

In the case of operationally-relevant medical research, the urgency of military medical requirements is hard to ignore. When this urgency is coupled with tangible products of medical research—products intended to directly support the provision of medical care in the field—their adoption is likely assured. Further, when these same products—which include knowledge that ultimately informs policies and practice patterns—have relevance to the larger nonmilitary health care industry, their likelihood of adoption, or put another way, their successful translation into standards of care, is almost certain. For this, we owe much to the military medical system, its clinicians in the field, its researchers in the Department of Defense research laboratories, and its medical educators at the Uniformed Services University who, faced with significant medical problems requiring urgent solutions, have truly made a difference for military and civilian medical care alike.

For policy makers and academic health center leaders, the challenge is to frame new innovations and medical knowledge into how they address urgent unmet needs in medicine, analogous to military operational needs, in an effort to spur the most effective translation possible. Clearly, the "Laboratory of War" can serve an important role in advancing and translating medical advances to the wider medical community [89].

Note: The views expressed are those of the authors and do not necessarily represent those of the Department of Defense.

REFERENCES

[1] Berwick DM. Disseminating innovations in health care. JAMA 2003;289:1969–75.

[2] Denis JL, Hebert Y, Langley A, Lozeau D, Trottie LH. Explaining diffusion patterns for complex health care innovations. Health Care Manage Rev 2002;27:60–73.

[3] Sussman S, Valente TW, Rohrbach LA, Skara S, Pentz MA. Translation in the health professions: converting science into action. Eval Health Prof 2006;29:7–32.

[4] Woolf SH. The meaning of translational research and why it matters. JAMA 2008;299:211–3.

[5] Zerhouni EA. Translational and clinical science—time for a new vision. New Engl J Med 2005;353:1621–3.

[6] Spinella PC, Holcomb JB. Resuscitation and transfusion principles for traumatic hemorrhagic shock. Blood Rev 2009;23:231–40.

[7] Kutcher ME, Kornblith LZ, Narayan R, Curd V, Daley AT, Redick BJ, et al. A paradigm shift in trauma resuscitation-evaluation of evolving massive transfusion practices. JAMA Surg 2013;148:834–40.

[8] Tieu BH, Holcomb JB, Schreiber MA. Coagulopathy: its pathophysiology and treatment in the injured patient. World J Surg 2007;31:1055–64.

[9] Borgman MA, Spinella PC, Perkins JG, Grathwohl KW, Repine T, Beekley AC, et al. The ratio of blood products transfused affects mortality in patients receiving massive transfusions at a combat support hospital. J Trauma 2007;63:805–13.

[10] Chandler MH, Roberts M, Sawyer M, Myers G. The US military experience with fresh whole blood during the conflicts in Iraq and Afghanistan. Semin Cardiothorac Vasc Anesth 2012;16:153–9.

[11] Ledgerwood AM, Blaisdell W. Coagulation challenges after severe injury with hemorrhagic shock. J Trauma Acute Care Surg 2012;72:1714–8.

[12] NIH. Fresh frozen plasma: indications and risks. NIH Consensus Statement Online, vol. 5. September 24–26, 1984. [cited 25.03.14]. pp. 1–12.

[13] Ferrara A, MacArthur JD, Wright HK, Modlin IM, McMillen MA. Hypothermia and acidosis worsen coagulopathy in the patient requiring massive transfusion. Am J Surg 1990;160:515–8.

[14] Hess JR, Holcomb JB. Transfusion practice in military trauma. Transfus Med 2008;18:143–50.

[15] Holcomb JB, Jenkins D, Rhee P, Johannigman J, Mahoney P, Mehta S, et al. Damage control resuscitation: directly addressing the early coagulopathy of trauma. J Trauma 2007;52:307–10.

[16] Holcomb JB, Wade CE, Michalek JE, Chisholm GB, Zarzabal LA, Schreiber MA, et al. Increased plasma and platelet to red blood cell ratios improves outcome in 466 massively transfused civilian trauma patients. Ann Surg 2008;248:447–58.

[17] Eastridge B, Jenkins D, Flaherty S, Schiller H, Holcomb JB. Trauma system development in a theater of war: experiences from operation Iraqi Freedom and operation Enduring Freedom. J Trauma 2006;61:1366–72.

[18] Eastridge B, Costanzo G, Spott M, Jenkins D, Wade C, West S, et al. Utilizing a trauma systems approach to benchmark and improve combat casualty care. J Trauma 2010;69:5S–9S.

[19] Joint Trauma System. United States Army Institute of Surgical Research. Available at: http://www.usaisr.amedd.army.mil/joint_trauma_system.html; 2014 [accessed 20.04.14].

[20] Clinical practice guideline for damage control surgery. United States Army Institute of Surgical Research. Available at: http://www.usaisr.amedd.army.mil/assets/cpgs/Damage%20Control%20Resuscitation%20-%201%20Feb%202013.pdf; 2013 [accessed 20.04.14].

[21] Cap AP, Spinella PC, Borgman MA, Blackbourne LH, Perkins JG. Timing and location of blood product transfusion and outcomes in massively transfused combat casualties. J Trauma Acute Care Surg 2012;73(Suppl. 1):S89–94.

[22] Nessen SC, Eastridge BJ, Cronk D, Craig RM, Berséus O, Ellison R, et al. Fresh whole blood use by forward surgical teams in Afghanistan is associated with improved survival compared to component therapy without platelets. Transfusion 2013;53(Suppl. 1):107S–13S.

[23] Kautza BC, Cohen MJ, Cuschieri J, Minei JP, Brackenridge SC, Maier RV, et al. Changes in massive transfusion over time: an early shift in the right direction? J Trauma 2012;72:106–11.

[24] Holcomb JB, Fox EE, Wade CE. The prospective observational multicenter major trauma transfusion (PROMMTT) study. J Trauma Acute Care Surg 2013;75(Suppl. 1):S1–2.

[25] del Junco DJ, Holcomb JB, Fox EE, Brasel KJ, Phelan HA, Bulger EM, et al. Resuscitate early with plasma and platelet or balance blood products gradually: findings from the PROMMTT study. J Trauma Acute Care Surg 2013;75(Suppl. 1):S24–30.

[26] Pragmatic, randomized optimal platelets and plasma ratios (PROPPR). Available at: ClinicalTrials.govhttp://www.usaisr.amedd.army.mil/joint_trauma_system.html; 2013 [accessed 20.04.14].

[27] Hunt BJ. Bleeding and coagulopathies in critical care. N Engl J Med 2014;370:847–59.

[28] Kragh JF, Swan KG, Smith DC, Mabry RL, Blackbourne LH. Historical review of emergency tourniquet use to stop bleeding. Am J Surg 2012;203:242–52.

[29] Jolly DW. Field surgery in total war. New York: Paul B. Hoeber, Inc; 1941. pp. 24–25.

[30] Watson-Jones R. Fractures and joint injuries. London: Livingstone; 1955. PP. 121–122.

[31] Wolff LH, Adkins TF. Tourniquet problems in war injuries. Bull US Army Med Dept 1945;87:77–84.

[32] Walters TJ, Wenke JC, Kauvar DS, McManus JG, Holcomb JB, Baer DG, et al. Effectiveness of self-applied tourniquets in human volunteers. Prehosp Emerg Care 2005;9:416–22.

[33] Beekley AC, Sebesta JA, Blackbourne LH, Herbert GS, Kauvar DS, Baer DG, et al. Prehospital tourniquet use in operation Iraqi Freedom: effect on hemorrhage control and outcomes. J Trauma 2008;64:S28–37.

[34] Kragh JF, Walters TJ, Baer DG, Fox CJ, Wade CE, Salinas J, et al. Survival with emergency tourniquet use to stop bleeding in major limb trauma. Ann Surg 2009;249:1–7.

[35] Welling DR, Burris DG, Hutton JE, Minken SL, Rich NM. A balanced approach to tourniquet use: lessons learned and relearned. J Am Coll Surg 2006;203:106–15.

[36] Lee C, Porter KM, Hodgetts TJ. Toruniquet use in the civilian pre-hopsital setting. Emerg Med J 2007;24:584–7.

[37] Callcut RA, Mell MW. Modern advances in vascular trauma. Surg Clin N Am 2013;93:941–61.

[38] Feliciano DV, Moore FA, Moore EE, West MA, Davis JW, Cocanour CS, et al. Evaluation and management of peripheral vascular injury. Part 1. Western Trauma Association/critical decisions in trauma. J Trauma 2011;70:1551–6.

[39] Sambasian CN, Schreiber MA. Emerging therapies in traumatic hemorrhage control. Curr Opin Crit Care 2009;15:560–8.

[40] Hoge CW, Castro CA, Messer SC, McGurk D, Cotting DI, Koffman RL. Combat duty in Iraq and Afghanistan, mental health problems, and barriers to care. N Engl J Med 2004;351:13–22.

[41] Bliese PD, Thomas JL, McGurk D, McBride S, Castro CA. Mental health advisory teams: a proactive examination of mental health during combat deployments. Int Rev Psychiatry 2011;23:127–34.

[42] Bliese PD, Adler AB, Castro CA. Research-based preventive mental health care strategies in the military. In: Adler AB, Bliese PD, Castro CA, editors. Deployment psychology. American Psychological Association; 2011. pp. 103–24.

[43] Riviere LA, Kendall-Robbins A, McGurk D, Castro CA, Hoge CW. Coming home may hurt: risk factors for mental ill health in US reservists after deployment in Iraq. Br J Psychiatry 2011;198:136–42.

[44] Andreasen NC. Posttraumatic stress disorder: a history and a critique. Ann N Y Acad Sci 2010;1208:67–71.

[45] Wells TS, Miller SC, Adler AB, Engel CC, Smith TC, Fairbank JA. Mental health impact of the Iraq and Afghanistan conflicts: a review of US research, service provision, and progammatic responses. Int Rev Psychiatry 2011;23:144–52.

[46] Ursano RJ, Benedek DM, Engel CC. Trauma-informed care for primary care: the lessons of war. Ann Int Med 2012;157:905–6.

[47] Kroenke K, Spitzer RL, Williams JB, Löwe B. The patient health questionnaire somatic, anxiety, and depressive symptom scales: a systematic review. Gen Hosp Psychiatry 2010;32:345–59.

[48] Hoge CW. Interventions for war-related posttraumatic stress disorder. JAMA 2011;306:549–51.

[49] Hoge CW, Castro CA. Preventing suicides in US service members and veterans. JAMA 2012;308:671–2.

[50] Olson-Madden JH, Brenner LA, Matarazzo BB, Sigoracci GM, Expert Consensus Collaborators. Identification and treatment of TBI and co-occurring psychiatric symptoms among OEF/OIF/OND veterans seeking mental health services within the state of Colorado: establishing consensus for best practices. Community Ment Health J 2013;49:220–9.

[51] Pompili M, Sher L, Serafini G, Forte A, Innmorati M, Dominici G, et al. Posttraumatic stress disorder and suicide risk among veterans. J Nerv Ment Dis 2013;201:802–12.

[52] Jordan NN, Hoge CW, Tobler SK, Wells J, Dydek GJ, Egerton WE. Mental health impact of 9/11 Pentagon attack: validation of a rapid assessment tool. Am J Prev Med 2004;26:284–93.

[53] Friedman MJ, Resick PA, Bryant RA, Strain J, Horowitz M, Spiegel D, et al. Classification of trauma and stressor-related disorders in DSM-5. Depress Anxiety 2011;28:737–49.

[54] North CS, Pfefferbaum B. Mental health response to community disasters: a systematic review. JAMA 2013;310:507–18.

[55] Ursano RJ, McKibben JBA, Resissman DB, Liu X, Wang L, Sampson RJ, et al. Posttraumatic stress disorder and community collective efficacy following the 2004 Florida hurricanes. PLoS ONE 2014;9:e88467.

[56] Ursano RJ, Goldenberg M, Zhang L. Posttraumatic stress disorder and traumatic stress: from bench to bedside, from war to disaster. Ann N Y Acad Sci 2010;1208:72–81.

[57] Milliken CS, Auchterlonie JL, Hoge CW. Longitudinal assessment of mental health problems among active and reserve component soldiers returning from the Iraq war. JAMA 2007;298:2141–8.

[58] Smith TC, Zamorski M, Smith B, Riddle JR, LeardMann CA, Wells TS, et al. The physical and mental health of a large military cohort: baseline functional health status of the millennium cohort. BMC Public Health 2007;7:340.

[59] Ursano RJ, Colpe LJ, Heeringa SG, Kessler RC, Schoenbaum M, Stein MB, et al. The army study to assess risk and resilience in servicemembers (Army STARRS). Psychiatry 2014;77:107–19.

[60] Zapor MJ, Erwin D, Erowele G, Wortmann G. Emergence of multidrug resistance in bacteria and impact on antibiotic expenditure at a major army medical center caring for soldiers wounded in Iraq and Afghanistan. Infect Control Hosp Epidemiol 2008;29:661–3.

[61] Kaspar RL, Griffith MJE, Mann PB, Lehman DJ, Conger NG, Hospenthal DR, et al. Association of bacterial colonization at the time of presentation to a combat support hospital in a combat zone with subsequent 30-day colonization or infection. Mil Med 2009;174:899–903.

[62] Hospenthal DR, Crouch HK, English JF, Leach F, Pool J, Conger NG, et al. Multidrug-resistant bacterial colonization of combat-injured personnel at admission to medical centers after evacuation from Afghanistan and Iraq. J Trauma 2011;71(Suppl. 1):S52–7.

[63] Murray CK, Wilkins K, Molter NC, Yun HC, Dubick MA, Spott MA, et al. Infections in combat casualties during operations Iraqi and Enduring Freedom. J Trauma 2009;66(Suppl. 4):S138–44.

[64] Murray CK, Wilkins K, Molter NC, Li F, Yu L, Spott MA, et al. Infections complicating the care of combat casualties during operations Iraqi Freedom and Enduring Freedom. J Trauma 2011;71(Suppl. 1): S62–73.

[65] Whitman TJ, Qasba SS, Timpone JG, Babel BS, Kasper MR, English JF, et al. Occupational transmission of *Acinetobacter baumannii* from a United States serviceman wounded in Iraq to a health care worker. Clin Infect Dis 2008;47:439–43.

[66] Visca P, Seifert H, Towner KJ. *Acinetobacter* Infection—an emerging threat to human health. IUBMB Life 2011;63:1048–54.

[67] O'Shea MK. *Acinetobacter* in modern warfare. Int J Antimicrob Agents 2012;39:363–75.

[68] Centers for Disease Control and Prevention (CDC). *Acinetobacter baumannii* infections among patients at military medical facilities treating injured U.S. Service members, 2002–2004. MMWR Morb Mortal Wkly Rep 2004;53:1063–6.

[69] Fighting superbugs: DoD's response to multidrug-resistant infections in military treatment facilities. Hearing before the Subcommittee on Oversight and Investigations of the Committee on Armed Services, House of Representatives. HASC No. 111–182. Available at: http://www.gpo.gov/fdsys/pkg/CHRG-111hhrg62994/pdf/CHRG-111hhrg62994.pdf; September 29, 2010 [accessed 30.04.14].

[70] Griffith ME, Gonzalez RS, Holcomb JB, Hospenthal DR, Wortmann GW, Murray CK. Factors associated with recovery of *Acinetobacter baumannii* in a combat support hospital. Infect Control Hosp Epidemiol 2008;29:664–6.

[71] Sutter DE, Bradshaw LU, Simkins LH, Summers AM, Atha M, Elwood RL, et al. High incidence of multidrug-resistant gram-negative bacteria recovered from Afghan patients at a deployed US military hospital. Infect Control Hosp Epidemiol 2011;32:854–60.

[72] Ake J, Scott P, Wortmann G, Huang X, Barber M, Wang Z, et al. Gram-negative multidrug-resistant organism colonization in a US military healthcare facility in Iraq. Infect Control Hosp Epidemiol 2011;32:545–52.

[73] Keen III EF, Murray CK, Robinson BJ, Hospenthal DR, Co E-MA, Aldous WK. Changes in the incidences of multidrug-resistant and extensively drug-resistant organisms isolated in a military medical center. Infect Control Hosp Epidemiol 2010;31:728–32.

[74] Hospenthal DR, Crouch HK, English JF, Leach F, Pool J, Conger NG, et al. Response to infection control challenges in the deployed setting: operations Iraqi and Enduring Freedom. J Trauma 2010;69(Suppl. 1):S94–101.

[75] Crouch HK, Murray CK, Hospenthal DR. Development of a deployment infection control course. Mil Med 2010;175:983–9.

[76] Hospenthal DR, Green AD, Crouch HK, English JF, Pool J, Yun HC, et al. Infection prevention and control in deployed military medical treatment facilities. J Trauma 2011;71(Suppl. 2):S290–8.

[77] Tribble DR, Lloyd B, Weintrob A, Ganesan A, Murray CK, Li P, et al. Antimicrobial prescribing practices following publication of guidelines for the prevention of infections associated with combat-related injuries. J Trauma 2011;71(Suppl. 2):S299–306.

[78] Hospenthal DR, Murray CK, Andersen RC, Bell RB, Calhoun JH, Cancio LC, et al. Guidelines for the prevention of infections associated with combat-related injuries: 2011 update. J Trauma 2011;(Suppl. 2):S210–34.

[79] Sandora TJ, Goldmann DA. Preventing lethal hospital outbreaks of antibiotic-resistant bacteria. N Engl J Med 2012;367:2168–70.

[80] Tribble DR, Conger NG, Fraser S, Gleeson TD, Wilkins K, Antonille T, et al. Infection-associated clinical outcomes in hospitalized medical evacuees after traumatic injury: trauma infectious disease outcome study. J Trauma 2011;71(Suppl. 1):S33–42.

[81] Lesho E, Gleeson T, Summers A, Kirkup B, Chahine M, Babel B, et al. Joint collaboration enhances infection control at home and abroad: the maiden voyage of the multidrug-resistant organism repository and surveillance network. Mil Med 2011;176:241–3.

[82] Lesho E, Craft D, Kirkup BC, Waterman P, Summers A, Vahey MT, et al. Surveillance, characterization, and preservation of multidrug-resistant bacteria. Lancet Infect Dis 2011;11:8–10.

[83] McGann P, Hang J, Clifford RJ, Yang Y, Kwak YI, Kuschner RA, et al. Complete sequence of a novel 178 Kb plasmid carrying *bla*NDM-1 in a *Providencia stuartii* strain isolated in Afghanistan. Antimicrob Agents Chemother 2012;56:1673–9.

[84] Mcgann P, Kwak YI, Summers A, Cummings JF, Waterman PE, Lesho EP. Detection of *qac*A/B in clinical isolates of methicillin-resistant *Staphylococcus aureus* from a regional healthcare network in the eastern United States. Infect Control Hosp Epidemiol 2011;32:1116–9.

[85] Waterman P, Kwak Y, Clifford R, Julius M, Onmus-Leone F, Tsurgeon C, et al. A multidrug-resistance surveillance network: 1 year on. Lancet Infect Dis 2012;12:587–8.

[86] Meyer WG, Pavlin JA, Hospenthal D, Murray CK, Jerke K, Hawksworth A, et al. Antimicrobial resistance surveillance in the AFHSC-GEIS network. BMC Public Health 2011;11 (Suppl. 2):S8.

[87] Lesho EP, Waterman PE, Chukwuma U, McAuliffe K, Neumann C, Julius MD, et al. The antimicrobail resistance monitoring and research (Armor) program: the Department of Defense's response to escalating antimicrobial resistance. Clin Infect Dis Augest 2014;59(3):390–7. http://dx.doi.org/10.6093/cid/civ319.

[88] EHR incentive programs. Centers for Medicare and Medicaid Services. Available at: http://www.cms.gov/Regulations-and-Guidance/Legislation/EHRIncentivePrograms/index.html?redirect=/ehrincentiveprograms/. [accessed 01.05.15].

[89] Elster E, Schoomaker E, Rice C. The laboratory of war: how military trauma care advances are benefiting soldiers and civilians. Available at: http://healthaffairs.org/blog/2013/12/; 2013 [accessed 01.05.15].

ABOUT THE AUTHORS

Charles L. Rice, MD was sworn in as the fifth president of the Uniformed Services University of the Health Sciences in 2005. Before assuming his present position, Dr Rice, a navy-trained surgeon and researcher, served as the vice-chancellor for health affairs at the University of Illinois, Chicago, from 1999 to 2004.

Kent E. Kester, MD is associate vice president and head, Clinical and Translational Sciences at Sanofi Pasteur, a position which followed his completion of a 24-year career in the US Army where he served in a variety of key positions directing malaria vaccine research, conducting clinical vaccine trials, and later, leading the Walter Reed Army Institute of Research.

Chapter 18

How Research Can and Should Inform Public Policy

Claire Pomeroy and Fred Sanfilippo

Never before has the potential for using research discoveries to improve health been greater, and yet we have much to learn about how research can and should inform health-related public policy. The need and opportunity for evidence to replace rhetoric in driving health policy is significant, and yet large gaps remain in the use of research-based evidence to inform health policy. The importance of applying research to policymaking has increased dramatically with the desire to optimize value received for health care expenditures. In the US, the impact of the Patient Protection and Affordable Care Act (ACA) [1] on many financial, regulatory, public health, and clinical practice policies has raised awareness of the importance of using evidence-based approaches in formulating and implementing policy.

The ultimate value of health-related research is to create new information and knowledge that can be translated to improve the care of patients and populations. Unfortunately, it is clear that much of what is done in an attempt to improve patient care and public health is not based on scientific evidence. Even when research provides new knowledge, providers, policymakers, and patients themselves may resist adopting and implementing changes that are based on evidence. Thus, a major challenge in clinical care and public health is that health policies are not adequately informed by the best available evidence, and that research findings may take too long to be utilized in the process of developing policy.

THE USE OF RESEARCH IN DEVELOPING HEALTH POLICY

Fields of Research Used

Many fields of research can be of value in the development and implementation of health policy. These range from basic science to translational and clinical research to the newer fields of health services research (HSR), comparative effectiveness research (CER), implementation and dissemination sciences, and knowledge transfer.

Basic Research

Basic research creates new understandings and technologies that form the basis for fundamental understanding of the mechanisms and manifestations of health and disease. Basic research has driven policymaking on issues ranging from stem cell research and cloning to ownership of genetic sequences. Scientific breakthroughs engendered by basic research provide the foundation for development of new diagnostics and therapeutics, and thus raise key questions about the appropriate application of these new discoveries.

Translational and Clinical Research

Translational research has a variety of definitions, but commonly is considered research that translates new information or knowledge that is created in one area to another application. There are generally two categories of translational research: T1 (basic to clinical) and T2 (clinical to population); however T1, T2, and T3 are often used to represent transitions between to patient-oriented, patient oriented to population-based, and basic to population-based research [2], respectively; a third usage has five categories, which include T0 (basic biomedical research), T1 (clinical translation), T2 (demonstrating efficacy), T3 (translation to practice), and T4 (translation to populations) [3,4].

Results from T1 and T2 research have impacted numerous policies related to bringing new diagnostics and therapeutics through clinical trials, while findings from T3 and T4 research have been effective in informing policy, especially regarding the models and financing of health care delivery in various populations.

Health Services Research

HSR is a growing area of research and training that focuses on the study of health care quality (e.g., outcomes, safety, and satisfaction), access, and cost [5]. It is also considered by many to represent T3 and T4 translational research. Academy Health, the largest professional HSR society, defines HSR simply as the scientific study of health care

The Transformation of Academic Health Centers. http://dx.doi.org/10.1016/B978-0-12-800762-4.00018-9

delivery—asking what works, for whom, at what cost, and under what circumstances [6]. Similarly, the Agency for Health care Research and Quality (AHRQ) defines the goals of HSR as finding the most effective ways to organize, manage, finance, and deliver high-quality care; reduce medical errors; and improve patient safety [7]. In addition to biomedical factors, HSR typically includes other factors that impact health—such as social determinants, the environment, personal behaviors, economics, human resources, and technologies.

Comparative Effectiveness Research

As defined by the Institute of Medicine (IOM), CER is an area of HSR focused on "the generation and synthesis of evidence that compares the benefits and harms of alternative methods to prevent, diagnose, treat, and monitor a clinical condition or to improve the delivery of care" [8]. CER has been important in helping policymakers make informed decisions to improve health care for specific cohorts as well as the general population. Although widely discussed in the US during the debates over the ACA in 2010, CER has been conducted for decades. Indeed, the American Recovery and Reinvestment Act (ARRA) stimulus legislation allocated more than $1 billion for CER in 2009 [9]. During the same year, the IOM issued a report which listed 100 priority research areas [8]. Of particular importance for developing health policy are "pragmatic trials." These studies are a subset of CER designed to measure effectiveness and define benefit; they go beyond the goals of many clinical trials that are focused mostly on whether a treatment works in the defined study population [10].

Implementation Research

Implementation research is the scientific study of methods to promote the uptake of research findings, and hence to reduce inappropriate care. It includes the study of influences on health care professionals' behavior and interventions to enable them to use research findings more effectively [11]. Biomedical research produces new findings, but these are too often not translated into practice or policy. Often, research projects focus on small-scale pilot studies or laboratory-based experiments and assume that findings can be generalized to roll out into practice and policy with few changes. Implementation research addresses this problem by developing effective interventions to translate research findings into "real-world" changes.

Knowledge Transfer

The interfaces between research and policy and practice are poorly understood; there are multiple barriers to translating knowledge to policy; that is, the "know-do" gap [12]. Knowledge transfer (KT) is a field of study dedicated to moving research into practice and policy [13]. As defined by the World Health Organization (WHO), KT is the "synthesis, exchange, and application of knowledge" to improve health [14]. Evidence-based medicine represents a subset of health-related knowledge transfer in which the results of scientific studies are used to make health care decisions, usually those involving the care of individual patients. Less widely discussed is the use of research and the resulting evidence to create broader health-related policies.

KT has most commonly focused on technical implementation of scientific findings in a context absent of political, economic, social, and cultural realities. These "classic linear models of KT are often poor predictors of research uptake" [15] into policy decisions because of other factors that frequently influence and occasionally overwhelm the scientific findings [16]. Policymakers must make decisions that affect diverse individuals and communities; thus, they are required to consider myriad influences in their interpretation and application of the scientific evidence.

Types of Policies That Can Use Research Evidence

The ways that research can and should inform health policy will depend upon the type of policy, and whether the policy is one of practice or law. Some have characterized guidelines, management, and business decisions as "small p" policies and formal laws, rules, and regulations as "big P" policies [17].

Evidence is used by provider institutions, payers, and professional societies to recommend "small p" policies for various patient cohorts, geographically defined populations, and/or society as a whole. Large hospitals, health maintenance organizations, and, more recently, accountable care organizations (ACOs) establish policies to manage quality, safety, and cost of services to the patient populations for whom they are responsible. While these policies may be based on expert consensus or other considerations, if they are based on research evidence reported in the literature they may be more widely accepted. For example, Infection Control Committees in many hospitals evaluate the research literature and make policy recommendations about requirements ranging from hand-washing to mandated vaccines to first-line antibiotic options. Pushback against such policies by hospital staff may be mitigated by providing evidence documenting the beneficial impact on patients. Similarly, many payers and provider institutions evaluate the literature and make recommendations about which medications, medical devices, and equipment should be used in the clinical care of patients for whom they are responsible.

The use of evidence to set clinical practice policy is generally done institution by institution and payer by payer in the US; however, in many European and Asian countries government or other oversight agencies make nation-wide decisions.

In the US, professional societies have recently advocated for the Choosing Wisely campaign [18]—a list of several procedures for which there is insufficient evidence to justify use that was distributed to health professionals and to the general public. While this generated important attention and is likely to change some clinicians' practices, the recommendations lacked the enforcement power of law or government policy.

"Big P" policies may be set at the local, state, or national level. Research evidence may be one of the factors used to make policy, but other considerations often result in significant variations in policy. For example, local governments have taken very different stances on the use of needle exchange programs—documented in many studies to decrease transmission of HIV [19]. Some cities and locales have enacted public health policies in support of such programs, while others have said that local considerations preclude them from endorsing this activity. Research evidence also serves as the basis for many US federal health policies, such as requirements for new drug approvals and determining which health services will be reimbursed by Medicare or provided by the Department of Veterans Affairs (VA). In practice, the reimbursement policy guidelines issued by the Centers for Medicare & Medicaid Services for Medicare (CMS) tend to be adopted by other payers and ultimately become general practice across the country.

The WHO has emphasized the importance of research as a basis for health system design and policymaking around the globe. WHO policy recommendations in response to wide-ranging health issues—from infectious disease outbreaks to maternal child health—attempt to use available research evidence as the basis for policy, while recognizing that the policies must be sensitive to local, cultural, financial, and logistical considerations.

Approaches to Using Research to Inform Health Policy

Research can be used by organizations to guide policies regarding their programs and recommendations, and by legislators to guide the processes of legislation, appropriation, and oversight (Table 1).

TABLE 1 Uses of Research Findings in Health Policy

- Define policy priorities and strategies
- Draft policies and legislation
- Evaluate proposed policies and legislation
- Monitor quality of programs created by policies and legislation
- Assess outcomes and impact of policies and legislation
- Determine cost-effectiveness of policies and legislation

The use of evidence in policymaking is quite different in countries with different political, economic, and cultural values and structures. In many European countries, a more centralized approach to health care policy, delivery, and reimbursement has led to the use of "health technology assessments," created by formally empowered entities and which serve as the basis for country-wide decisions about publicly financed health interventions. In contrast, in the US a more decentralized approach [20] often leaves health policy decisions to multiple, potentially nonaligned groups. As some have said, in the US, "policy is made through ad-hoc issue specific coalitions" [21]. In this case, research is often used as "legitimation or ammunition to justify policy decisions and defend them against opponents' criticism" [21]. Indeed, recent federal legislation forbids Medicare from using CER to make cost-based decisions about health care services [22]. The state by state differences in coverage for US Medicaid services highlight this as well. Finally, in less-developed nations, lack of resources and technical capacity may limit the ability to create, access, and utilize evidence in the creation of health policy. The WHO created EVIPNet (Evidence Informed Policy Network) to foster a systems approach to better link policy and research, especially in developing countries.

Defining Policy Priority Areas

Researchers may be unaware of the areas in which policymakers would desire additional evidence. To better understand the knowledge gaps which research could help fill, interaction between researchers and policymakers enhances insight into the priority areas for policy. Bidirectional discussions characterized by mutual trust and respect can better align research projects with policy priorities.

Systematic Reviews

A major advance in using evidence as a basis for making health policy decisions was the recognition of the importance and power of systematic reviews. Understanding that policymaking based on evidence from a single study is fraught with problems, meta-analysts review the literature to combine findings from multiple studies. For example, the development of the Cochrane Reviews in the 1990s provided impartial compilations of evidence that are widely appreciated as facilitators of evidence-based decision-making [23]. The US Council of State Governments has issued a State Policy Guide on using research in public health policymaking [24]. They emphasize to their constituency of policymakers that the "strongest evidence is obtained from summaries of multiple research studies called systematic reviews" [24].

Of course, systematic reviews do not solve all problems in defining the "best" evidence. Optimal use of systematic reviews by policymakers requires that the systematic review itself is well done and effectively explained. The review's effectiveness is dependent upon the quality and reliability of the data being collated. Attention must be paid to the credibility of the people doing the systematic reviews and the processes they use. Systematic reviews summarizing trials that are flawed or address the wrong questions are likely to reinforce the wrong messages. Unless meta-analysts are cognizant of the potential problems in study design, the methods for the generation of data, and the accuracy and completeness of the publication of results, they may end up further amplifying biased evidence [25]. For example, if the literature is dominated by studies comparing surgical approaches to treat a specific condition, they may fail to reveal that the ideal treatment is a medical or lifestyle intervention. To help address this, a move from summative reviews to interpretive synthesis has been advocated [26].

"Clearing Houses" to Synthesize and Disseminate Research Evidence

Once high-quality evidence has been generated and synthesized, there must be unbiased and effective "user-friendly" mechanisms for policymakers to receive the results. The amount of data available to policymakers can be overwhelming, and it is essential that it be presented in a way that is understandable with reasonable time and effort. One example of such a "receptor" organization for health research is the National Institute for Health and Clinical Excellence (NICE) in the UK [27,28].

To help ensure that research evidence optimally informs health policy, broad discussion, debate, analysis, and access to the study design and its findings are all important. In some cases, this occurs through the policy side—as with the Federal Register; in other cases, it is through the research side—as with professional journals; and in others, it is by direct engagement of the public through mainstream and social media. All of these may play "clearing-house" roles, with varying levels of quality and oversight.

Assessing the Impact of Research on Policy

To optimize the effectiveness of evidence in policymaking, it is important to understand how research does and does not affect policy, including its development and implementation [29]. Most experts agree that assessment is most likely to show an impact when analyzed forward; that is, by starting with the research and looking at the effect on policy, rather than by trying to start with policy and tracing backward to the research [12]. To truly assess the impact of research on policy, it is also necessary to research the processes of policy design and implementation [17].

CONSIDERATIONS IN USING RESEARCH TO INFORM POLICY

Researcher versus Policymaker Perspectives

One of the challenges in using research to inform policy is that researchers and policymakers come from different professional cultures, operate in different venues, and bring distinct styles and values to their decision making. Some have discussed the "practical-operational orientation" of policymakers for whom evidence is "context-based and defined less by its quality and more by its relevance, applicability, or generalizability to a specific context" [30]. This contrasts with researchers who often have a more academic, theoretical, and/or discipline-driven approach, referred to as a "philosophical-normative orientation" [30].

Researchers and policymakers are likely to have very different perspectives about what evidence is needed and how that evidence should be used. In quantitative medical research, problems are typically framed with a focused hypothesis and explicit parameters, while the policymaker must work in a "messier" and more complex environment [31]. It can be very difficult to "generalize" from the researcher's focused and parameter-defined area of study to the more context-dependent nature of a policymaker's decision making. For researchers, a well-defined issue with a testable hypothesis and answers that can be reproduced are desirable. In contrast, policymakers work in a world of complicated agendas and diverse determinants of feasibility and acceptance.

One way to address these differences is to move from purely researcher-driven processes, which summarize research, to "co-production processes" [26] that allow policymakers and researchers to interpret implications together. To accomplish this, it will be important to invest more resources and time into bridging the cultural and intellectual gaps between policymaking and research. Sustained interaction between scientists and policymakers is the key to "unleashing the value of science for the policy process" [27].

Evidence-Based versus Evidence-Informed Health Policy

The term "evidence-based medicine" has been used to describe the application of evidence to individual patient management decisions. It is clear that the development of policy also requires evidence, but, in addition, information is needed about the social, political, cultural, and economic context of the issues under consideration [21]. The complexity of the policymaking process requires that policymakers evaluate competing desires and outcomes in ways that are accountable to the range of individuals and constituencies. Health policy should not only

TABLE 2 Factors to Consider in Health Policy Making in Addition to Evidence

- Social and cultural context
- Political realities
- Ideology
- Traditions
- Economic situation
- Feasibility
- Public acceptance
- Media
- Community priorities
- Available resources
- Community attributes and capacities
- Trade-offs
- Timing

use evidence focused on quality, access, and/or cost, but also must be informed by other key determinants such as available resources, community needs and priorities, perceived value, timing, culture, and feasibility (Table 2). In other words, effective health policy will be "evidence-informed," taking into account evidence in a particular context and community, rather than exclusively based on the results of scientific research [32].

To state that only research-derived evidence should guide policy decisions risks a dogmatic approach blind to the complex realities of those being impacted. It has been pointed out that adoption of oversimplified concepts of evidence-based policy without consideration of local realities is not conducive to good governance [21].

Limitations and Issues in Using Research to Inform Health Policy

Movement of research findings into policy can be slow, and evidence-informed policies face a number of challenges and barriers. Problems with the evidence itself and challenges created by the processes with which the evidence is used must be overcome to achieve the goal of ideal evidence-informed health policy.

Quality, Reproducibility, and Generalizability

Wide variations in the quality of research clearly exist. Pressures to publish in academia and "bottom-line" pressures in industry have raised ongoing concerns that poorly-designed studies and nonreproducible results are not always identified prior to publication or even after. Recent reports indicate that a disturbing number of studies contain data that is not reproducible [33].

The difficulty in generalizing or extrapolating research findings to populations beyond defined study groups is a major challenge for both researchers and policymakers. While high-quality evidence may be generated using individuals who meet the researchers' exclusion and inclusion criteria, it may be difficult for policymakers to discern if and how the evidence applies to their communities. Well-documented instances of interventions that work in academic settings but not in "real-world" settings have appropriately prompted the call for more community-based research. Furthermore, high-quality evidence of effectiveness in one geographic locale, racial or ethnic group, or other subpopulation cannot be automatically applied to all groups. Yet, it is also not feasible to reproduce studies with every combination of parameters. Researchers can help address this by including diverse subjects in their study population, but policymakers will nevertheless be left with the need to interpret the implications of the research evidence for the "less controlled" populations that they serve.

Statistical Caveats

The flawed use of statistical analysis in many research studies, whether due to lack of researcher knowledge or deliberate misrepresentation, is a source of significant concern. Policymakers may have difficulty understanding specific statistical methodologies and thus in determining the appropriateness of using the study as a basis for policy. Unfortunately, tools such as randomization do not necessarily ensure generalizability. Even when significant differences are found between groups, the results may not apply to every member of that group due to individual variation. Thus, policy decisions based on research evidence may have unintended and potentially deleterious impacts on some individuals.

Flawed Use of the Evidence

Many factors influence the ways in which evidence is used in health policy [33]. Inappropriate influence can be due to selective use of evidence and conflicts of interest [34]. Selective use of evidence is a distressingly common flaws. Specific groups may seek to influence policy in a manner that is advantageous to their constituency, even if it is potentially harmful to others [35]. Unfortunately, outright manipulation or fabrication of results has been linked to conflicts of interest. Research and evidence from special interest groups should be subject to sufficient scrutiny to ensure that bias in design, performance, and/or interpretation of study results has not occurred.

Objectives of the wider political agenda may influence the ways in which evidence is used. Political pressures may encourage selective use of evidence to support predetermined policy choices or ideological positions. Personal agendas, cultural values, and nationalist perspectives can bias selection or interpretation of evidence used for health policy. Evidence will not change strong ideological support for or opposition to policy positions, but it can be used by the open-minded to make more informed decisions.

ROLES OF ORGANIZATIONS IN ADVANCING EVIDENCE-INFORMED HEALTH POLICY

Given the multiple challenges facing evidence-informed health policy, there has been considerable interest by many segments of the research and policymaking sectors to enhance and facilitate research that can better inform policy. On the policy side, this includes government agencies and foundations with interest in policy, while on the research side this includes consulting organizations and academia that conduct contract or independent research.

Government Agencies

Government agencies at the federal, state, and local levels have significant involvement in setting health policy decisions. On the national level in the US, these range from the approval of diagnostics and therapeutics through the Food and Drug Administration (FDA), to health standards issued through the Centers for Disease Control and Prevention (CDC), the Environmental Protection Agency (EPA), the Occupational Safety and Health Administration (OSHA), and the US Department of Agriculture (USDA), to the criteria for reimbursement of health care-related activities through CMS, the Department of Defense (DOD), the Health Resources and Services Administration (HRSA), and the VA. Each of these agencies support research in the areas in which they have regulatory authority. Moreover, a number of other agencies—such as AHRQ, the National Institutes of Health (NIH), and the National Science Foundation (NSF)—support research, much of which can be used to inform health policy decisions [36].

Support for Health Research

Funding by government agencies for biomedical research has been flat or in decline, raising concerns that evidence may not be available for future policy decisions [37]. Support for research in health related services has been identified as a priority at some federal agencies. For example, in 2005 the NIH created a centralized online system for health services research (HSR Information Central) that provides information on HSR activity and funding opportunities across all government agencies [5]. More recently, the NIH created the National Center for Advancing Translational Research (NCATS), which is directing more of its efforts in translational research and support of Clinical and Translational Science Awards (CTSAs) to health policy related studies. The VA has also developed an extensive program in HSR, with the creation of more than 20 Health Services Research & Development (HSR&D) centers across the country.

As part of the ACA, the CMS Center for Medicare and Medicaid Innovation (CMMI), and the independent Patient Centered Outcomes Research Institute (PCORI, see below) were created with new and significant funding streams. Both organizations support research that impacts policy at several levels. For CMMI, the research goals are to identify practices and consequent policies that could be scaled to reduce costs and increase value for CMS. For PCORI, the research goals focus on patient-centered outcomes and are directed toward identifying better quality and access to health care. While intended to avoid questions of costs, the knowledge gained from PCORI research will ultimately help inform many health policies.

Support for HSR Education

In addition to the increase in government funding of HSR, there is also an increasing awareness of the importance of federal support from the NIH and VA to train scientists to develop expertise in these areas [38]. Currently, 18 institutions have NIH T32 grants supporting pre- and postdoctoral training programs in HSR, and 60 institutions have clinical research training programs supported through their CTSA grants. Likewise, the VA supports 15 fellowship programs through its HSR&D centers.

Translation of Findings to Policy

In addition to supporting research that can be used to inform health policy, government agencies play an important role in translating evidence to regulations. One important role of government in this regard has been to facilitate access to data. For example, legislators have mandated that CMS provide researchers with ways to access and utilize information available in Medicare and other databases. The development of the Multi-Payer Claims Database facilitates access to information for research involving a broader array of patient groups [22]. At the state level, access to public health and Medicaid claims data vary, and unfortunately in some cases is quite difficult or expensive to acquire.

Recognizing that there needs to be a balance between evidence and the urgent need for patients to receive new therapies, the US government has instituted new procedures that allow "fast tracking" of drug approvals in cases of promising evidence and high potential benefit to patients without other options [39]. Highlighted in the early days of the HIV epidemic, patients with diseases for which no established therapy is effective, are willing to try new drugs before all the traditional requirements of evidence for approval is available.

NGOs, Consulting Groups, and Foundations

A wide range of private organizations have interests in the research and/or policy aspects of health, and some are specifically focused on the issues around developing evidence-informed policy. These include nongovernmental organizations (NGOs) that support studies and activities to help provide evidence-informed health policy, consulting companies that engage in contract as well as in independent research in policy areas, and foundations and professional societies that support policy-related research as a primary function.

Institute of Medicine

The IOM is an independent, nonprofit organization that works outside of government to provide unbiased and authoritative advice to decision makers and the public. Many of the studies that the IOM undertakes begin as specific mandates from Congress; still others are requested by federal agencies and independent organizations [40]. The IOM utilizes its standing committees, as well as forums, roundtables, and expert consensus committees to facilitate discussion, discovery, and critical, cross-disciplinary thinking in order to provide evidence-based advice. In addition, the IOM often plays the critical role of an impartial clearinghouse to provide quality control of research results and the process of synthesizing those results.

Patient-Centered Outcomes Research Institute

PCORI is a public/private entity charged with identifying research priorities, establishing an agenda for, and funding appropriate patient-centered CER studies. The ACA created PCORI to promote and fund HSR and CER to ensure that research addressed and improved real-world health care decisions. CER became highly politicized during the debate over the ACA. Indeed, limitations were imposed by the Congress on PCORI's ability to influence federal payment policies. Overall, the ACA reflects political resistance to binding policy too closely to CER [41]. The political controversy over CER appears to stem from fundamental ideological differences in what is considered the appropriate role of government, but also in part from a recognition that studies could challenge the usefulness of common treatments, as well as the income of drug companies, device firms, and provider groups [42]. Moreover, while the public understands that research helps identify the relative effectiveness and efficiency of different treatments, they are also concerned that using CER to mandate clinical practices or resource allocation would limit choices or threaten the doctor–patient relationship.

Consulting Groups

Government, corporate, and other parties interested in a particular policy issue will often not have the expertise or capacity to conduct the necessary studies to help inform the policy decision-making process. In such cases, for-profit consulting groups are often engaged, typically through RFA processes with contracts to conduct defined, targeted studies. Many such firms have focused expertise on narrow areas of research and/or health policy, while others have significant multidisciplinary capacity and a broad portfolio of interests with a multinational scope.

Foundations and Professional Societies

A number of foundations and professional societies have a major interest in health and support research studies that can help inform health policy. In addition to supporting research through grants, foundations such as the Robert Wood Johnson Foundation, Kaiser Family Foundation, and others provide extensive information daily on topics related to health care delivery and health policy through a wide range of media. Professional societies such as Academy Health and the American Medical Association (AMA), and institutional member organizations such as the Association of Academic Health Centers (AAHC), American Hospital Association (AHA), Association of American Medical Colleges (AAMC), and University Health System Consortium (UHC) are focused on health care and actively engage in generating and utilizing evidence to help inform legislators on policy issues.

Universities and Academic Health Centers

Universities, in general, and academic health centers (AHCs) in particular, are well-positioned to play key roles in advancing evidence-informed health policy (Table 3). As respected thought leaders, AHCs can act as conveners to

TABLE 3 Role of Universities and AHCs in Advancing Evidence-Informed Policy

- Convene researchers and policymakers
- Link the public to research and to policy
- Provide multi-sector expertise to address complex policy issues
- Review the research portfolio of AHCs to identify areas of expertise useful to policy makers
- Enhance ability of AHCs media relations to communicate research evidence to policy makers
- Encourage government relations staff to facilitate exchange between researchers and policy makers
- Use evidence to design and implement health policies within their own institution
- Encourage professional societies/membership organizations to support evidence-informed health policy

bring together researchers and policymakers. By virtue of their role in the community, AHCs can facilitate connections between the general public, researchers, and policymakers to learn more about community priorities for health research and policy and to develop networks to facilitate acceptance and uptake of policy decisions [43].

Universities and AHCs are in a unique position to bring together the multi-sector expertise necessary to do the research needed to inform health policy [44]. AHCs often have a range of health sciences faculty in medicine, public health, allied health, nursing, pharmacy, and other health professions. The university provides an environment where these experts in the health sciences can be brought together with experts in law, social sciences, education, industrial and systems engineering, business, and other disciplines, to provide a multidisciplinary perspective and collaboration to address complex policy decisions. In some cases, universities have institutes or centers of policy that can also act as an organizing entity for such multidisciplinary collaboration.

AHCs can provide connections between research and policy issues involving government, NGOs, foundations, and health care providers. A good example has been the Durham Community Health Network (DCHN), a public–private partnership that was developed by Duke University, Durham County, and the state of North Carolina to provide care management for Medicaid and CHIP enrollees. Several studies and projects of the DCHN, such as the Medicaid In-Home Aide Service, have led to improved clinical practice, with state mandated changes in policy and substantial financial savings [45].

Another good example of AHC engagement in policy has been in recent studies and policies regarding nursing education and scope of practice. In 2008, the IOM and the Robert Wood Johnson Foundation partnered to study the evidence and produce a report on the future of nursing [46]. The report concluded that nurses should practice to the full extent of their education and training, achieve higher levels of education in programs that promote seamless academic progression, and be full partners in multi-professional, team-based approaches to redesigning US health care. The report also emphasized the importance of better data collection and information infrastructure to enable nursing workforce planning and policymaking. Many changes have occurred in nursing training programs as a result of the report. In addition, several states have been prompted to review their laws and regulations related to nursing, especially as regards to the scope of practice.

CASE EXAMPLES: EVIDENCE-INFORMED HEALTH POLICIES

Patient Care

Thalidomide

The story of thalidomide highlights the importance of using research evidence to make good decisions about new therapeutics [47]. Developed by a German pharmaceutical company, thalidomide was marketed as a panacea for a variety of conditions—including morning sickness in pregnant women—after limited animal testing and no clinical trials. Even after reports of devastating birth defects, including phocomelia in the babies of women who had taken the drug, the company continued to market the drug. When application was made to the FDA to expand its use in the US, one young official, Dr. Francis Kelsey, deemed the lack of research evidence to be unacceptable and resisted multiple attempts by the company to get approval. As a result, cases of phocomelia in the US were limited and Dr. Kelsey subsequently received the 1962 President's Award for Distinguished Federal Civilian Service. The thalidomide tragedy prompted the US and other countries to strengthen the rules for testing drugs in clinical trials prior to receiving marketing approval.

However, the memories of the devastating effects of thalidomide also resulted in delays in studies that have demonstrated potential beneficial effects of the drug. Today, thalidomide is approved for treatment of some cancers and is being studied as an immune-modulatory drug in patients with HIV and with autoimmune diseases. The painful lessons of thalidomide highlight the importance of using research evidence in making policy decisions about new drug approvals, as well as the challenges of ensuring that regulatory requirements do not interfere with or prevent the development of beneficial drug uses.

Improved Outcomes and Equitable Access in Organ Transplantation

The success of cadaveric organ kidney transplantation in the US rapidly led to a significant shortage of donor organs by 1980, as well as a significant debate as to the criteria that should be used for organ sharing. Studies at the time clearly showed several factors had detrimental effects on graft outcome, such as ABO and HLA (Human Leukocyte Antigens) incompatibility, warm ischemia time of the donor kidney, and immune sensitization from prior graft loss [48]. To address these issues, several groups of transplant centers coordinated organ sharing with common criteria, while many individual centers used their own criteria for sharing organs.

To improve the organ matching and the placement processes, the US Congress passed the National Organ Transplant Act (NOTA) in 1984. The Act established the Organ Procurement and Transplantation Network (OPTN) to maintain a national registry for organ matching [49]. The Act also called for the network to be operated by a private, nonprofit organization under federal contract with HRSA. Medicare reimbursement was then tied to compliance with OPTN rules, and in 1987 all solid organ transplant programs were mandated to provide donor and recipient data to the Scientific Registry of Transplant Recipients (SRTR).

These data have been used extensively for routine review of clinical outcomes by comparing predicted to actual outcomes, which in turn has informed policies on reimbursement. The results of these studies also showed disparate waiting times for some patient cohorts, especially African-American patients, based on a variety of factors—including factors grounded in the distribution algorithms [50]. These findings have led to ongoing changes in organ distribution policies, which are now continually updated based on the available evidence and input from patients, providers, scientists, and policymakers.

Screening Mammograms

Mammograms have become a mainstay of preventive care for women in the US, and proponents include both medical experts and women who tell stories of their lives being saved by the diagnostic modality. Yet, recurrent controversies have arisen about the age recommendations for screening and the true effectiveness of the procedure [51].

For many years, it was recommended that annual screening start at age 40 years. In 2009, the US Preventive Services Task Force revised their guidelines and did not advocate routine screening for women aged 40 to 49 years, suggesting that the harm–benefit ratio did not justify routine mass screening in this age group. There was an immediate outcry from advocacy groups and especially from patients whose cancer had been diagnosed by mammography performed when they were younger than 50 years. While the Task Force experts tried to respond that their recommendations resulted from the evidence and did not preclude individual women from deciding in consultation with their physician to be screened, their voices were not often heard over the simplified headlines.

Most recently, studies have questioned the value of nearly all mass screening by mammography [52]. A Cochrane review in 2006 concluded that it was not clear whether screening does more good than harm and recommended that women should be fully informed before electing to undergo the procedure.

Different interpretations of the data regarding mammogram screening and the conflicting conclusions can be very confusing [53]. The ability of government, payers, advocacy groups, and professional organizations to formulate policy recommendations is made very difficult by the lack of consensus. The controversies around the age limits and effectiveness of screening mammography highlight the importance of reproducible data, the key role of systematic reviews, the critical position of clearing houses for evidence, and the challenges in communicating evidence to policymakers, as well as professionals and the public.

Vaccines/Autism

Awareness of autism has skyrocketed in recent years, and parents desperately search for explanations when their child is diagnosed. In 1998, a study reported by Dr Andrew Wakefield posited a link between the measles–mumps–rubella vaccine and autism in a group of 12 children. Although the paper was subsequently retracted by the journal, and an investigation concluded that there had been deliberate fraud and manipulation of the data [54], many parents continue to believe that vaccines play a causal role in autism and demand changes in policies that mandate vaccination, such as those for school entry.

The CDC, the IOM, the American Academy of Pediatrics, the UK National Health Service, and the WHO have all concluded that there is no link between vaccines and autism. Yet, many parents still do not allow their children to be vaccinated because of the fear of autism. Despite the multiple subsequent studies which failed to show a connection between vaccination and autism, children are being put at risk of preventable infectious disease by well-meaning parents who avoid vaccines.

This controversy highlights the challenges in the role that traditional and social media play in providing information about complex health issues. As quoted on one Web site [55], "In the absence of any answers from the scientific community, any scintilla of suggestion is going to get magnified by the social process of talking it out." As another Web site [56] cautioned "There are passionate people on each side…. Not all of their research is of equal quality, however, so it's important to go beyond the media…and look for credible, well-designed research studies."

This example highlights the negative effects that poor quality and/or fraudulent research can have on public opinion and the problem that it poses for policy. The attempts by respected review groups to provide accurate interpretation have been important, but are often not as influential as a media-savvy celebrity opinion leader's stories. The controversy also emphasizes the importance of ongoing research to provide evidence that is accurate and to explore alternative explanations for the causes of autism.

Public Health

Antismoking Initiatives

Tobacco remains the single most important cause of preventable death worldwide, and evidence-informed policy has resulted in decreased cigarette smoking rates in the US, which has saved millions of lives. The US Surgeon General's 1964 report decisively linked smoking to cancer and other diseases, prompting multiple antismoking policy initiatives. New laws mandate higher tobacco taxes, restrictions on cigarette advertising to minors, warning labels on cigarette packages, nonsmoking workplaces and public settings, and public funding for programs that support smoking cessation. Hospitals no longer sell cigarettes, and recent announcements by pharmacy chains that they will discontinue cigarette sales have emphasized that cigarettes should not be part of the business plan for providers of health-related products and services. Fifty years after the Surgeon General's report, the smoking rate in the US has decreased significantly.

The multitude of lives saved by antismoking policies based on research evidence is a dramatic example of the importance of evidence-informed policymaking. But there is still work to be done. Many Americans still live in areas with smoke-free laws for workplaces and restaurants. Many young people choose to smoke despite the evidence. Disparities in smoking rates due to race, ethnicity, education, and income persist.

Disturbingly, global inequities are also growing with disproportionate increases in cigarette smoking rates in less-developed countries. For example, China is home to about a third of the world's smokers, with smoking rates among men at nearly 50%. A recent study estimated that almost 13 million lives could be saved by 2050 in China if the country implemented the comprehensive tobacco control recommendations from the WHO [57]. Some countries aspire to become tobacco free, for example New Zealand by 2025 and Finland by 2040.

Folate-Enriched Foods to Reduce Neural Tube Defects

Neural tube defects (NTDs), including spina bifida and anencephaly, are devastating congenital abnormalities that incur great medical costs to society and personal suffering to affected patients and families [58]. Policies now require folate enrichment of flour and other grain foods, and there has been a consequent dramatic decrease in the incidence of this condition in the US [59].

A 1991 study reported that women who had previously borne a child with a NTD experienced a 72% risk reduction when they took periconceptual folate supplements [58]. Very shortly thereafter, the CDC advised high-risk women in the US to take folic acid before conception and through the first trimester. By 1992, the CDC recommended supplementation for all women of reproductive age. This guideline was also endorsed by a number of medical professional societies, foundations, and the IOM.

However, it was clear that folate was needed early, even before the woman knew she was pregnant. As a result, in 1996 the FDA mandated folate fortification of enriched breads, cereals, flour, and other grain products, with full implementation completed by 1998. Since then, NTDs in the US have decreased by 20–30%, preventing an estimated 1000 NTDs each year.

This public health policy demonstrates the power of research to inform public policy. It also demonstrates that policymakers must go beyond the evidence and consider human behavioral factors when designing policy. While the evidence showed that periconceptual folic acid was of benefit, the practical considerations of women taking the supplements before they knew they were pregnant required a more creative response. Fortification of grain products accomplished this goal.

Fluoridation of Water

By the 1930s and 40s, research had established the beneficial effect of fluoride in preventing tooth decay. Subsequent investigations documented that protection could be achieved through topical application. On the basis of this research, in 1951 the US Public Health Service adopted a policy supporting fluoridation of community water supplies. By 1960, fluoridation was widely used in the US, significantly reducing the incidence of tooth decay and cavities. The fluoridation of public water has been cited by the CDC as one of the top health achievements of the twentieth century [60].

Despite the evidence, controversy has characterized this intervention since its initial adoption. Early on, some claimed that fluoridation was a communist plot to ruin the water supply and adversely affect health. This argument was particularly effective in the McCarthy era when theories of Communist conspiracies were frequent. Others object to fluoridation because they view it as a government-mandated medical treatment which violates their autonomy and to which they have not assented. Still others consider fluoride to be a poison and claim that exposure violates their rights. While exposure to very high levels of fluoride can have adverse effects, the research evidence documents that the levels used in the US are safe.

In Europe, fluoridation of water supplies is no longer used in most countries, and substantial declines in childhood cavities have nevertheless occurred. Some have speculated that this is due to the provision of topical fluoride treatments in government-sponsored child dental programs, which are not widely available in the US. This highlights the need for public health policy decisions to take into account the larger context of the community and emphasizes the importance of asking the "right" question. In other words, it is important to not only investigate the effectiveness of water supply fluoridation, but also to do comparative studies of other potential interventions, such as widely accessible topical fluoride treatments.

RECOMMENDATIONS TO ENHANCE EVIDENCE-INFORMED HEALTH POLICY

To maximize the translation of research evidence to health policy and to overcome the barriers to evidence-informed health policy, attention should be paid to the needs of researchers and policymakers, as well as to the systems that support policymaking. Both policymakers and researchers should understand that coming together in formal and informal settings can develop relationships that help increase mutual respect and trust. Several recommendations to enhance the effective use of research evidence in developing and implementing health policies are provided in Table 4.

TABLE 4 Approaches to Enhancing Evidence-Informed Health Policy

- Strengthen demand from policymakers for evidence
- Enhance knowledge of policymakers about research methods and processes
- Facilitate and enhance dialogue between researchers and policymakers
- Support systematic reviews and clearing house functions to validate credibility
- Create infrastructure to receive, review, and disseminate research evidence and policy makers
- Increase researcher awareness of evidence needed for health policy
- Increase ability to "package" research results in accessible form useful for policy making
- Support research of health policy implementation processes
- Celebrate successful evidence-informed health policy

Recommendations for Policymakers

Strengthening demand from policymakers for research evidence is the key. By creating processes and venues for policymakers and their staff to interact with researchers, better communication and understanding can occur, helping to overcome differences in jargon or context. More interaction with researchers can also help build mutual trust and foster interest from policymakers in research results that might guide their policy decisions.

In order for policymakers to take full advantage of research-generated evidence, opportunities should be provided to policymakers and their staff to learn about research methods and processes [61]. By enhancing the technical knowledge of policymakers to understand and interpret research, including its potential benefits and limitations, policymakers will be able to make the most appropriate use of scientific evidence. Policymakers will be best served if they are able to obtain data from multiple sources and from researchers without personal or political agendas.

Policymakers should establish relationships with researchers that are stewarded over time, so that when evidence is needed quickly to address policy questions, it can be provided in a most timely manner possible. Policymakers should then attempt to obtain data as early as possible in the formulation of policy and use it to define their positions, rather than searching for data that support decisions that have already been made.

Recommendations for Researchers

Researchers who are interested in contributing to evidence-informed policy must work to understand the issues that are of importance to policymakers. While evidence can generate interest in a new focus area for policymakers, most often the issue is already a priority for the policymaker. By understanding these priorities and agendas, researchers can explore issues that are most important to the policymaker and increase the likelihood that the evidence they generate will be used to guide the policymaking. A collaborative approach with policymakers in setting the research agenda will increase the likelihood that evidence will be used to shape policy.

Researchers must present the results of their work in a way that it is accessible and easy to understand. Researchers should avoid use of jargon and summarize key "take-home" points. In addition, researchers need to understand that evidence needs to be provided quickly to meet the timelines of policymakers.

Researchers need to strive to ensure that the body of evidence is of the highest quality, free of biases and inappropriate influences. Results need to be reproducible, and to the greatest extent possible, generalizable to all the populations served by the policymaker.

Recommendations for Health Research Funders

Good evidence to inform health policy will only be available if research programs are adequately funded. Government agencies—such as NIH, NSF, CDC, AHRQ, PCORI, and others—need adequate federal funding that supports the investigations necessary to generate evidence. Funding from philanthropy and from industry are additional important sources that will facilitate the generation of evidence. Funding of scientific inquiry should not be subject to political restrictions or biased agendas.

In order for policymakers to take full advantage of research evidence, improved systems for generating, processing, and sharing information from research studies need to be supported. Investments in creating useful and accessible data repositories, as well as the mechanisms that support systematic analyses and provide clearing-house functions, will make evidence more accessible and easy to use for policymakers, and thus more likely to help guide policymaking. Further, established systems to receive and analyze evidence will make it more likely that policymakers will trust and utilize the research results.

Recommendations for Academic Health Centers

Universities and AHCs play central roles in the generation of evidence that can be used to inform health policy. They should take lead roles in overseeing research quality, so that reproducibility and accuracy build trust with policymakers and the public. AHCs should act as conveners of policymakers and researchers to help build relationships and create venues for an exchange of ideas and understandings. AHCs should review their research portfolios to identify areas of expertise that are desired by policymakers, as well as areas where evidence is currently insufficient and needed by policymakers.

AHCs have a responsibility to connect with their communities (local to global) so that they can best define the areas in which evidence is needed to guide policy. They should enhance media and government relations expertise within their institutions to ensure that research findings are communicated and disseminated to both policymakers and the general public. In addition, as members and influencers of professional societies, AHCs should support advocacy for evidence-informed policymaking and for funding of the research needed to provide that evidence.

Finally, AHCs also have the opportunity to role model the development and implementation of evidence-informed health policies, from employee policies to the practice of evidence-informed care in their clinical facilities.

REFERENCES

[1] The Patient Protection and Affordable Care Act of 2010. Pub No. 111-148, 124 Stat. 119.

[2] Rubio DM, Schoenbaum EE, Lee LS, Schteingart DE, Marantz PR, Anderson KE, et al. Defining translational research: implications for training. Acad Med 2010;85:470–5.

[3] ITHS. T-phases of translational health research. https://www.iths.org/about/translational [last accessed 22.10.14].

[4] Khoury MJ, Gwinn M, Yoon PW, Dowling N, Moore CA, Bradley L, et al. The continuum of translation research in genomic medicine: how can we accelerate the appropriate integration of human genomic discoveries into health care and disease prevention? Genet Med 2007;9:665–74.

[5] NIH. http://www.nlm.nih.gov/hsrinfo/ [last accessed 22.10.14].

[6] Academy Health. What is HSR? http://www.academyhealth.org/About/content.cfm?ItemNumber=831&navItemNumber=514 [last accessed 22.10.14].

[7] NIH. Translation. http://obssr.od.nih.gov/scientific_areas/translation/index.aspx [last accessed 22.10.14].

[8] Institute of Medicine. Initial national priorities for comparative effectiveness research. Washington, DC: The National Academies Press; 2009.

[9] American Recovery and Reinvestment Act of 2009. Pub No. 111-5, 123 Stat. 115.

[10] Roland M, Torgerson DJ. Understanding controlled trials: what are pragmatic trials? BMJ 1998;316:285.

[11] Walker AE, Grimshaw J, Johnston M, Pitts N, Steen N, Eccles M. PRIME – process modelling in implementation research: selecting a theoretical basis for interventions to change clinical practice. BMC Health Serv Res 2003;3:22.

[12] Hanney SR, Gonzalez-Block MA. Yes, research can inform health policy; but can we bridge the 'Do-knowing it's been done' gap? Health Res Policy Syst 2011;9:23.

[13] AHRQ's Knowledge Transfer Program. Agency for healthcare research and quality. 2014. Rockville, MD. http://www.ahrq.gov/cpi/centers/ockt/index.html.

[14] World Health Organization. Bridging the "Know-Do" gap: Meeting on knowledge translation in global health; 2005.

[15] Macintyre S. Evidence in the development of policy. Public Health 2012;126:217–9.

[16] Murphy K, Fafard P. Taking power, politics, and policy problems seriously: the limits of knowledge translation for urban health research. J Urban Health 2012;4:723–32.

[17] Nilsen P, Stahl C, Roback K, Cairney P. Never the twain shall Meet?—a comparison of implementation science and policy implementation research. Implement Sci 2013;8:63.

[18] Cassell CK, Guest JA. Choosing wisely: helping physicians and patients make smart decisions about their care. JAMA 2012;307:1801–2.

[19] Shalala DE. Needle exchange programs in America: review of published studies and ongoing research. Report to the Committee on Appropriations for the Department of Labor. Health and Human Services Education and Related Agencies; February 18, 1997.

[20] Van Herck P, Annemans L, Sermeus W, Ramaekers D. Evidence-based health care policy in reimbursement decisions: lessons from a series of six equivocal case-studies. PLoS One 2013;8:e78662.

[21] Liverani M, Hawkins B, Parkhurst JO. Political and institutional influences on the use of evidence in public health policy. A systematic review. PLoS One 2013;8:e77404.

[22] Mohr P. Looking at CER from Medicare's perspective. J Manag Care Pharm 2012;18(4 Suppl. A):S5–8.

[23] Starr M, Chalmers I, Clarke M, Oxman AD. The origins, evolution and future of the Cochrane database of systematic reviews. Int J Technol Assess Health Care 2009;25(Suppl1.):182–95.

[24] Council of State Governments. State policy guide: using research in public health policymaking; 2008.

[25] Tricco AC, Tetzlaff J, Sampson M, Fergusson D, Cogo E, Horsley T, et al. Few systematic reviews exist documenting the extent of bias: a systematic review. J Clin Epidemiol 2008;61:422–34.

[26] Lomas J. Using research to inform healthcare managers' and policy makers' questions: from summative to interpretive synthesis. Health Policy 2005;1:55–71.

[27] Hanney SR, Gonzalez-Block MA. Evidence-informed health policy: are we beginning to get there at last? Health Res Policy Syst 2009;7:30–4.

[28] National Institute for Health and Care Excellence. http://www.nice.org.uk/ [last accessed 22.10.14].

[29] Pellini A, Serrat O. Enriching policy with research. Washington DC: Asian Development Bank; 2010.

[30] Hyder AA, Corluka A, Winch PJ, El-Shinnawy A, Ghassany H, Malekafzali H, et al. National policy-makers speak out: are researchers giving them what they need? Health Policy Plan 2011;26:73–82.

[31] Overseas Development Institute. Helping researchers become policy entrepreneurs. London: ODI; 2009.

[32] Humphreys K, Piot P. Scientific evidence alone is not sufficient basis for health policy. BMJ 2012;344:e1316.

[33] Prinz F, Schlange T, Asadullah K. Believe it or not: how much can we rely on published data on potential drug targets? Nat Rev Drug Discov 2011;10:712.

[34] Stamatakis E, Weiler R, Ioannidis JP. Unique industry influences that distort healthcare research, strategy, expenditure and practice: a review. Eur J Clin Invest 2013;43:469–75.

[35] Jernigan DH. Global alcohol producers, science, and policy: the case of the International Center for Alcohol Policies. Am J Public Health 2012;102:80–9.

[36] Fielding JE, Briss PA. Promoting evidenced-based public health policy: can we have better evidence and more action? Health Aff 2006;25:969–78.

[37] Pomeroy C. A call to action: invest in biomedical research now to secure a healthier future. Surg Oncol 2014;23:1–4.

[38] Ricketts TC. Preparing the health services research workforce. Health Serv Res 2009;44:2227–41.

[39] Reichert JM, Rochon SL, Zhang BD. A decade of the fast track programme. Nat Rev Drug Discov 2008;7:885–6.

[40] http://www.iom.edu/.

[41] Garber AM. How the patient-centered outcomes research institute can best influence real-world health care decision making. Health Aff (Millwood) 2011;30:2243–51.

[42] Gerber AS, Patashnik EM, Doherty D, Dowling CM. Doctor knows best: physician endorsements, public opinion, and the politics of comparative effectiveness research. J Health Polit Policy Law 2014;39:171–208.

[43] Michener L, Cook J, Ahmed SM, Yonas MA, Coyne-Beasley T, Aguilar-Gaxiola S. Aligning the goals of community-engaged research: why and how academic health centers can successfully engage with communities to improve health. Acad Med 2012;87:285–91.

[44] VanLare JM, Conway PH, Rowe JW. Building academic health centers' capacity to shape and respond to comparative effectiveness research policy. Acad Med 2011;86:689–94.

[45] Cook J, Michener JL, Lyn M, Lobach D, Johnson F. Community collaboration to improve care and reduce health disparities. Health Aff 2010;29:956–8.

[46] Institute of Medicine. The future of nursing: leading change, advancing health. Washington, DC: The National Academies Press; 2010.

[47] Stephens T, Brynner R. Dark remedy: the impact of thalidomide and its revival as a vital medicine. Cambridge, Massachusetts: Perseus Publishing; 2001.

[48] Sanfilippo F, Vaughn WK, Spees EK, Light JA, LeFor WM. Benefits of HLA-A and HLA-B matching on graft and patient outcome after cadaveric-donor renal transplantation. N Engl J Med 1984;311: 358–64.

[49] US Dept. of Health and Human Services. Organ procurement and transplantation network. http://optn.transplant.hrsa.gov/ [last accessed 22.10.14].

[50] Sanfilippo F, Vaughn WK, Peters TG, Adams PL, Lorber MI, Williams GM. Factors affecting the waiting time of cadaveric kidney transplant candidates in the United States. JAMA 1992;267:247–52.

[51] Reynolds H. The big squeeze: a social and political history of the controversial mammogram. IRL Press/Cornell University Press; 2012.

[52] Biller-Andorno N, Jüni P. Abolishing mammography screening programs? a view from the Swiss Medical Board. N Engl J Med 2014;370:1965–7.

[53] Pace LE, Keating NL. A systematic assessment of benefits and risks to guide breast cancer screening decisions. JAMA 2014;311: 1327–35.

[54] The Editors of the Lancet. Retraction – Ileo-lymphoid-nodular hyperplasia, non-specific colitis, and pervasive developmental disorder in children. 2010;375(9713):445.

[55] Downs M. Autism-vaccine link: evidence doesn't dispel doubts. WedMD. http://www.webmd.com/brain/autism/searching-for-answers/vaccines-autism [last accessed 05.05.14].

[56] Rudy LJ. What is the Vaccine/Autism controversy about? About. com. http://autism.about.com/od/whatisautism/a/vacci/nequestion. htm [last accessed 05.05.14].

[57] Levy D, Rodriguez-Buño RL, Hu TW, Moren AE. The potential effects of tobacco control in China: projections from the China Simsmoke simulation model. BMJ 2014;348:g1134.

[58] Pitkin R. Folate and neural tube defects. Am J Clin Nutr 2007;85: 2855–85.

[59] Centers for Disease Control. Recommendations for the use of folic acid to reduce the number of cases of spina bifida and other neural tube defects. MMWR Morb Mortal Wkly Rep 1992;41:1.

[60] Burt BA, Tomar SL. Changing the face of America: water fluoridation and oral health. In: Ward JW, Warren C, editors. Silent victories: the history and practice of public health in twentieth-century America. Oxford University Press; 2007. p. 307–22.

[61] Jewell CJ, Bero LA. Developing good taste in evidence: facilitators of and hindrances to evidence-informed health policymaking in state government. Milbank Q 2008;86:177–208.

ABOUT THE AUTHORS

Claire Pomeroy, MD, MBA was appointed president of the Albert and Mary Lasker Foundation in 2013. Her prior positions include that of CEO and Vice Chancellor of the Health System and Dean of the School of Medicine at the University of California, Davis. An expert in infectious diseases, Dr Pomeroy is a long-time advocate for patients, especially those with HIV/AIDS, and public health.

Fred Sanfilippo, MD, PhD is the Director of the Emory-Georgia Tech Healthcare Innovation Program, which serves to accelerate innovation in healthcare research, education, and service. Prior to his current position, he served as Emory University Executive VP for Health Affairs, CEO of Emory's Woodruff Health Sciences Center, and Chairman of Emory Healthcare.

Preparing for Health System Change

Population Health and the Patient

Jay A. Perman, C. Daniel Mullins and Robin Newhouse

INTRODUCTION

Health care providers and administrators of academic health centers (AHCs) must appreciate and respond to the reality that the health status for each patient is impacted not only by the individual's genetic make-up, but also the individual's wealth, education level, and access to resources, such as reliable transportation, adequate housing, and grocery stores with nutritional food [1]. There are five key social determinants of health (SDH)—genetics, social circumstances, environment, behavioral choices, and access to medical care—that determine, or at least strongly influence, the ability to achieve and maintain good health throughout one's life span [2]. Enhanced understanding and implementation of interventions that address SDH and holistic approaches to health care can improve the research, practice, and service mission of AHCs. It is important that every clinician and staff member at AHCs address SDH in a nonjudgmental manner.

SOCIAL DETERMINANTS OF HEALTH

SDH are factors regarding how and where a person lives, works, or interacts with others that influence one's health status. SDH suggest that an individual's health is not only impacted by whether one can access quality health care, but is also shaped by the ability to access healthy foods, as well as the safety and sanitary conditions of one's environment. A patient's cultural norms, health literacy, and competing needs—such as the availability of transportation or child care—govern individual actions and can influence others' actions. Physicians, nurses, pharmacists, and others associated with health care delivery and research should have an understanding of the causal pathways that predict health in terms of proximate factors (immediate health risks such as environmental hazards like air pollution for children with asthma) and protective factors (those that mitigate or reduce risks, such as social supports and access to regular health care for children with asthma) in order to facilitate multiple approaches to health interventions [3].

IMPLICATIONS FOR AHCs

Education

Improving the quality and equity of care requires that all health professions work together toward a common goal of health for the populations being served. For example, when considering the care of children with asthma: physicians, physician assistants, or advanced nurse practitioners diagnose and treat; pharmacists counsel on inhaler use; social workers assist families in locating resources; and nurses provide patient and family self-management education or a home environment evaluation. All disciplines are focusing on a common goal to reduce asthma exacerbations. Because interprofessional teams are essential to promote quality care, interprofessional education (IPE) will be a core element in AHC curriculum to assure that clinicians are "practice ready" at the completion of their training [4].

Academic training of all health care professionals needs to include cross-cultural competence. Communication skills of the provider can be enhanced if the patient's social and cultural background is taken into consideration at the initial point of the medical encounter and, more importantly, before the provider offers to manage a health condition in which cultural traditions put medical treatment at odds with patients' health beliefs. It is also important for the academic health institution to take into account the effect of the social environment (e.g., poverty) on health outcomes [5].

Complex economic relationships play a critical role between people seeking care, their providers, and the ability to pay for or access care (through insurance, federal or state programs, or self-pay mechanisms). These SDH make a difference in care provided (or not), and lead to some of the greatest disparities, requiring the health care team to develop a high level of cost consciousness. While AHCs cannot always address all SDH, they can improve the number of services available to assist patients and reduce barriers to care.

Medical education curriculum for resident internists, developed by the Alliance for Academic Internal Medicine and the American College of Physicians, has recently

The Transformation of Academic Health Centers. http://dx.doi.org/10.1016/B978-0-12-800762-4.00019-0

addressed a gap in training by including ten case-based interactive sessions that teach high-value and cost-conscious principles that can be performed in clinical practice [6]. Future plans are to offer this type of training to established physicians, since they may operate their practices under the former more is always better work code. The framework [7] includes: (1) consider the benefits, harm, and relative costs of treatment; (2) remove treatment that offers no benefit or may cause harm; (3) consider comparative and cost-effectiveness data to reduce harm and costs while increasing benefits; (4) develop a care plan that includes patient values and concerns; and (5) check for system-level opportunities to improve outcomes, minimize harms, and reduce health care waste. The first three steps highlight the importance of comparing and selecting the right treatment. The fourth incorporates individual patient attributes, while the fifth obligates providers to seek effective resources for current and future patients and populations.

Nurse education traditionally has included SDH in the curriculum at the point of addressing community health. The Affordable Care Act (ACA) includes multiple mandates with the goal to improve population health, necessitating inclusion of the SDH at multiple points and across programs in the nurse education curriculum [8]. It is no longer adequate for nurse training to simply identify health risks; rather, nurses need to be trained to become health care advocates who partner with interdisciplinary teams to advance research and ultimately transform public policy [9]. Nurse training needs to also be aligned with the Health Resources and Services Administration's (HRSA) funding mechanism, which has now been mandated by the ACA to include SDH as one of the requirements to receive grant funding. Therefore, the full long-range educational curriculum must include training of nurse scientists [9].

Similar to nurse training and the Nurse Practice Act (which stresses the role of being an advocate for community-based clients), medical education programs are providing service-learning opportunities for residents to learn firsthand how SDH impact a patient's ability to obtain and regain health. For example, the Oregon Health and Science University four-week resident social medicine curriculum partners with the Central City Concern, a Portland community-based facility that gives support to the homeless and those with addiction [9]. While the training sessions were constrained to fit within the existing curriculum, a multi-phase program was devised with the support of the community organization. The two-phase program gave participants a real-world opportunity to talk with local patients and garner understanding that SDH are not merely theoretical concepts, but that real people live in poverty and have inadequate food and housing.

The attitudes of residents regarding care for the underserved were recently studied [10]. If doctors are to engage patients with living situations and health literacy issues that are dissimilar to the resident, it is suggested that volunteering in the local community may help to improve attitudes and break down barriers working with the underserved populations. In order to measure these attitudes and behaviors, Wieland et al. administered a cross-sectional, multi-institutional survey to 18 residency programs in the United States with a response rate of 52% [11]. Medical school students' attitudes and actual rate of volunteerism were in concordance with those who did not volunteer; however, as the medical student transitioned to the role of resident physician, the attitude remained high while the rate of volunteering diminished. Some suggested reasons for this reduction include time constraints due to young families often being formed during residency and the burden of a residency curriculum, as well as a sense of lower community integration. It is interesting to note that older physicians had an even lower rate of volunteering services. This may be because free medical services were already being offered in the clinical setting and the need to seek additional volunteering was not on the practicing physician's agenda. It is recommended that the medical resident curriculum be revised to offer a formal opportunity to volunteer.

Patient Care

The health care that people receive is too often suboptimal, and frequently dependent on who they are and where they live [11]. Minority and low-income groups are at the greatest risk of suboptimal care and too often use emergency departments because they do not have or cannot access a primary care provider. SDH impact the quality care that many minority patients receive. For example, non-Hispanic Whites receive better care than Blacks or Hispanics for approximately 40% of quality measures [11]. The quality measures align with the 2011 National Strategy for Quality Improvement in Health Care goals for better care, healthy people/healthy communities, and affordable care [12]. The report contains core measures that are important to quality care and supported scientifically. SDH impact all individuals, yet patients with multiple chronic conditions, minority patients, and those with low health literacy skills represent a significant subset of priority populations for examining SDH.

Consider the case of pediatric asthma. There are 7 million children under the age of 18 with asthma [13]. The total cost for treating asthma is estimated at $56 billion [14]. Unfortunately, the percentage of people with asthma who take their medication daily or almost daily has decreased over time for all age categories, except for those over 65 years of age. These troublesome trends exist despite national evidence-based guidelines published by the National Institutes of Health in 2007 [15]. These guidelines focus on control of asthma by reducing impairment (e.g., symptom management) and reducing risk (e.g., pharmacotherapy with minimal or no adverse effects). The national asthma control initiative seeks to translate evidence and action through mobilizing champions, convening and activating local leaders, developing a communication infrastructure to share resources, demonstrating best practices, and monitoring progress [16].

Taking best practices to scale is another area where the health of populations can be affected. For example, a randomized, controlled three-armed trial tested the efficacy of home-based asthma education plus adherence monitoring compared to education with a home visit (asthma basic care) and usual care with standard education for inner-city children [17]. Results indicate that both home-based interventions demonstrate a faster decrease in emergency department use and courses of oral steroids. However, there was no difference in asthma symptom frequency or hospitalizations. These study results indicate that there is a modest but significant effect when home visits are conducted. The message is that we need to go to the community as part of the plan to improve care for children with asthma.

AHCs are well situated to lead practice initiatives to improve care for some of the most vulnerable populations. They are the locus for the training of medical professions, have some of the greatest medical care available locally on staff representing all professions, and have some of the greatest resources with connections to communities in need. They hold the capacity to generate evidence to solve some of the most difficult problems, test interventions through pragmatic designs, and communicate results through professional and clinical venues. It is these AHCs that have the capacity and responsibility to solve some of the nation's most difficult health care problems.

Research

There are two major implications for research conducted at AHCs. The first is to include priority areas of health disparities in which SDH interact significantly with the intervention to impact patient outcomes. The second is the incorporation of methods needed to include the lens of the patient, caregiver, or family in the conceptualization, design, delivery, analysis, and dissemination of research results.

The need to identify the root cause of a health disparity and test interventions to deliver equitable high-quality health care at the patient, system, and population levels of the ecological framework is paramount in the generation of new knowledge. New advances in geographic information systems are providing data to easily identify hot spots (populations at risk for poor outcomes) so that interventions can be targeted toward modifiable risk reduction. Research aims can then be focused on building infrastructure and health equity in line with funding agencies' SDH priorities, professional organizations (such as the American Public Health

Association), or state priorities that help to build capacity in targeted areas [18,19]. Table 1 includes selected examples of funding agencies with research priorities aligned with SDH.

County and state-level evaluations are needed to assess opportunities to reduce disparities and target areas where SDH can be leveraged locally. For example, in Maryland's Prince Georges County, a multimethod design used interviews and surveys to collect data from residents, leaders, and policy-makers to inform a new county system design to improve health for the county's residents [20]. Prince Georges is Maryland's most diverse county, with 80% of the population representing minority groups (65% Blacks, 15% Hispanics, 15% Whites). Results indicate that chronic diseases (diabetes, heart disease, hypertension, asthma, and cancer) were the key health priorities (ethnic and racial differences identified); these chronic diseases are the most amenable to improvement in a new health system. In addition, primary care needs were identified, including 61 physicians (13% more) and 31 dentists (7% more) to meet the minimum recommendation ratios of providers to population.

To fully engage a multidimensional research platform, Hardy et al. (2013) embedded evidence-based strategies in an existing community research project—Rapid Assessment, Response, and Evaluation (RARE)—which was partnered with academic faculty and a nonprofit agency. The project had the goal of identifying strategies to prevent childhood obesity in a southwest United States city [18]. According to the author, "RARE is a time-efficient (six- to eight-week) community-engaged research process to address SDH" [17]. Local community experts were paid and trained in methods of performing data collection. The research team did not disrupt existing community projects, but embedded the research strategy into the already existing community projects. Secondary data were collected by a research student to match local statistics with the results obtained by the trained data collection staff. A triangulation method was then used to sift out important items that were top-priority items for the community. A steering committee guided the work of the team, and the community had buy-in from the start of the project. Some of the findings of this particular research study identified specific steps to prevent childhood obesity: implement structured recess to involve students in sports, establish after-school activities to include educational family fun nights, and initiate a school-based obesity prevention clinic.

Research results can have more impact and be more meaningful for patients if they are engaged to assure that "the patient's voice" is heard throughout the research process. A patient-centered emphasis throughout the research process helps to ensure that it remains focused on problems that people care about and that the research questions, design, comparators, and outcomes are more applicable to

TABLE 1 Examples of Agencies that Fund Research Related to Social Determinants of Health

Agency	Funding Program/ Agenda	Priorities	Link
National Institute on Minority Health and Health Disparities (NIMHD)	Community-Based Participatory Research Initiative	• Enhance community capacity by supporting equal community participation in research for which community members will directly benefit • Establish sustainable programs that improve health behaviors and health outcomes in health disparity populations • Accelerate the translation of research findings to health disparity communities by designing effective and culturally-tailored interventions	http://www.nimhd.nih.gov/programs/extra/cbpr.html
All NIH Institutes and Centers (ICs)	Basic Behavioral and Social Science Opportunity Network (OppNet)	• Advance basic behavioral and social science research through activities and initiatives that build a body of knowledge about the nature of behavior and social systems	http://oppnet.nih.gov/
Office of Behavioral and Social Sciences Research (OBSSR)	Social and Cultural Factors in Health	• Develop knowledge related to racial/ethnic, social class, and rural-urban health disparities influenced by behavioral and social factors, their specific causes and by participating in finding solutions	http://obssr.od.nih.gov/scientific_areas/social_culture_factors_in_health/index.aspx
HRSA	Improve Health Equity	• Reduce disparities in quality of care across populations and communities • Monitor, identify and advance evidence-based and promising practices to achieve health equity • Leverage programs and policies to further integrate services that address the social determinants of health • Partner with diverse communities to create, develop, and disseminate community-based health equity solutions for populations with the greatest health disparities	http://www.hrsa.gov/about/strategicplan.html
PCORI	Addressing Disparities	• Identify potential differences in prevention, diagnosis or treatment effectiveness, or preferred clinical outcomes across patient populations, and the healthcare required to achieve best outcomes in each population	http://www.pcori.org/research-we-support/priorities-agenda/
Robert Wood Johnson Foundation	RWJF Vulnerable Populations Portfolio	• Vulnerable populations are a major area of priority focusing on finding innovative and effective ways of addressing the many factors that influence the health vulnerable people. • Supported research and dissemination of *A New Way to Talk About the Social Determinants of Health*	http://www.rwjf.org/en/our-work.html

real-world patients [21]. When considering studies related to SDH, patient engagement becomes paramount. Studies need to: engage patients and stakeholders related to the problem of interest; select study participants who are representative of the target population; and use patient-reported outcomes when possible [22].

PRIORITY POPULATIONS

SDH impact all individuals, yet patients with multiple chronic conditions and others who have been underrepresented in research and underserved in clinical practice represent priority populations for examining SDH.

Multiple Chronic Conditions

Sixty-eight percent of Medicare beneficiaries have multiple chronic conditions [23], with 14% experiencing six or more. A person has multiple chronic conditions if they have two or more conditions that last a year or more, limit activities of daily living, and require medical attention [24]. Other definitions add that the nature of the conditions are not self-limited and are related to reoccurring health problems that last months and years (as opposed to days and weeks) [25,26]. The health burden of living with multiple chronic conditions is high, with a disproportionate share of Medicare cost allocated to treatment [24].

Central to the Health and Human Services Strategic Framework on multiple chronic conditions [27] are four interdependent goals:

1. Health care providers need to break out of the silo mentality and direct care toward evidence-supported multidisciplinary care management—led by improved patient-provider communication. Specifically, there is a need to maintain patient functional status while reducing unnecessary hospitalizations. Additionally, stakeholders need to consider reimbursement reform, especially for non-physician providers. Health systems need to develop strategies that foster the meaningful use of electronic health records. Patients should be encouraged to participate in disease prevention programs.
2. Health care providers need to recognize the importance of family and other caregivers who assist patients by providing evidence-based self-care management to prevent falls, reduce depression, and oversee medication management. Education aimed toward patients and care providers is needed to reduce adverse drug interactions and improve medication adherence.
3. Educators of health care professional trainees must be provided additional and specific education to improve cultural competence, understand the role of family and other caregivers, and, when necessary, address the need for chronic physical illness care in conjunction with mental health care or substance abuse interventions.
4. Health care professionals should encourage clinical trial participation for people who are currently underrepresented in studies and for those in which disparities exist. These include the priority populations with multiple chronic conditions—who should be included in research on health disparities and all aspects of patient-centered health research.

TRANSFORMING AHCs THROUGH INFUSION OF SDH

From an AHC's perspective, a high-performing health care team focuses not just on diagnosis and treatment of a patient's disease, but also on SDH that affect the patient's social, family, and environmental context. As stated by Korenstein, "clinicians must change the way they approach every patient in every setting" [28].

From a public health perspective, academic training needs to include specific methods and guidance to establish partnerships with local stakeholders so that a blending of scientific training and buy-in from community action leadership organizations becomes routine [29]. However, awareness of SDH for individuals is only one aspect of committing to the change in health inequality.

Transforming Patient Care at AHCs to Address SDH

The real transformation of the health care workforce will begin when patient care is moved from a disease-cure emphasis to a disease-prevention approach, and when an understanding of SDH is embedded in the care delivered by the health care team. Linkages across medical specialties, the use of electronic health records, and outreach programs to the community to assist patients who manage their medical and social needs require an educated workforce attuned to the local social environment. Some of this transformation will take place through continuing education programs for established health care workers, while other means of providing this new population-based emphasis will be provided by AHCs as they update their curriculum to include interprofessional training at both campus and community-based practice settings [30].

The role of interprofessional alignment for better patient health care is best described by the Robert Wood Johnson Foundation publication, *Time to Act: Investing in the Health of Our Children and Communities* [31]. As primary care providers for the most vulnerable populations, i.e., those who do not have a medical home, medical centers functioning in silos are not acceptable. Today's medical professionals need to incorporate new vital signs that impact their patients' health, i.e., employment, education, health literacy, and safe housing. Now more than ever, partnerships and coordination with local community organizations is imperative. It is no longer acceptable for a physician to suggest that a patient eat more nutritional foods without taking into consideration the patient's financial status, much less their physical ability to get to the store.

The ability to see patients as members of the local community will require additional training of health professionals who work in medical facilities. AHCs need to train their workforce to first identify and subsequently address the realities of patients' resources so that their three-year needs assessment, as required by the ACA, can aid AHCs with strategic planning and subsequent investment in community support initiatives. This additional training will connect the physician and patient in an enduring way. It will build a bridge between the patient and the medical facility that will directly impact the patient's and community's health outcomes, as well as the AHC's downstream costs.

Success Stories from Maryland and University of Maryland Baltimore

The Maryland Department of Health and Mental Hygiene (DHMH) partnered with University of Maryland Baltimore (UMB) to establish the Institute for a Healthiest Maryland [32]. Start-up funds to establish the Institute are derived from a portion of a five-year $9.5 million federal Community Transformation Grant from the Centers for Disease Control and Prevention (CDC). The focus is on three wellness areas: obesity prevention, tobacco prevention, and appropriate management of hypertension and high cholesterol.

The Institute for a Healthiest Maryland guides policy, systems, and environmental changes in local communities by providing evidence-based resources from higher-education experts. The Institute's advisory board is cochaired by UMB President Jay A. Perman, MD, and DHMH Secretary Joshua M. Sharfstein, MD. The board consists of leaders from academia, public health, public education, hospitals, providers, and community organizations, and addresses policy, program, and quality assurance needs of community partners.

The Community Health Resources Commission (CHRC) and the DHMH established the Health Enterprise Zone (HEZ) Initiative, which was called for by the 2012 Maryland Health Improvement & Health Disparities Reduction Act [33]. This initiative is a four-year pilot program with a budget of $4 million per year. The purposes of the HEZ Initiative are to: (1) reduce health disparities among racial and ethnic minority populations and among geographic areas; (2) improve health care access and health outcomes in underserved communities; and (3) reduce health care costs and hospital admissions and readmissions.

One of the five HEZs is centered at Bon Secours Baltimore Health System, West Baltimore Primary Care Access Collaborative. The focus of this HEZ is to expand access to primary and preventive care while creating a community health infrastructure for residents in West Baltimore. This HEZ will recruit 18 new primary care professionals and 11 community health workers and assist with providing access to healthy food retail stores and exercise facilities.

In Baltimore, asthma rates are as high as 20% (national prevalence of 7.5%). The Breathmobile is a pediatric asthma and allergy clinic that travels to schools and has provided care to 3,500 patients (more than 7,000 visits), including asthma care to children, since 2002 [34]. The Breathmobile has improved asthma care for children by reducing lost school days and emergency department visits, and increasing use of preventative asthma medicines. These improvements in health are immediate and local, as well as boost children's health in a city of high asthma prevalence.

PATient-centered Involvement in Evaluating effectiveNess of TreatmentS (PATIENTS) is a five-year, five million dollar grant from the Agency for Healthcare Research and Quality (AHRQ)[1] awarded to researchers at the University of Maryland, Baltimore and the university itself to lead partnerships with health care systems, patient communities, and advocacy groups to improve health care research and outcomes [35]. PATIENTS builds, conducts, expands, disseminates, and implements patient-centered outcomes research in partnership with diverse local, regional, and national communities and health care systems to: (1) improve researchers' cultural competency and patients' health outcomes; (2) improve patient decision-making; and (3) improve strategies based upon research to impact health care providers and health care systems design. The community partnership includes: Association of Black Cardiologists; Bon Secours; Center for Medical Technology Policy; Mt Lebanon Baptist Church; Patients Like Me; Riverside Health System; University of Maryland Medical Center; and Westat.

Challenges in Addressing Social Determinants of Health

What is the cost to AHCs in transforming health care to address SDH? First, these centers are safety nets to individuals who have complex medical conditions and have little resources available to them in terms of their economic, educational, and social environment. These factors place the AHC in the position of offering undercompensated care. A need exists to develop a business case for AHCs to address SDH [36]. Accountable care is at the heart of a business model for getting it right the first time, particularly when there are reduced financial payments for avoidable readmissions. However, financial levers are not sufficient; there is a need for resources to help practitioners and administrators of AHCs to understand and address SDH among the patients they serve. Some challenges include lack of training, long work shifts, insufficient transitions of care models, and silos between health delivery systems and government systems that impact the health and welfare of individuals.

CONCLUSION

AHCs have high potential and accompanying responsibility to address some of health care's most pressing problems. The academic enterprise can and should be transformed through infusion and incorporation of SDH in education, practice, and research. We propose three main recommendations. First, with their unique position in training an interprofessional health care workforce, AHCs can instill an expectation and culture of interprofessional care delivery that incorporates SDH in the assessment and delivery of treatment options. Second, research conducted should

---

[1] Grant Number: R24 HS22135-01.

incorporate SDH to inform high-quality, equitable, and cost effective practice models that reduce health disparities among people and settings. Third, the voice of the patient should be incorporated in AHC education, patient care, and research. Forming community partnerships to understand health care needs, train clinicians to incorporate the assessment of SDH within the care processes, and generate research questions will be one of the most transformative influences as AHCs improve the health of populations.

REFERENCES

[1] Commission on Social Determinants of Health (CSDH). Closing the gap in a generation: health equity through action on the social determinants of health. Final report of the commission on social determinants of health. Geneva: World Health Organization; 2008.

[2] McGinnis JM. In: Koop CE, editor. "United States," in critical issues in global health. San Francisco: Jossey-Bass; 2001. p. 80–90.

[3] Link BG, Phelan J. Social conditions as fundamental causes of disease. J Health Soc Behav 1995;35. Spec No:80–94.

[4] Interprofessional Education Collaborative Expert Panel. Core competencies for interprofessional collaborative practice: report of an expert panel. Washington, DC: Interprofessional Education Collaborative; 2011. Available at: https://www.aamc.org/download/186750/data/core_competencies.pdf [accessed 26.04.14].

[5] Green AR, Tan-McGrory A, Cervantes MC, Bentancourt JR. Leveraging quality improvement to achieve equity in health care. Jt Comm J Qual Patient Saf 2010;36(10):435–42.

[6] Smith CD. Alliance for academic internal medicine–American college of physicians high value; cost-conscious care curriculum development committee. Teaching high-value, cost-conscious care to residents: the alliance for academic internal medicine–American college of physicians curriculum. Ann Intern Med 2012;157(4):284–6.

[7] Owens DK, Qaseem A, Chou R, Shekelle P. Clinical guidelines committee of the American college of physicians. High-value, cost-conscious health care: concepts for clinicians to evaluate the benefits, harms, and costs of medical interventions. Ann Intern Med 2011;154:174–80.

[8] Mahony D, Jones EJ. Social determinants of health in nursing education, research, and health policy. Nurs Sci Q 2013;26(3):280–4.

[9] Iles-Shih M, Sve C, Solotaroff R, Bruno R, Gregg J. Health and illness in context: a pragmatic, interdisciplinary approach to teaching and learning applied public health within an urban safety net system. J Public Health Manag Pract 2011;17(4):308–12.

[10] Wieland ML, Beckman TJ, Cha SS, Beebe TJ, McDonald FS. Underserved care curriculum collaborative. Residents' attitudes and behaviors regarding care for underserved patients: a multi-institutional survey. J Grad Med Educ Sep 2011;3(3):337–44.

[11] AHRQ. National health care quality report. 2012. Available at: http://www.ahrq.gov/research/findings/nhqrdr/nhqr12/highlights.html [accessed 26.04.14].

[12] National Strategy for Quality Improvement in Health Care. 2011. Available at: http://www.ahrq.gov/workingforquality/nqs/nqs2011-annlrpt.pdf [accessed 23.05.14].

[13] Bloom B, Cohen RA, Freeman G. Summary health statistics for U.S. Children: national health interview survey. Vital Health Stat 2010;10(250):2011. Available at: http://www.cdc.gov/nchs/data/series/sr_10/sr10_250.pdf [accessed 22.04.14].

[14] Barnett SB, Nurmagambetov TA. Costs of asthma in the United States: 2002-2007. J Allergy Clin Immunol 2011;127(1):145–52.

[15] U.S. Department of Health and Human Services, NIH. National asthma education and prevention program. Expert panel report 3. Guidelines for the diagnosis and prevention of asthma. 2007. Available at: http://www.nhlbi.nih.gov/guidelines/asthma/asthsumm.pdf [accessed 26.04.14].

[16] National Asthma Control Initiative. The national heart, lung, and blood institute launches new effort to put what works into action. 2010. NIH Publication No. 10-7541. Available at: http://www.nhlbi.nih.gov/health/prof/lung/asthma/naci/pubs/naci-factsheet.pdf [accessed 26.04.14].

[17] Otsuki M, Eakin MN, Rand CS, Butz AM, Hsu VD, Zuckerman IH, et al. Adherence feedback to improve asthma outcomes among inner-city children: a randomized trial. Pediatrics 2009;124(6):1513–21.

[18] Hardy LJ, Bohan KD, Trotter 2nd RT. Synthesizing evidence-based strategies and community-engaged research: a model to address social determinants of health. Public Health Rep 2013;128(Suppl. 3):68–76.

[19] APHA Advocacy & Policy. Priorities. Available at: http://www.apha.org/advocacy/priorities/ [accessed 23.05.14].

[20] University of Maryland School of Public Health. Transforming health in Prince George's county, Maryland: a public health impact study. 2014. Available at: sph.umd.edu/princegeorgeshealth [accessed 26.05.14].

[21] Mullins CD, Abdulhalim AM, Lavallee DC. Continuous patient engagement in comparative effectiveness research. JAMA 18 April, 2012;307(15):1587–8.

[22] PCORI (Patient-Centered Outcomes Research Institute) Methodology Committee. The PCORI methodology report. pcori.org/research-we-support/research-methodology-standards. 2013. Available at: http://www.pcori.org/assets/2013/11/PCORI-Methodology-Report.pdf [accessed 30.04.14].

[23] Centers for Medicare and Medicaid Services. Chronic conditions among medicare beneficiaries, chartbook, 2012 edition. 2012. Available at: http://www.cms.gov/Research-Statistics-Data-and-Systems/Statistics-Trends-and-Reports/Chronic-Conditions/Downloads/2012Chartbook.pdf [accessed 23.05.14].

[24] HHS, "GOV HHS initiative on multiple chronic conditions," Available at: http://www.hhs.gov/ash/initiatives/mcc/ [accessed 23.05.14].

[25] Thrall JH. Prevalence and costs of chronic disease in healthcare system structured for treatment of acute illness. Radiology 2005;235(1):9–12.

[26] Goodman RA, Posner SF, Huang ES, Parekh AK, Koh HK. Defining and measuring chronic conditions: imperatives for research, policy, program, and practice. Prev Chronic Dis 2013;10:120239.

[27] Parekh AK, Goodman RA, Gordon C, Koh HK. HHS interagency workgroup on multiple chronic conditions. Managing multiple chronic conditions: a strategic framework for improving health outcomes and quality of life. Public Health Rep 2011;126(4):460–71.

[28] Korenstein D, Kale M, Levinson W. Teaching value in academic environments: shifting the ivory tower. JAMA 2013;310(16):1671–2.

[29] Israel BA, Coombe CM, Cheezum RR, Schulz AJ, McGranaghan RJ, Lichtenstein R, et al. Community-based participatory research: a capacity-building approach for policy advocacy aimed at eliminating health disparities. Am J Public Health 2010;100(11):2094–102.

[30] Interprofessional Education Collaborative Expert Panel. Core competencies for interprofessional collaborative practice: report of an expert panel. Washington, DC: Interprofessional Education Collaborative; 2011. Available at: https://www.aamc.org/download/186750/data/core_competencies.pdf [accessed 26.04.14].

[31] RWJF Commission to Build a Healthier America. Time to act: investing in the health of our children and communities. 2014. Available at: http://www.rwjf.org/en/research-publications/find-rwjf-research/2014/01/recommendations-from-the-rwjf-commission-to-build-a-healthier-am.html [accessed 28.03.14].

[32] Institute for a Healthiest Maryland, "About," Available at: http://www.healthiestmaryland.org/about [accessed 26.04.14].

[33] "Press release: Lt. Governor Brown announces Maryland's first five health enterprise zones". Available at: http://www.governor.maryland.gov/ltgovernor/pressreleases/130124.asp [accessed 26.04.14].

[34] University of Maryland Medical Center, "Children's hospital. Breathmobile," Available at: http://umm.edu/programs/childrens/services/breathmobile, Updated October 8, 2013 [accessed 26.04.14].

[35] University of Maryland, MPowering the state. PATIENTS, Available at: http://patients.umaryland.edu/ [accessed 26.04.14].

[36] Knettel A. The business case for academic research centers addressing environmental, behavioral and social determinants of health. Washington, DC: The Association of Academic Health Centers; 2011.

ABOUT THE AUTHORS

C. Daniel Mullins, PhD is a Professor and Chair of the Pharmaceutical Health Services Research Department at the University of Maryland School of Pharmacy. His research and teaching have focused on comparative effectiveness research, patient-centered outcomes research (PCOR), pharmacoeconomics, and health disparities research. He directs the University of Maryland PATient-centered Involvement in Evaluating effectiveNess of TreatmentS (PATIENTS) Program.

Robin Newhouse, PhD, RN, NEA-BC, FAAN is a Professor and Chair of the Department of Organizational Systems and Adult Health, and Co-Director of the Center for Health Outcomes Research at the University of Maryland School of Nursing. Dr. Newhouse was appointed to the Methodology Committee of the Patient-Centered Outcomes Research Institute by the Comptroller General of the U.S. Government Accountability Office and is currently serving as the committee's chair.

Jay A. Perman, MD was appointed President of the University of Maryland, Baltimore in 2010. This marked a return to the UMB campus where he chaired the Department of Pediatrics in the School of Medicine from 1999 to 2004. Previously, he served as the Dean and Vice President for Clinical Affairs at the University of Kentucky College of Medicine from 2004 to 2010.

Chapter 20

The Changing Delivery of Patient Care

Katy A. Stevenson, Sarah E. Peyre, Katia I. Noyes and Bradford C. Berk

EXECUTIVE SUMMARY

Like many academic health centers (AHCs) in the United States, the University of Rochester Medical Center (URMC) is currently embarking on a transformative strategic plan to maintain its leadership position in the changing health care environment. Academic health care enterprises are evolving into high-value, integrated, transparent systems that effectively compete for a market share and can impact the health of large populations. Patient care is increasingly moving out of fixed institutional and office settings and into wherever the patient happens to be. This chapter will explore how the changing dynamics of patient care, including large data sets and by teams, are impacting AHCs. Topics covered include the emerging delivery system science, the role of faculty, interprofessional education, creative partnerships, and payment models that will change the nature of AHCs' care delivery models.

HISTORY OF ACADEMIC HEALTH CENTERS IN THE LANDSCAPE OF THE US HEALTH CARE SYSTEM

The major AHCs are often called the crown jewels of health care in the United States. These are conglomerations of privately owned health care facilities—including hospitals, cancer centers, outpatient clinics, and research organizations that also either include, or are closely affiliated with, a medical school. While their national and international reputation is usually based on their state-of-the-art specialty medical services and research discoveries, AHCs serve as regional health care hubs and typically provide a full spectrum of medical services—including prevention and screening, primary care, imaging, emergency medical services, and palliation.

Not surprising, the challenges that the US health care system is facing also directly affect AHCs. Numerous attempts at health care reform led by some twentieth-century US presidents—Harry Truman, Richard Nixon, Jimmy Carter, and Bill Clinton—were made against the fierce opposition from stakeholders, such as the American Medical Association, businesses, and the insurance industry. Fragmented political institutions and Americans' skepticism about government, enabled opponents to scare the public with the specter of socialized medicine and tales of horrors in foreign health systems [1].

The Obama administration and congressional Democrats learned from their predecessors' mistakes and sought to neutralize any stakeholder opposition, such as the Pharmaceutical Research and Manufacturers of America and the American Hospital Association. The administration negotiated deals with health industry groups to support reform in exchange for millions of newly insured-patients to treat [1]. While the future impact of the 2010 Patient Protection and Affordable Care Act (ACA) is unknown, it is becoming increasingly obvious that the role and function of AHCs must undergo fundamental changes [2].

CURRENT CHALLENGES

The American health care system faces mounting pressures for cost containment, quality improvement, and public accountability. Health care workforce needs are increasing as baby boomers age and chronic disease grows. Our society is becoming increasingly diverse, intensifying the need for cultural awareness in health care. As society becomes more consumer and technology savvy, expectations of how health care is delivered and at what cost are changing. Health care reform is not just legislation to extend coverage to tens of million more Americans; it is a seismic shift in the way providers practice medicine, the way clinicians train, the way scientists discover cures, and the way health insurance is provided.

In the emerging clinical environment, the health care system will demand more clinical integration, more risk for providers, and more transparency for quality, outcomes, and cost. The 2010 ACA has initiated new payment structures, through both Medicare and Medicaid, shifting incentives from volume to value (quality/cost) and driving care models to highly integrated, team-based approaches to care. Improved information technology infrastructures will provide patient data to help these clinical teams manage large populations with an emphasis on prevention and coordinated care of the chronically ill. Managing care of unpredictable,

The Transformation of Academic Health Centers. http://dx.doi.org/10.1016/B978-0-12-800762-4.00020-7

complex events will also be essential. Over time, reimbursement models will likely transition from fee-for-service to episodic, bundled payments to population-based health delivery systems that are reimbursed on per member, per month basis. These payment models can present risk-sharing opportunities.

With each transition, the emphasis on quality and cost incentives will become stronger; and the need for a large patient population will become paramount. To reach larger patient populations, AHCs are developing a robust care continuum, comprised of low-to-high acuity services in a broad array of outpatient and inpatient settings. This continuum is a mechanism to provide timely, appropriate care in a suitable setting, ultimately managing health care costs by reducing utilization.

As all health systems compete for market share, AHCs can capitalize on their broad tertiary and quaternary care, community engagement and patient catchment areas, and their access to capital in order to attract new health system partners and regional referrals. Most AHCs possess the expertise and resources necessary to navigate through changes initiated by health care reform. As payment models shift from fee-for-service to value-based bundled reimbursement, AHCs can leverage their academic and clinical resources to provide the highest level of service for commodity illnesses (e.g., pneumonia, urinary tract infections) and create centers of excellence for expensive core services such as cancer, cardiac surgery, and pediatric subspecialties. As health care shifts to a population health model, emphasis on primary care as a patient gateway approach, as well as wellness models to keep the population healthy, will grow.

Through local and national recruitment efforts, AHCs can directly place their graduates into new and current practices, thus filling increasing skills gaps as older providers retire and demand for new skills grows. However, reform also exposes some long-standing AHC vulnerabilities. AHCs' often fragmented and siloed leadership can produce inefficiencies and barriers to increased integration. As the emphasis on volume in the current fee-for-service model disappears, focus will shift to providing value through resource use. Components of resource use include utilization management, administration, evidence-based care models, and maximizing clinical team members' scope of practice. With this shift, AHCs must successfully reconceptualize the value of their academic components and adopt multidisciplinary models of care delivery. For example, URMC has successfully implemented several clinical service lines—starting with the heart/vascular and pediatric service lines established in 2013. In April of 2014, a cancer service line was added. The service line has been organized into 10 patient-centered groups, each focusing on a specific disease area, including brain and spine; breast cancer; chronic lymphocytic leukemia/lymphoma; gastrointestinal; genitourinary; head, neck, and thyroid; leukemia; myeloma,

bone marrow transplant; lung/thoracic; and melanoma and sarcoma. Augmenting the patient-centered groups listed above, six other groups provide additional representation to the service line including cardiac, geriatrics, palliative care, pathology, pediatrics, and survivorship/integrative services.

Success in this new environment will require a fair amount of teamwork and willingness to change. Health providers, including both physicians and advanced care partners (ACPs), are facing fairly substantial changes in their job descriptions. For example, URMC is currently expanding its system by developing and growing collaborations with regional physicians, hospitals, and other facilities. Through this approach, URMC providers will see more local patients for primary and secondary care while managing an influx of regional referrals for tertiary and quaternary care. To successfully manage these shifts, URMC providers are increasing their technology competencies (such as eRecord and telemedicine), improving access and communication pathways and policies with community providers and facilities, and increasing internal throughput and efficiencies through "Lean" processes. This approach allows patients and their families to receive care in their local community as appropriate. Rethinking individual members' scope of practice to maximize the overall clinical team effectiveness will be an essential cultural shift in clinical care. As our society increasingly integrates technology into day-to-day activities, patients' expectations of how their providers use technology will evolve. Understanding when and how to use a technological solution in providing patient care will become an important component of a health care provider's skill set.

As today's health care delivery becomes increasingly complex and integrated, a new educational model must appropriately address the current and emerging learning needs of students, trainees, and practitioners. In the modern health care system, society needs practitioners who demonstrate clinical and research excellence, work within an integrated health care team, develop and utilize evidence-based methodologies, are technologically competent, have high patient safety standards, focus on patient engagement, and practice efficiencies. Health education programming should prepare learners to function and lead in a diverse, heterogenous society.

This transformation at AHCs will be dramatic and require intensive coordination across the tripartite missions. Long-engrained workplace cultures must evolve. Previously disparate areas of AHCs must enhance efforts in collaboration to become more efficient and competitive. How AHCs engage with outside partners will also be essential to their success. While health care systems must increase their partnership with traditional partners—payers, health care facilities, governmental agencies—new, perhaps nontraditional partners who help the clinical health system deliver high-quality, convenient care to patients outside of a traditional office setting can be useful to reach patients. With insurance products placing more cost burden on individuals and technology changing the way society interacts and communicates, consumerism in health care will increase. Patients will gravitate toward low-cost, convenient health care options when possible. While the quality of care is important, patients' perceptions of quality can greatly differ from a clinician's perception.

Health care reform's heightened focus on value—as well as emerging technologies—is also increasingly changing biomedical research portfolios. By leveraging the research mission, AHCs have both the opportunity and challenge to lead in developing new knowledge in areas necessary to developing high-performing health care systems. Implementation science—the study of methods to promote the integration of research findings and evidence into health care policy and practice—helps clinicians reduce waste and improve quality. Implementation science seeks to understand the behavior of health care professionals and other stakeholders as a key variable in the sustainable uptake, adoption, and implementation of evidence-based interventions. Grant funding is also shifting to encourage implementation science research (e.g., Patient-Centered Outcomes Research Institute and Center for Medicare and Medicaid Innovation programs).

As biomedical science increasingly becomes information science, a medical center's need for biomedical informatics will grow in order to effectively use biomedical data, information, and knowledge for scientific inquiry, problem solving, and decision-making. In response to these challenges, AHCs will have to balance budget constraints with the need to invest in emerging and essential research areas.

THE FUTURE OF CARE DELIVERY

To successfully navigate the changing dynamics of patient care, AHCs' should focus resources on modifying and developing their patient care. Areas that will require close consideration include the emerging delivery system science, the role of faculty, multidisciplinary care teams, creative partnerships, and future payment models that will change the nature of AHCs' care delivery models.

Multidisciplinary Care Teams

Multidisciplinary teams are widely thought to be a more effective way to deliver health care. By bringing together many professionals, teams often produce improved adherence to evidence-based guidelines, better treatment decisions, and better clinical outcomes. While these benefits are well documented, the economic benefits and organizational aspects of team care are not well understood. Also less known are the characteristics that benefit and

hinder a team, which may vary across settings and patient types (Taylor, 2012). Finally, emerging team training and educational models are insufficient to meet the needs of all teams in all settings.

Historically, teams have been better utilized and most effective in nursing homes, pediatrics, and breast cancer management. However, many more areas appear to be ripe to shift from the traditional model where a patient is expected to follow up with multiple specialists—one for each organ system or a new diagnosis—to a team-based, coordinated model with providers sharing access to electronic medical records and appointment schedules, utilizing care managers or nurse coordinators, and delivering care in a more patient-friendly setting (multidisciplinary outpatient clinic or colocated providers or medical groups). Patient-centered medical homes[1], multidisciplinary cancer service lines[2], and accountable care organizations[3] are some examples of multidisciplinary teams formed around AHCs and supported by aligned financial incentives.

To thoughtfully approach these opportunities to improve care and science, URMC has developed the Institute for Innovative Education. This URMC group of interprofessional educational leaders develops projects that collaboratively prepare future and current providers to work effectively in this increasingly collaborative environment.

The shift toward multidisciplinary care also highlights the importance of ACPs in the success of modern day AHCs. Active involvement of ACPs—such as nurse practitioners, physician assistants, and social workers—in vertical patient care teams allows medical centers not only to reduce overall cost of care, but improve quality and patient satisfaction. The additional benefit of reduced time in the waiting room, allows for less-rushed provider–patient interactions, and hence, more opportunities for shared decision-making and better treatment satisfaction. More research on the social structure, relationships, and status of health professionals should help us understand how successful teams produce superior health outcomes.

Multidisciplinary teams of providers will replace the sacrosanct one-to-one doctor–patient relationship. These teams will have to fully leverage their collective expertise and skills to provide high-quality, low-cost care. As clinical margins decrease, figuring out efficient, effective models to gain the most value from these teams will be key. This will require a concerted effort to change behavior and embedded culture. Disciplines currently train separately, yet will be expected to behave as a team on the hospital floor, with implicit understanding of everyone's skills, expertise, and specific strengths. This is tantamount to pulling together

---

[1] http://pcmh.ahrq.gov/.

[2] http://sites.mc.rochester.edu/faculty/faculty-focus/top-stories/ts_140403_cancerserviceline.aspx.

[3] http://www.cms.gov/Medicare/Medicare-Fee-for-Service-Payment/ACO/.

individual baseball players and expecting them to play well together in games without any team practice. This bumpy transition from student to provider can be alleviated through interprofessional education. This is an opportunity to instill the team mind-set early in training and help grow an appreciation of the various skills and expertise of a team. Scope-of-practice regulations will be reviewed with careful consideration of the new delivery system.

In sum, it is becoming more commonly accepted that multidisciplinary care delivery leads to better patient outcomes and has a potential for reducing costs. To further implement this idea of collaboration in future generations of health care providers, medical education needs to shift from its current teaching style that emphasizes individualist to one that encourages teamwork. If professionals are taught to use all of their available resources, including the disciplines of their peers, they will likely be more welcoming of teams in their careers and have higher motivation to work in these teams.

Emphasis on Specialization and Standardization

Advances in biomedical science over the past few decades have made it possible to successfully treat many chronic and acute conditions, significantly improving survival for millions of patients. At the same time, efficient delivery of state-of-the-art medical care has become increasingly complex, technologically challenging, and resource intensive. As a result, many community hospitals, rural centers, and single providers can no longer accomplish or financially sustain such services and remain financially viable.

Evolution in cancer care is by far the most striking illustration of shifts in the highly-specialized and standardized care paradigm. Since the 1980s, the synergy of five main principles have led to dramatic reductions in rates of local recurrence, increases in disease-free and overall survival, and reductions in permanent stoma rates in colorectal cancer [3–8]. In countries that have implemented standardized care pathway programs based on these principles, cancer-specific outcomes for colorectal cancer have improved dramatically. The standardization of services involves: (1) a multidisciplinary team approach that identifies, coordinates, delivers, and monitors the optimal treatment on an individual patient-by-patient basis; (2) appropriate stage- and site-specific imaging techniques identifying those patients at high risk of local recurrence; (3) raising quality of surgery by using organ-preserving techniques and minimizing permanent stomas; (4) the use of newer, more effective neo-adjuvant and adjuvant therapies, including radiotherapy and chemotherapy; and (5) use of pathology assessment, along with other time-sensitive metrics. In most areas, these resources are concentrated under the umbrella of an AHC and generally are unavailable to community and rural providers.

One way to overcome this geographic maldistribution and improve access to high-quality recommended care for patients residing far from AHCs is to develop regional standardized pathways of care and cost-effective regional systems. While there is not extensive literature on clinical pathways in the United States, existing data show that implementation of regional clinical pathways and centers of excellence is an effective method to improve quality, reduce variability, and subsequently decrease costs. A report by Neubauer et al. [9] demonstrated a 35% decrease in outpatient costs with equivalent outcomes for patients with nonsmall cell lung cancer treated according to clinical pathways, compared to those receiving nonpathway treatment. A study by Hoverman et al. [10], which investigated the cost-effectiveness of a clinical pathway for colon cancer, confirmed decreased cost, a lower rate of chemotherapy-related hospital admissions, and comparable survival with the use of the pathway.

The success of these pathways has been largely through alignment of incentives among academic and community providers for pathway compliance and improved outcomes. Using financial incentives for compliance, one collaboration in Michigan was able to obtain pathway compliance from cancer care providers of 95% within the first year [11]. Geisinger Health System, a nonprofit, integrated delivery system in Pennsylvania caring for over 290,000 patients, has reported considerable success with the use of biannual incentive payments based on a combination of fee-for-service with additional metrics for high quality, outcome improvement, and innovation. This strategy has led to sustained reductions in costs through improved patient outcomes and a drop in complications and readmissions [12].

Big Data

The ongoing transition to the electronic health record (EHR), combined with rapid increases in computing power and rapid declines in the cost of generating massive amounts of health-related data (exemplified by the 5-$\log_{10}$ decrease in the cost of sequencing a human genome over the past 14 years[4]), have created an unparalleled—and exponentially growing—amount of health data. This has, in turn, created new opportunities to apply data science to the improvement of health and health care. For example, health is determined by multiple factors—including genetics, behavior, social circumstances, and environmental influences[5]; data science allows us to aggregate these disparate information sources and to analyze the resulting data to make more accurate predictions of disease risk at both the population and individual level—and, to develop more effective approaches to care for patients, their families, and their communities.

---

[4] http://www.genome.gov/sequencingcosts/.
[5] http://www.nejm.org/doi/full/10.1056/NEJMsa073350.

Big data is also changing the patient–provider relationship, by expanding access to care while also enhancing the ability of patients to take charge of their own health. This is occurring, for example, through the expansion of mobile devices—such as FitBit—in home care services, as well as via greater patient access to their own health information (e.g., Epic's MyChart). At URMC, faculty are piloting "virtual house calls" for Parkinson's disease with the twin goals of increasing patient access to care while also reducing costs. Another application of big data is in "precision medicine"—or the use of genetic information to diagnose or treat disease. This is likely to have a major impact on cancer care by allowing clinicians to match treatments to the specific mutations present in an individual patient's tumor. While most genetic information is presently not clinically actionable, this is likely to change over the coming years—and as it does, the need for education of both providers and patients in the science of genomics will become increasingly important.

A particular challenge to the analysis of health information is the fact that the EHR is dominated by unstructured data (such as clinician's notes) that do not lend themselves to simple computational analysis. As a result, natural language processing has become a critically important need because it can enable computers to efficiently extract meaning from human language. Cognitive computing systems, such as IBM's Watson platform, can offer a powerful solution to this problem because they are able to build knowledge over time, thereby improving their performance in much the same way as do people. Such systems can rapidly analyze vast amounts of unstructured information—including physician's notes, lab reports, and published clinical literature—and then present hypotheses (or treatment options) based on that analysis. Consequently, Watson is being used to help community physicians provide better treatment options for their cancer patients.

Fully optimizing data aggregation and analytics can help achieve the triple aim of improving quality while driving down cost and enhancing patient outcomes. For example, data analytics can identify the drivers of cost in specific health conditions and unwarranted variations in delivery of care for those conditions—thereby allowing health systems to deliver higher value care at reduced cost. This also has implications for reducing health disparities across patient populations and improving community health—a major focus at URMC through its Center for Community Health and its new Rochester Center for Health Informatics. Other "low hanging fruit" include the use of health analytics to predict risk of hospital admission or readmission among patients with chronic medical conditions—such as heart failure and hypertension—and to then use that information to target interventions (such as home visits by nursing professionals) that can stabilize such at-risk individuals. As a result, one of the three major focus areas for the University

of Rochester's newly established Institute for Data Science is in predictive health analytics.[6]

A significant constraint on realizing value from big data may be in the shortage of talent—particularly of people with in-depth expertise in statistics and machine learning and the managers and analysts who know how to operate systems by using insights from big data.[7] There is also a dearth of clinicians and scientists with expertise in informatics and data science—and the development of new training and education programs in bioinformatics is a major unmet need across AHCs. At the University of Rochester, the Institute for Data Science is actively collaborating with the URMC's Clinical and Translational Sciences Institute, and its nascent Center for Biomedical Informatics, to provide such training programs, with an emphasis on interdisciplinary education that combines a core curriculum in computer science and statistics with track-specific specialization in specific clinical/biomedical domains. Individuals trained in this manner will be essential to meeting health care needs—and the health care systems that are successful at recruiting, retaining, and developing their analytics talent will have a clear competitive advantage.

Related to this is the ability to leverage institutional computational resources for training and research/analytical purposes. In the last five years, the University of Rochester has invested heavily in high-performance computing infrastructure and expertise through its Center for Integrated Research Computing (CIRC) and its partnership with IBM—which has led to the establishment of the new Health Sciences Center for Computational Innovation (HSCCI). The HSCCI is presently one of the five most powerful university-based supercomputing sites in the nation, and is home to IBM's latest Blue Gene/Q supercomputer. In addition, a new state-of-the-art data visualization laboratory is about to come online with the goal of further-enhancing institutional capacity around big data. Combined university and medical center resources can help attract a broad pool of talent to collectively unlock big data's potential.

Interprofessional Education

Changes in care delivery are driving changes in health professions education as dramatically as the Flexner report shaped medical education in the 1910s. The Flexner report called for a greater education for physicians in science and to anchor medical education in our universities. Before that time, physicians largely relied on their bedside manner and relationships with patients to "treat" their illness. In the 1950s, an emphasis on physicians becoming proceduralists and interventionists placed importance on integrating clinical skills in medical education. Most recently, the

---

[6] http://www.rochester.edu/data-science/about/index.html.

[7] www.mckinsey.com/mgi.

awareness of the need to develop the skills to lead and be a member of a health care team has become the greatest influence on medical education. Our understanding of the levels of learning in medical education has evolved from informative (information and skills), to formative (socialization and values), and, finally, to transformative learning—placing an emphasis on leadership attributes and physicians as change-agents in complex health care systems (Lancet). Teaching and developing core competencies in collaborative care are now the drivers in health professions education. This includes teaching and defining values and ethics for interprofessional practice, roles and responsibilities, interprofessional communication, and elements of teams and teamwork [13].

Interprofessional education is a driving influence in the call for change in health professions education. With the assumption that interprofessional education facilitates collaborative care, all health professions education programs are working to create opportunities for students to learn "with, from, and about each other" in meaningful ways. This includes, at a prelicensure level, shared curriculum on patient safety, ethical decision-making, and observation skills. Within the clinical setting, we are observing interdisciplinary teaching rounds, interprofessional and interdisciplinary grand rounds, conferences, and lectures. In a 2009 Accreditation Council for Graduate Medical Education (ACGME) multispecialty resident survey, results positively correlated formal team training with nonphysicians to resident satisfaction and self-reported serious medical errors [14]. To train and retrain health care teams, providers need to learn and grow together, both within the clinical environment and within professional development educational programming. Recent changes in accreditation standards, both at the student and continuing education levels, reflect this shared value and vision for interprofessional education.

At URMC and its affiliated Wilmot Cancer Institute, interprofessional team-based care delivery is becoming a new standard. In addition to services lines, care plans for cancer patients are discussed at weekly tumor boards that bring together all providers involved in cancer care including medical and radiation oncologists, surgeons, radiologists, pathologists, and nursing. Furthermore, specific multidisciplinary standardized care pathways have been developed and implemented to reduce the risk of surgical site infections and blood clots in cancer patients.

Challenges for health professions education include the financial and culture barriers that remain from an apprenticeship model of education. Currently, funding for education is largely siloed and/or driven by the academic schools that have varying partnerships with clinical health systems. In theory, this would support increased collaboration and shared resources, but the culture of siloed training prohibits, in many cases, team-based education. An ideal model for team-based education would be to align educational initiatives anchored in patient care around service lines that represent the continuum of learners as well as all related specialties and disciplines.

Implementation Science

As technology in health care allows for greater mobility, gaining a better understanding of which technologies are effective in patient care is essential to health systems' success. The rise of implementation science—the study of how to effectively implement clinical processes and regimens into practice—will play a significant role. Driven by AHCs, this research can have far-reaching impact across the health care system. In 2008, URMC opened the Saunders Research Building (SRB), which serves as the hub of clinical and translational research for both URMC and a network of researchers across the state. Through colocation, multidisciplinary teams of scientists and physicians can collectively tackle tough problems in human health and educate the next generation of translational researchers. The URMC Center for Research Implementation and Translation (CRIT, often referred to as the Implementation Center) encourages and supports research translation through the development of implementation science. The Center's scientists conduct rigorous, hypothesis-driven studies in which clinical research findings are tailored, tested, and systematically incorporated in clinical practice, and novel research methods are developed for the field.

As the use of technology in medicine grows, implementation science will be essential to determine what interventions and measures are and are not effective. Providers can perform exams and communicate with patients remotely through the use of telemedicine devices. Patients can also be monitored remotely (e.g., home health care telemonitoring) and provided in-person care as needed. Patients have greater access to their health records as well as health maintenance tools (e.g., blood pressure cuffs at shopping centers, health and wellness apps). Programs, such as the National Healthy Mothers, Healthy Babies Coalition's Text4Baby, allow pregnant mothers to receive timely texts throughout their pregnancy with health and wellness advice. Understanding what is effective may prove to be a challenge when you consider how long a research study can take compared to how quickly technology evolves. For example, it may take 7 years to conduct a research study and publish the results. Technology capabilities will advance more rapidly in that same time frame [15]. Similarly to the technology development process, implementation science focused on technology in health care must be rapid-paced, iterative, adaptive, and transparent. With appropriate analytics tools to pull information from a variety of sources, rapid collection and analysis of relevant data can be assessed and compared to determine the most effective pathways and settings. This culture shift from more traditional biomedical research will utilize research resources more effectively while yielding key findings necessary to the evolution of health care delivery and communication.

Often, once a best practice or process is determined, integrating it into clinical practice is slow and challenging. A variety of reasons could explain these barriers: lack of knowledge or education, engrained attitudes or behaviors, organization resources, and decision support systems. Health care educational institutes play a huge role in educating scientists on how to develop or evaluate health interventions or policies as well as increasing emerging clinicians' understanding of the effects of implementation science and how it can positively impact the health care system. Graduating clinicians must not only expect, but demand, changes that continually improve the delivery system throughout their career [16].

Role of Faculty

As systems change, providers need to develop their own competencies in team-based and collaborative care; and to do that they need faculty at all levels who can model and deliberately teach change process. Faculty are the essential component for change; however, many of our current clinical educators need retraining in the core competencies for interprofessional teamwork cited earlier. Traditional health professions education, including continuing education, is based on the faculty as the expert model. Because collaborative care delivery is an emerging framework, we do not have the expert panel that is found within the clinical silos for clinical care, such as disease management and treatment.

The faculty role must evolve to be facilitative and reflective around group process in learning, thus modeling process change. It is important to consider how faculty can serve as advocates and change-agents within these adaptive systems; and a continuous quality improvement process will drive education. As learners can pursue knowledge-based education through online learning, the faculty role will center on systems-based teaching and shared decision-making [17]. Along the continuum of learning, faculty's role will continue to shift dominantly on connecting theory to practice. Faculty can critique and analyze practice gaps, lead shared decision-making, and promote metacognitive approaches to learning that support lifelong learning (reflection, error detection, pattern recognition).

Most importantly, the definition of who are faculty is also evolving. Collaborative practice and team training provide tools for scope of practice and, as a result, provide teaching and learning from multiple sources within integrated teams. In essence, teaching and learning can occur in a 360 approach between physicians, mid-level providers, nursing, allied health, and patients and their families. This supports the World Health Organization definition of interprofessional education as: "When students from two or more professions learn about, from and with each other to enable effective collaboration and improve health outcomes" [18]. Faculty will need to be interprofessional as well, supporting interprofessional education that supports collaborative practice.

Creative Partnerships

Academic medicine is currently undergoing a metamorphosis. Biomedical research is becoming a more data-driven science. Research funding is threatened. Educational learning environments, tools, and curriculum are evolving. Clinical reimbursements are changing to incentivize prevention and wellness. Health systems are developing risk contracts with payers and employers. Hospital mergers are rapidly consolidating markets. Capital and information technology demands weigh on hospital budgets. As this environment changes, AHCs that can be nimble and form creative, strategic partnerships and collaborations will ultimately fare better.

As health systems compete for market share, creative, strategic partnerships that help enhance the quality, accessibility, and outcomes of clinical care will grow in value. An example is the myriad of partnerships that hospitals are forming with health and wellness centers. Hospitals are engaging with health and wellness center members—those individuals and families who already are motivated to regain or maintain their health, thereby reducing their level of risk essential in population health management. Many health and wellness centers offer a range of one-stop shopping services such as fitness, diet and nutrition counseling, day care, and laundry services. Adding primary care practices or wellness-centric medical services (e.g., biometric screening, orthopedic services) is valuable as it strengthens the fitness center's membership offering. Cobranding of these partnerships allows for the medical partner to be associated with community-based health organizations.

An example of this partnership is the Healthy Living Center™ in Des Moines, Iowa. This partnership formed in 2009 between the Mercy Medical Center–Des Moines, LADCO Development, and the YMCA of Greater Des Moines. The YMCA and Mercy have a joint operating agreement to run the Healthy Living Center™ on hospital campus grounds. The YMCA provides daily management and Mercy provides clinical services and medical oversight, and each party shares 50% responsibility for the center's financial performance. Serving a membership comprised mainly of health seekers who were referred to the YMCA by their physician, the Healthy Living Center runs an array of medical-based fitness programs—including physical therapy, cancer survivor wellness, weight loss, pain management, cardiac rehabilitation, stroke recovery, neurological wellness, and diabetes wellness. The cross-directional referrals between Mercy and the YMCA have created new margins while building both Mercy's patient and the YMCA's membership base. The benefits to Mercy and the YMCA are numerous—competitive advantages, increased referrals and membership, increased physician alignment, enhanced reputation in community as a health and wellness provider, and risk mitigation.

One of the keys to the success of the Des Moines Y Healthy Living Center is the rich referral network developed with local physicians and other health professionals. Patients referred by a physician to the Y Healthy Living Center tend to remain members. While some join other Y branches more convenient to them, over 50% of referrals continue with membership at the Healthy Living Center. In addition to being an effective source for new members, referring physicians are content having wellness-related services for their patients [19]. As care-delivery changes and margins constrict, creative partnerships like the Healthy Living Center can contribute to a medical center's long-term stability.[8]

With our local YMCA network, URMC is developing a collaboration modeled after the Y Healthy Living Center. Initial efforts to develop a clinical wellness presence at the largest YMCA local branch include a full-service primary care office, physical and occupational therapy, disease management and biomedical screenings, as well as wellness-focused education. URMC is working to grow this collaboration across the local YMCA network. Especially with a primary care presence, these partnerships represent a new avenue to grow URMC's patient base.

Payment Models

AHCs are undergoing a period of great fiscal uncertainty. Hospital care delivery and payment models are changing and markets are consolidating. To be competitive, academic health systems must decrease costs and increase quality to provide value. Under the ACA of 2010, the clinical mission (the economic engine of most AHCs) will not be able to achieve its current margin levels in the near-future due to mandated cuts in Medicare and Medicaid, as well as decreases in the commercial market, secondary to the development of health exchange insurance.

As reimbursement models transition from fee-for-service to episodic, bundled payment to possibly population-based, and capitation models, the emphasis on quality and cost will become stronger in addition to the need for a larger population to spread risk. To support population health management, more flexible financing models are needed. They must empower provider networks to choose where in the system they need to direct resources to achieve the best possible outcomes. As health systems develop, their low-cost facility offerings (e.g., community hospital, outpatient settings) will help systems provide medically appropriate and cost-effective care. Patients who want

accessible, nonhospital-based care will gravitate toward these facilities. Better integration of remote monitoring and other technologies, such as telemedicine, will also appeal to patients while allowing health systems more flexibility to develop effective, efficient processes. To facilitate this monumental shift, URMC is developing a population health management program that will manage URMC and affiliate planning, directing, organizing, and controlling of services needed for population health management. This new program will streamline population health management activity by consolidating previously disparate committee efforts.

URMC is growing its footprint through a local primary care expansion strategy as well as a regional effort to increase tertiary and quaternary referrals. While it is unclear when it is best to transition from fee-for-service to population health management, URMC is laying the groundwork with the help of a local provider network and its faculty group to expand its services to new populations. Through the new provider network, members will participate in emerging payment plans that reward hospitals and doctors for better results and improved efficiency. The overall strategy uses a myriad of approaches—acquiring hospitals and practices, building new collaborations, negotiating employer and payer contracts, and improving care systems to efficiently accommodate referrals and new patients. These payer and employer contracts allow URMC to gain population with policies and standards in infrastructure in place to help manage their risk. URMC is rapidly growing into a regional hub and model that allows patients to receive high-quality primary and secondary care supported by URMC in their community, with access to referrals for URMC's tertiary and quaternary services as needed.

CONCLUSION

As the US health care model changes, it must evolve to meet societal expectations for accessibility, cost, quality, and transparency. The health systems that do so successfully will thrive in the new health care environment. This transformation will take tremendous effort, requiring change in education, clinical care, research, reimbursements, technology, partnerships, and culture. While the initial efforts to be patient-centered have yielded some success, the coming rise of consumerism will drive markets to be even more competitive. Like many AHCs, URMC is working to successfully evolve into its own unique system that meets the needs of its patients.

REFERENCES

[1] Oberlander J. Long time coming: why health reform finally passed. Health Aff 2010;29(6):1112–6.

[2] Fuchs VR. Health reform: getting the essentials right. Health Aff (Millwood) 2009;28(2):w180–3. http://dx.doi.org/10.1377/hlthaff.28.2.w180. Epub 2009 Jan 16.

---

[8] http://www.ymcahealthylivingcenter.org/.

[3] Elkin EB, Bach PB. Cancer's next frontier: addressing high and increasing costs. JAMA 2010;303(11):1086–7.

[4] Khani MH, Smedh K. Centralization of rectal cancer surgery improves long-term survival. Colorectal Dis 2010;12(9):874–9.

[5] Khoo CK, Vickery CJ, Forsyth N, Vinall NS, Eyre-Brook IA. A prospective randomized controlled trial of multimodal perioperative management protocol in patients undergoing elective colorectal resection for cancer. Ann Surg 2007;245(6):867–72.

[6] Morris E, Haward RA, Gilthorpe MS, Craigs C, Forman D. The impact of the Calman-Hine report on the processes and outcomes of care for Yorkshire's colorectal cancer patients. Br J Cancer 2006;95(8):979–85.

[7] Siegel R, Naishadham D, Jemal A. Cancer statistics, 2013. CA Cancer J Clin 2013;63(1):11–30.

[8] Wille-Jørgensen P, Sparre P, Glenthøj A, Holck S, Nørgaard Petersen L, Harling H, et al. Result of the implementation of multidisciplinary teams in rectal cancer. Colorectal Dis 2013;15(4):410–3.

[9] Neubauer MA, Hoverman JR, Kolodziej M, Reisman L, Gruschkus SK, Hoang S, et al. Cost effectiveness of evidence-based treatment guidelines for the treatment of non-small-cell lung cancer in the community setting. J Oncol Pract 2010;6(1):12–8.

[10] Hoverman JR, Cartwright TH, Patt DA, Espirito JL, Clayton MP, Garey JS, et al. Pathways, outcomes, and costs in colon cancer: retrospective evaluations in 2 distinct databases. Am J Manag Care 2011;17(Suppl. 5):SP45–52.

[11] Feinberg BA, Lang J, Grzegorczyk J, Stark D, Rybarczyk T, Leyden T, et al. Implementation of cancer clinical care pathways: A successful model of collaboration between payers and providers. Am J Manag Care 2012;18(5):e194–9.

[12] Lee TH, Bothe A, Steele GD. How Geisinger structures its physicians' compensation to support improvements in quality, efficiency, and volume. Health Aff (Millwood) 2012;31(9):2068–73.

[13] IPEC reference: Interprofessional Education Collaborative Expert Panel. Core competencies for interprofessional collaborative practice: report of an expert panel. Washington, DC: Interprofessional Education Collaborative; 2011.

[14] Baldwin Jr DC. Fostering interdisciplinary teamwork in graduate medical education. Presentation. In: Sixth annual AAMC physician workforce research conference. Alexandria, VA, 2010.

[15] Glasgow RE, Phillips SM, Sanchez MA. Implementation science approaches for integrating eHealth research into practice and policy. Int J Med Inform 2013. http://dx.doi.org/10.1016/j-ijmedinf.2013.07.002.

[16] Josea Kramer B, Howe JL. The role of education in implementation science. Gerontology Geriatrics Educ 2014;35:1–3.

[17] Frenk J, Chen L, Bhutta ZA, Cohen J, Crisp N, Evans T, et al. Health professionals for a new century: transforming education to strengthen health systems in an interdependent world. Lancet 2010;376(9756):1923–58. http://dx.doi.org/10.1016/S0140-6736(10)61854-5. Epub 2010 Nov 26.

[18] World Health Organization (WHO). Framework for action on interprofessional education & collaborative practice. Geneva: World Health Organization; 2010. Retrieved April 11, 2011 from http://whqlibdoc.who.int/hq/2010/WHO_HRH_HPN_10.3_eng.pdf.

[19] The YMCA Healthy Living Center, http://www.ymcahealthyliving-center.org/, Retrieved April 17, 2014.3k.

ABOUT THE AUTHORS

Bradford C. Berk, MD, PhD is the Senior Vice President for Health Sciences at the University of Rochester and CEO of the University of Rochester Medical Center (URMC). Dr. Berk joined URMC in 1998 as Chief of the Cardiology Division. Prior to this, he served on the faculties of Harvard Medical School, Emory University, and the University of Washington.

Katia I. Noyes, PhD, MPH is Professor of Surgery and Public Health Sciences and Scientific Director of Surgical Health Outcomes and Research Enterprise (SHORE) at the University of Rochester School of Medicine. She is co-Director of Comparative Effectiveness Research Key Function Committee for the University of Rochester Clinical and Translation Science Institute, and co-Director of Center for Research Implementation and Translation (CRIT) at the URMC.

Sarah E. Peyre, EdD is the Assistant Dean for Interprofessional Education, an Associate Professor of Surgery, and an Assistant Professor of Nursing at the University of Rochester Medical Center. She serves as the Director of the Center for Experiential Learning, a centralized simulation and educational support center for the educational programs within the School of Medicine and Dentistry, School of Nursing, Eastman Institute of Oral Health, Strong Hospital and the Faculty Practice Group.

Katy A. Stevenson, MPP joined the Strategic Planning Department at the University of Rochester Medical Center (URMC) in 2010. As a Manager for Strategy, Planning and Development, she focuses on institution-wide strategic efforts that especially impact the research and education missions.

Chapter 21

Making the Patient Paramount

James Merlino

A physician colleague once asked: "why are we focusing on making patients happy? We have more important things to do like practicing medicine" [1]. Others have pointed out "a patient will never be happy with a new diagnosis of cancer." However, patient experience and patient-centered care is not about just making patients happy, it is about how health care professionals can better deliver the best care—ensuring and providing safe, high-quality health care in an environment of patient centeredness.

Patients will never be "happy" with a diagnosis of cancer. As health care providers and caregivers, we have an obligation to ensure that the news is delivered in a compassionate, humanistic, and caring manner so that patients believe—in fact know and feel—that they have a partner in what will be a very difficult journey for them and their families. Doing this correctly is not only the right thing to do, but it is the way we would want ourselves or our families to be treated. Health care providers are all part of a health care ecosystem—not because we work in health care or a field that impacts health care, but because someday, perhaps right now, we or our family will be patients. We should all demand that care be more centered on the patient. The physician colleague noted above did not understand that you cannot separate human caring from the medical care that we deliver.

CHALLENGES TO PATIENT-CENTERED CARE

We are seeing many changes in health care delivery. The Affordable Care and Accountability Act of 2010 set in motion health care reform that has never been experienced in the United States. Among many of the new initiatives—and for the first time ever—performance in safety, quality, and patient experience metrics has been linked to hospital Medicare reimbursement. This linkage is successfully changing the way providers and systems do business. Medicare, using its position as the largest payor in the United States and its power as the hospital regulator, has successfully linked the two things that hospitals and physicians care about most (exclusive, of course, of taking care of patients)—reputation and money— through public reporting of data (data transparency) and its link to hospital reimbursement.

The linkage of performance data and reimbursement is redefining the relationship between physicians and hospitals and the way they deliver care to patients. Traditionally valued physician independence and autonomy is declining as standardization of practice and regulation of performance and outcomes are increasing. Since hospitals are the most at risk for reimbursement penalties, it is forcing them to pay close attention to physician performance. Physicians are also being held more accountable for their behavior and their ability to communicate effectively with patients. Patient-reported physician communication performance is part of Medicare's Consumer Assessment of Healthcare systems and Providers survey program. Hospitals and physicians are being forced to find ways to work together as they must rely on each other in order to be successful to meet these new challenges and be successful in this ever-changing and complex health care environment.

Other issues that are contributing to the burning platform of change in health care delivery include cost and price transparency. The services we deliver cost too much and, in many circumstances, we do not understand how much something costs or why things cost what they do. The amount of money that we charge—or the price of our services—is inconsistent and poorly designed and understood. From the patient perspective, health care financing is a black box. As pricing variations become more available and transparent, it is clear that the same test that costs $450.00 at one site can cost as much as a $1000 at a different site operated by the same organization.

These changes are helping to fuel consumerism in health care, which is also increasing as insurance plans require greater out-of-pocket payment responsibility. How much something costs, as well as performance on safety and quality metrics, will impact how consumers choose their health care—and they will want to pay less. If a patient's out-of-pocket deductible per year is $2000 at the beginning of January and the cost difference of a diagnostic test between two different health care systems is $250–$750.00, the patient will choose the organization with the lower cost. The government, private payors, and private companies are forcing greater transparency in pricing so that consumers can become more invested in their health care, helping them to make decisions that are centered on out-of-pocket cost.

The Transformation of Academic Health Centers. http://dx.doi.org/10.1016/B978-0-12-800762-4.00021-9

In addition to these changes, systems and practices are challenged by a need to continually grow and see more patients.

Other challenges are upon us as well. For the next 18 years, 8000 baby boomers will turn 65 every day in the United States [2]. One in eight Americans over 65 years and half of Americans over 85 years have Alzheimer disease [3]. As our population ages and grows—and as medical technology helps us to continue to push life expectancy—the number of people with this disease and other chronic health issues, such as heart disease and diabetes, will also grow. We will need to transform our model of care delivery not only to take better, more efficient care of patients, but also to educate patients and their families to be more engaged in the care process.

These reforms and others have changed both the way hospitals manage their operations and the way providers care for patients. Some will argue that many of the regulatory changes are placing increasing stress on health care delivery. But what is often not discussed is that many of these reforms are necessary to improve a system that has done an inadequate job of monitoring quality, enacting standards, and controlling costs. An Institute of Medicine (IOM) report highlighted the risks and vulnerabilities of our health care system and helped to ignite this burning platform that is necessary to drive reforms [4].

ADDRESSING THE BARRIERS TO PATIENT-CENTERED CARE

To meet these challenges head on, we need to think differently about how we deliver care. Just reducing costs by ordering fewer tests or cutting our operations expenses will not be enough. We need to drive greater standardization of care to reduce variability and increase quality; push "health" education out to our communities to reduce the drivers of chronic disease, such as sedentary lifestyles, smoking, and obesity; and develop and drive new strategies and tactics to better engage our patients and their families. We need new health care partnerships with patients where we work better together to deliver care and assure health.

More important than transforming the care we deliver is the need to transform the way we teach and develop our caregivers. Our goal to meet the changing demands of the health care environment is to better train people to think in terms of a health care system. Health care workers are typically taught to solve problems, cure diseases, and take care of sick people. Physicians are presented with a patient problem and challenged to fix it. Nurses are traditionally branded as those assisting with the treatment of disease. These professionals, and others, are taught in silos. The way we teach our professionals and the way we think of their roles need to evolve. We need to recognize the skills and promote the competencies that different health care professionals bring to the people we serve—e.g., making sure we promote and support nurses operating at the top of their license.

Academic health centers are poised to lead health care to define and teach the way health professionals will deliver care in the future. They can teach providers to accommodate and thrive in this reform environment, to develop and drive the standard by which patients should be cared for, and how we provide health to the populations we serve. To achieve this transformative approach, one of the first tactics that we have to employ is to place a renewed focus on why we do what we do and recognize that we exist to care for our patients and their families. Focusing on patients and their families by placing them at the center of our strategic planning and operational execution of all our care delivery will allow our caregivers to meet the challenges of the future.

PATIENT CENTEREDNESS

Patient centeredness, as defined by the IOM, "encompasses quality of compassion, empathy, and responsiveness to the needs, values, and expressed preferences of the individual patient" [4]. It picks up the "humanistic" side of patient care that is so often ignored. Note that safety and quality are not mentioned—but they are "givens" and represent the basic foundation of effective medical care. Patient centeredness is the patient experience. This definition should be taken to a more operational level so that health care leaders can clearly understand what we are trying to achieve. Often leaders are inundated with a variety of metrics and are responsible for delivering on numerous goals for the organization. Just introducing a concept such as patient experience without defining it and placing it in context will put the strategy of patient centeredness at risk of dilution when focusing on many other competing priorities in health care.

At Cleveland Clinic, we define our patients first. Patient experience strategy is: first, the delivery of safe care; second, high-quality care; and third, care in an environment of satisfaction. This definition and framework for thinking about patient centeredness allows us to think about operationalizing patient-centered care in a clinical environment that must meet the priorities and demands of safety and quality first. Thinking about satisfaction within this framework not only appropriately prioritizes patient centeredness but also places it relative to important initiatives around safety and quality.

Patient centeredness and experience is not about making patients happy. It is about paying better attention and improving the process of care delivery. For example, communication is a critical element of patient-centered care that drives the patient experience. There are nine questions on the Hospital Consumer Assessment of Healthcare Providers and Systems (HCAHPS) [5] survey that relate to how well providers communicate with patients; three questions relate to how well doctors communicate, three to how well nurses communicate, and three to communication about new medications. Certainly if measuring patient happiness was the primary goal, we would not need nine questions to measure it. But when nurses at the bedside communicate more effectively with patients, medication errors, falls, and pressure ulcers are reduced. These are safety issues. When physicians communicate more effectively with patients, compliance with treatment increases. When physicians communicate more effectively and partner with nurses, coordination of care improves. Collectively, these drive improved quality. There is no doubt that when caregivers communicate more effectively with patients, they and their families are more satisfied. Recognizing these relationships should remind us that when we tackle best practices that touch safety, quality, and satisfaction together, we actually improve effectiveness, efficiency, and value in health care. That is what the patient experience and being patient centered is about and how it is connected to health care.

Build Strategies with Patients at the Center

Why care about putting patients at the center of all that is done in health care? For starters, it is the right way to approach the health care industry. Patient centeredness defines purpose. We health care providers are in the health care delivery business to care for people, and we need to work to align our cultures to deliver our product and our service offerings to meet that goal.

Alignment of the organizational culture supports the business of medicine, as well, and can deliver a strong return on investment. Forrester Research keeps an annual customer experience index of some of the top customer-centered organizations in the world. The top 10 performing companies are labeled customer experience leaders, and the bottom 10 are labeled customer experience laggards. Watermark Consulting analyzes these 20 companies against the S&P 500 index. Companies that were considered leaders generated an average of performance three times higher than those considered laggards [6]. Thus, companies that maintain an absolute internal and external customer-centric approach do better than those that do not. Customer-centric organizations build strategies and execute operations to support customer needs.

This analogy is applicable to health care. Dr Delos "Toby" Cosgrove, chief executive officer and president of Cleveland Clinic, related a story about when he spoke before a group of Harvard Business School students. A young MBA student told him that her father was a physician in North Carolina and, needing mitral valve heart surgery, researched the various options available to him. He had read about Dr Cosgrove's pioneering clinical reputation in the field and the Cleveland Clinic's reputation as the No. 1 heart center in the United States, but he chose to go to a competitor because the Cleveland Clinic did not teach its doctors empathy. She asked bluntly: "Dr. Cosgrove, do you teach your doctors empathy?" Stunned, Cosgrove's only response could be "no we don't." What this student had done with one clever anecdote was frame the return on investment (ROI)

on patient-centered care to an organization like Cleveland Clinic that depends on brand and reputation to drive referral business. Nearly 50% of the Clinic's heart business derives from patients outside Ohio. The demographics of these patients represent those that have choice. This student's father had rationalized that he would be fine from a clinical standpoint at any top center, but had differentiated on the experience he believed he would receive [7].

There are those who scoff at the renewed focus on placing the patient at the center and aligning a health care culture around the patient. Some might say "of course we are patient-centered, it is what we do. We are doctors." An argument can be made that this is actually less true. We have become efficient at our trade of medical care. Nurses take care of multiple, complicated cases every day, collecting hundreds of data points, synthesizing observations, recording their findings, and delivering care. Physicians are taught to rapidly zero in on a problem and "get the diagnosis in three questions or less"—not wasting time on things that do not matter to the ultimate problem or solution. While medicine may have been more patient centered 100 years ago, the rapidly expanding armamentarium of medical technology and delivery has resulted in a movement away from our primary mission of treating the patient holistically. Health care providers are taught in medical, nursing, and other health-related professions' schools to keep the patient at the center, but as we progress through the frenetic pace of care delivery and work hard to hone our skills, we lose perspective on why we went into health care in the first place. We—doctors, nurses, and other health care providers—become experts at the product of care we deliver but not in the delivery of that care in a patient-centric fashion.

Before You Begin, Understand What Patients Want

First, we need to understand what is important to patients. Leaders in medicine believe that they know what patients want because they are the professionals and have also personally been patients—therefore they must know. These assumptions can derail a better approach. While health care professionals are very good at telling a patient what they think he or she wants to know, often they fail to listen and provide the patient what he or she wants. At Cleveland Clinic, we have conducted studies to try to better understand what it means to be on the other side of health care. One of the most important elements that stands out is that patients do not want to be treated like patients; they want to be treated like individuals. Being a patient is a very depersonalizing experience. Patients want their health care providers to remember that they are people too, and they often believe that if we caregivers know more about them personally and are invested in them, we may be less likely to make errors that affect their care. William Osler may have been wrong when he suggested we need to be detached to remain objective [8]. It is insulting to believe that we lose our professionalism if we care too much. Developing an emotional connection to a patient does not necessarily mean that we lose our objectivity, but it does signify to the patient that we care more. It is no different than how we would want ourselves treated or our family members treated.

Patients want to be kept informed "in the moment" about what is happening. Just the overall plan of care is not good enough. Patients and families want details—almost like a play-by-play, up-to-the-minute report of what is happening with their care. If a physician walks into a patient's room at 7:00 am and orders a chest X-ray, patients do not want to wait all day to find out what it demonstrated—they want to know immediately. Think of the mind of a patient. If sent down early in the morning for an X-ray, a patient believes it must be for an important reason. If no one talks to them about the results until 4:00 pm in the afternoon, they may wonder: did the physician forget to check it? And if it was necessary that they go down immediately, how could their doctor forget? Or they may think something is wrong with the film and the physician is trying to decide what to do with the information. Patients sit in that bed all day long with nothing to do but worry about their condition and think about what is happening. They do not want to wait; waiting drives anxiety and fear. They want to be kept up to date.

Patients want empathy—they deserve empathetic caregivers! They want to believe that we, as health care providers, understand what it is like to be on the other side of health care. Teaching empathy is difficult. It is what is known as a latent construct—we believe we know it when we see it, but we often have difficulty defining it and much more difficulty teaching it. Several different definitions of empathy exist. Spiro defines it as the ability of a person to "experience an emotional connection with someone and their experience" [9]. Hojat defines it as "an ability to understand the patient's inner experiences and perspective and a capability to communicate this understanding" [10]. The ability to understand the other side will make us care more and be more compassionate toward the people we serve. In 2013, we conducted an internal study of 1000 patients sampled from across the United States. We learned that patients want physicians and other care providers to be able to understand the position they are in. Physicians who have empathy are more likely to receive a patient's benefit of the doubt when they have to suffer long wait times or if medical errors occur. Physicians who demonstrate empathy are viewed as being more compassionate.

Finally, caregivers must remember that no one wants to be a patient. This should be intuitive but we do not think about it. Patients walk through our doors with anxiety, fear, and often terror. Many have never experienced being sick before. Suddenly, they may find that their first hospitalization is for something that is life threatening. Some equate being a patient in a hospital to being in prison. Think about it: you go somewhere you do not really want to be, they take your clothes and give you a hospital gown, you have to wear a wrist band, you are often placed into a room with a total stranger, and people are always telling you what to do. If you have experienced that, then you understand what it is like—nobody chooses this.

Understand Treatment Flow—*The Patient's Perspective*

Traditionally, patients are treated in episodes. A patient comes in with a problem and receives attention. But this is not the model of the future. Our challenge will be based more on managing populations of people along a continuum of care; the reality is that people flow through a health care delivery system, and that flow must encompass the activities before and after the acute encounter. The responsibility of health care delivery begins with the patient before they experience an encounter. What we do to impact patients before they interact directly with an episode is as important as the care we deliver at the episode. For instance, helping people ensure that they have access to the right provider at the right place and right time is important. Helping people understand how to take care of themselves in their home and how to leverage nonprofessional, home-based caregivers (family members, friends, neighbors) is as important as providing care during an acute episode in a hospital or clinic.

Finally, how we move people out of the acute episode is equally important. Preparing patients for discharge from the hospital, or ensuring that they have a complete understanding of their needs and access to follow-up after they leave an ambulatory environment, is just as important as delivering the actual care. Our responsibility becomes that of a team of providers focused on preventing illness, managing access when care is needed, and getting patients back to their healthy or baseline state. This is a challenge of managing the "360" of health care. Our responsibility to health care professionals of the future is to make sure they understand that responsibility for the care of patients begins before they have an acute episode, during that acute episode, and in returning them to normal or their starting point. This is patient-centered "health" delivery.

Train Medical, Nursing, and Other Health Professional Students Together

To achieve an appropriate level of systems thinking and success with what we believe is a new model of health care, we have to transform the way we teach health care professionals. Moving toward greater interprofessional education will be critical. Interprofessionalism is defined as "two or more professions working together as a team with a common purpose, commitment and mutual respect" [11]. Generally, physicians, nurses, and other professionals work adequately together. Hospitals today spend time and resources trying to find ways to drive and improve high-performance teamwork among providers. To meet the needs of the future, we need professionals entering their careers already trained in high-performance team work.

Doctors, nurses, and other professionals will need to work more closely together as partners in care delivery. American medical teaching and training is siloed and hierarchical. Most medical students and nursing students never interact with each other while obtaining their professional education. These separations continue when nursing students graduate and begin their first job and medical students become house staff. The team-learning components often come into play when new house staff interacts for the first time with nurses and others—it is on the job experience with no training. Professionals are left to figure it out for themselves.

Doctors can learn from nurses. Medical students graduate believing that medical delivery is an individual sport; autonomy and independence are valued. This mentality is against what we know is now an important reality—which is that patients are taken care of by teams of people. Nurses are schooled more formally in team-based care delivery; however, they are rarely taught with physicians. These separations in training are a barrier to obtaining the building of highly functioning teams necessary to take care of complex patients. Academic health centers are now working to develop programs that promote better teamwork between professionals, but these programs are a Band-Aid approach to changing a dysfunctional culture. More is needed to transform the culture of patient care.

We will need to evolve our thinking about the role of nursing. Nursing is a profession that is responsible for much more than just "acute care delivery," but it is a profession that is often inappropriately branded in the public eye as those who take care of patients in the hospital. In fact, the role and scope of nurses is broader and more encompassing. Components of their primary mission includes providing for the health and welfare of our population. In addition to providing acute care, they lead patient education, health delivery, and maintenance of individual health outside of the acute health care delivery. Our responsibility is to manage the 360° of care—and the foundation of that delivery is the nurse. Nurses are the backbone of health care delivery, ensuring that the care received in the acute care environment transitions to the health maintenance strategy outside in the population. They are responsible for the navigation of patients with chronic medical conditions outside of the clinic and hospital. Evolving the traditional thoughts about the health professions will take a generation, but the starting point is in our academic health centers.

Thinking about Execution

How do we refocus our organizations on the patient? After setting patient centeredness as a top strategic priority,

execution becomes paramount. To successfully execute an improvement strategy, three critical areas must be addressed: process, people, and patients.

Hospitals and health care delivery systems are full of processes. A primary responsibility is to ensure that the processes necessary for effective patient care are working properly. There is no service-excellence strategy or interdisciplinary teamwork training that can fix a broken hospital process. If people wait for hours in the waiting room due to inadequate physician capacity, then the capacity of appointments needs to be addressed first. If patients are on eating restrictions and waiting in hospital rooms for procedures to take place and the procedure is canceled at the end of the day without the patient or the nursing team aware, these process failures will drive staff unhappiness, create inefficiencies, and waste money. Scheduling procedures and managing patients need to be fixed first. Once we have the required processes fixed and working effectively and efficiently, the next challenge is what other best practices can we layer on to improve what we are trying to do?

One best practice in the inpatient environment is nurse hourly rounding. The nursing literature has demonstrated that when a nurse goes into a patient's room every hour and runs a checklist that includes questions about pain, need to use the bathroom, patient positioning, and proximity of personal belongings, then medication errors, falls, and pressure ulcers are reduced. We conducted a study in our institution that looked at consistent hourly rounding compared to sporadic rounding. We asked the patients to tell us on their hospital inpatient survey if they witnessed the rounding taking place. When patients experience a rounding that regularly occurs, each nursing HCAHPS domain achieved the 90th percentile performance. If patients experienced less rounding, the scores declined significantly. Nurse hourly rounding is a best practice that should be implemented in every hospital [12].

Aligning staff around the North Star of patients and creating an environment of service delivery is critical. When evaluating the number of caregivers who provide service to patients in the hospital, the importance of teamwork and the vital role of every individual in care delivery is highlighted. At Cleveland Clinic, we assigned everyone the title "caregiver" to signify their individual and important role in care delivery. This was a tough argument for some, as many believed that if you were not a nurse or doctor you could not be a caregiver. We held firm on the belief that if you support the Cleveland Clinic mission, which is basically to deliver care to patients, then—regardless of the role—you are a caregiver.

Service-excellence training is another important component of cultural development. Teaching every health care worker how to interact with patients and families is critical for ensuring that the first impression that is created is affirmed and sustained throughout the health care journey. To ensure that patients are treated with respect and dignity, employees must have a framework about how to interact with patients and fellow caregivers. Equally important to how we treat people is how we recover service when something goes wrong. Patients are in a captured environment—often for many days—and there is ample opportunity for service failures to occur. Teaching caregivers how to apologize not only helps to ensure that the appropriate actions are taken to correct the problem but also can provide an opportunity to avoid a potential patient grievance.

Treat Employees Like Patients

If we agree that the focus on the patient and family is paramount, then we must also understand that developing and caring for our employees is just as important. Organizations need to implement strategies to drive engagement. Many health care organizations do not measure employee engagement scores, but this is an important first step.

Our personal experience impacts the work we do and the care we provide. This is called emotional intelligence. Popularized by Ronald Goleman, emotional intelligence "refers to the ability to perceive, control and evaluate emotions" [13]. In essence, it is being smart with your own and others' emotions. Emotional intelligence has four primary pillars: self-awareness, self-management, social awareness, and relationship management. The first two relate to our ability to understand our emotions and manage how they impact what we do. The second two relate to our ability to understand others and manage our relationships with other people better.

Some have argued that having sound emotional intelligence is a much better predictor of leadership success than intellectual intelligence. Health care as a people business requires frontline caregivers, as well as managers and leaders, to better manage relationships. Empathy is not just for patients, it is for our coworkers as well. Working in health care and taking care of patients is stressful and hard work.

PATIENTS AS PARTNERS

We have seen an evolution in terminology used to describe how health care professionals impact patients' understanding of disease and health care. Historically, providers have been tasked with educating patients about their disease. Caregivers then worked to empower patients to take more responsibility for their care. In recent years, the term patient engagement has been introduced, which defines "actions individuals must take to obtain the greatest benefit from the health care services available to them" [14]. More recently, patient activation has become the new focus. An activated patient is an "Individual [who] understands their role in the care process, and has the knowledge, skill and confidence to carry it out" [15].

These are important elements, but a better way to frame this issue would be to ask how can we form better patient partnerships. Partnership is defined as "one that is united with another in an activity of common interest." There is no greater connection, relationship, or partnership than that between a caregiver and patient. Patients share their deepest concerns with providers and reach a level of vulnerability and trust that is seen in no other relationship. Providers make decisions and recommendations that impact the health, welfare, and life of people. Caregivers often form similar relationships with patient families.

There are many ways that we can form better partnerships with patients. If every patient and family member understood the importance of infection risk and hand washing, we could potentially raise the awareness of this important issue so that every time a caregiver walked into a patient's room and did not wash their hands, the patient or family would remind them. At our organization, we developed an Ask 3, Teach 3 program [16] as a tactic to better educate patients about the medications they are receiving. For every medication that patients are given, they are taught to ask what it is, what it is treating, and what the side effects are. Then the nurse or pharmacist is educated to provide the same three points back to the patient. This is a form of dialogue that ensures the patient's response matches the caregiver response and is a tool to help reduce medication errors. It helps to create a partnership between the patient and the caregiver around the delivery of their medication.

We also want patients to help us educate them by working to educate themselves. The availability of health information today on the Internet is nearly unlimited. Many clinicians complain that patients are often misled by information they find on the Internet; I disagree. Patients should be encouraged to research their disease so that they become better prepared for discussion. Research demonstrates that consumer willingness to use the Internet to obtain health information reaches nearly 68% across all demographics, including senior citizens.

The partnership, however, goes both ways. Our responsibility is to treat disease; patients have to take greater responsibility for their health. Lifestyle choices, such as smoking and lack of exercise, can trigger and exacerbate chronic disease. Compliance with treatments is often compromised, and patients and their family members have an obligation to drive compliance.

LEADING TO TRANSFORM PATIENT-CENTERED CARE

Health care delivery is changing and academic health centers are poised to capitalize on this opportunity to redefine and teach how we deliver care to our citizens. Executing on a strategy of patient centeredness requires health care organizations to set it as a strategic priority. Because the scope of this initiative is so broad, success will only be achieved if the top person—the leader of the organization—owns this as his or her initiative.

Transformation starts with placing the patient and their families at the center of everything we do and making patient experience improvement a strategic priority for the organization. Successfully meeting this challenge will require us to tackle an execution strategy that articulates how to operationalize this change, which will include better ways to develop our culture. Patient centeredness does not belong to a single stakeholder group, such as nursing or operations; it is the responsibility of everyone in the organization. The need for high-performing teamwork among groups of professionals will be necessary and will only be achieved by our ability to tackle interprofessional education.

Our collective challenge is not to try to predict the future, but to position our organizations and health care professionals to be able to adapt to any change that is necessary in order to meet the goal of delivering safe, high-quality, and patient-centered health care. That is the patient experience; it is the right thing to do, and it is the way we would want care provided for us and our families.

REFERENCES

[1] Merlino JI. Speaking from experience. Mod Healthc November 24, 2012.

[2] http://www.aarp.org/personal-growth/transitions/boomers_65/ [last accessed 28.10.14].

[3] http://www.alz.org/downloads/facts_figures_2012.pdf [last accessed 28.10.14].

[4] Institute of Medicine. Crossing the quality chasm: a new health system for the 21st century. Washington, DC: National Academy Press; 2001.

[5] http://www.hcahpsonline.org/surveyinstrument.aspx [last accessed 28.10.14].

[6] Manning H. Outside in 2012 (lecture, 3rd Annual Patient Experience: Empathy & Innovation Summit). Cleveland, Ohio, May 20–22.

[7] Cosgrove T. The Cleveland clinic way: lessons in excellence from one of the world's leading health care organizations. New York: McGraw-Hill; 2013. p. 110.

[8] Bryan CS. "Aequanimitas" redux: William Osler on detached concern versus humanistic empathy. Perspect Biol Med 2006;Summer;49(3): 384–92.

[9] Spiro H. What is empathy and can it be taught? Ann Intern Med 1992;116(10):843–6.

[10] Mohammadreza H, Gonnella JS, Nasca TJ, Mangione S. Physician empathy: definition, components, measurement, and relationship to gender and specialty. Am J Psychiatry 2002;159:1563–9.

[11] Dunston R. Interprofessionalism health education in Australia. Australia: University of Sydney; 2009.

[12] Hancock K, Merlino JI. Nurse hourly rounding: a critical driver of HCAHPS scores. Group Pract J 2012:30–9.

[13] Goleman D. Emotional intelligence. New York: Bantam; 1995.

[14] Center for Advancing Health. A new definition of patient engagement. 2010. http://www.cfah.org/file/CFAH_Engagement_Behavior_Framework_current.pdf [last accessed 28.10.14].

[15] Hibbard J. Patient activation and engagement for ACOs (PowerPoint Presentation). University of Oregon; 2008.

[16] Woicehovich L, Rivera ML, Merlino JI. Ask 3/Teach 3: improving medication communication scores and patient safety. Group Pract J 2013:20–8.

ABOUT THE AUTHOR

James Merlino, MD is the Chief Experience Officer of the Cleveland Clinic health system, leading the Office of Patient Experience, and is a practicing staff colorectal surgeon in the Digestive Disease Institute. He is also the founder and current president of the Association for Patient Experience.

Chapter 22

Positioning Academic Health Centers for Quality, Safety, and Patient Empowerment

Carolyn M. Clancy and Arthur Garson Jr.

INTRODUCTION: CASE EXAMPLE

"JK" is a 60-year-old white male who has been experiencing increased fatigue and occasional mild chest discomfort over the past 4–6 weeks. Increased exercise and more sleep have not improved his symptoms. Upon referral by his primary care physician, JK undergoes a cardiac stress test. The test results are "wildly positive" and he receives cardiac catheterization the same day. At the time, he is admitted to a community hospital, which has recently become part of an extended network with a renowned academic health center (one that is frequently ranked in the upper tiers of *US News and World Report's* annual Best Hospitals list). JK is then informed that his catheterization has indicated several major blockages. Fortunately, there is an available slot at the same hospital for a coronary artery bypass surgery the next morning. JK contacts a physician colleague with expertise in quality assessment and inquires about the specific surgeon. His colleague reports that no information is publicly available. His family also wonders if JK would be better off if he elected to have the procedure done at the major hospital center nearby. They turn to the Internet, but a quick Web search turns up little information about what it means to have a procedure performed at a hospital associated with a major academic institution. Would it be safer? Or, would JK be better off having the procedure performed at the major hospital with which they are more familiar? Lacking the desired information, but anxious to address JK's immediate health problem, JK and his family make arrangements for the bypass to be performed at the community hospital.

The preceding case reflects a patient's experience—a patient who is a highly-educated professional with little detailed knowledge of clinical care. Despite the growing amount of information about quality and safety that is readily available (especially on the Internet), JK and his family were unable to find information needed for a time-sensitive decision—i.e., the safety and quality of care provided at a hospital affiliated with an academic health center (AHC).

In addition, the case raises important questions about the role of AHCs as implementation of the 2010 Patient Protection and Affordable Care Act (ACA) continues to move forward. How do these facilities measure up in terms of quality and patient safety? With organizational structures and practices based on reimbursement structures rooted in 1960s policy, are AHC providers working in systems designed to provide the best care? Does the organizational and leadership structure of a given AHC align with the way new doctors should be taught about patient safety and quality? Additionally:

- Lacking complete information, is choosing an AHC—when confronting imperfect information—a practical, fail-safe approach?
- What are the quality and patient safety implications for community hospitals that are part of an AHC's extended network?
- With health care organizations aspiring to deliver patient-centered care and provide evidence-based information to help guide patient decision-making, what can AHCs do to close the gap between patients' expectations for salient information and the current reality?

To help address these questions, this chapter provides an overview of US health care quality and patient safety. It then examines several unique characteristics, which may need to be revisited if AHCs aspire to leadership roles post-ACA, and reviews patient empowerment in this context. The chapter concludes with eight recommendations for consideration.

QUALITY AND PATIENT SAFETY OVERVIEW

The many successes of biomedical science and public health in the past century have resulted in substantial

The Transformation of Academic Health Centers. http://dx.doi.org/10.1016/B978-0-12-800762-4.00022-0

increases in life expectancy—Americans are living much longer. Among the more notable achievements are dramatic reductions in cardiovascular diseases, the transformation of an HIV diagnosis from "death sentence to "highly manageable chronic condition," and significant reductions in the percentage of Americans who smoke.

However, numerous reports indicate there is much greater room for improvement, and along many dimensions. For example, precisely because previously life-threatening conditions have been so successfully confronted, addressing chronic illnesses now presents the greatest opportunity for improvement. Multiple studies in the past decade confirm a sizable gap between the best possible care and care that is routinely provided, as consistently demonstrated by the US Agency for Healthcare Research and Quality's (AHRQ) annual reports to Congress on the state of health care quality [1]. While AHRQ has found statistically significant increases in quality across all settings and populations since 2003 (the year it began reporting), the magnitude of improvement reported most often has been modest and the disparities in care associated with individuals' race/ethnicity, age, education, income, and other factors have remained pervasive. Other studies also have confirmed that the United States has a long way to go [2].

At the international level, studies comparing health and health care in the United States with other countries are sobering. A 2014 Commonwealth Fund study of 11 countries (Australia, Canada, France, Germany, the Netherlands, New Zealand, Norway, Sweden, Switzerland, the United Kingdom, and the United States) found that the United States ranked dead last—a position it also held in previous studies for 2010, 2007, 2006, and 2004 [3]. Further, an analysis by the World Health Organization of 191 countries found that the United States spent a higher proportion of its gross domestic product on health than any other country, yet ranked 37th on performance [4].

At the most fundamental level, numerous studies examining the process and outcomes of care have shown substantial variations in clinical practice (itself an indicator that quality may be questionable or at least not consistent) and have resulted in a movement to develop better methods for determining the relationship between care processes and outcomes. These efforts have also revealed underuse, overuse, and misuse of services, as well as a substantial time lapse for new scientific findings to be translated into practice—a problem that persists to this day. In addition, the past 15 years have witnessed increased attention to preventing avoidable harms that occur as a result of receiving care, such as health care-associated infections (HAIs), surgical complications, and errors in prescribing and dispensing medications. There has also been increasing recognition of how factors external to direct care delivery—including reimbursement, organizational structure, and leadership—influence safety and quality.

Nearly a quarter of a century ago, quality was defined by the Institute of Medicine (IOM) as "the degree to which health services for individuals and populations increase the likelihood of desired health outcomes and are consistent with current professional knowledge" [5]. Implicitly, this definition covers both individuals and patient groups, including those who seek care and those who do not. Further, the definition was intended to focus on outcomes or end results important to individuals and to recognize that medical knowledge evolves. In 1999 and 2001, the IOM published two landmark reports that galvanized the public's attention and challenged all health care delivery professionals to improve care, and even change how they worked. The first report, *To Err is Human*, focused on patient safety, estimating that between 44,000 and 98,000 preventable deaths occurred each year due to avoidable harms from medical care itself [6]. Two years later, *Crossing the Quality Chasm* offered a more expansive assessment of all factors that facilitate or impede provision of care that is safe, timely, effective, equitable, efficient, and patient-centered [7]. While neither report focused explicitly on AHCs, both included extensive recommendations that resulted in major changes to the health care system at large, as well as external demands for transparency.

Turning to the issue of performance transparency, contemporary efforts to promote such transparency in hospitals and other care settings commenced in the late-1980s when the Medicare program published hospital mortality rates, adjusted for patient risk and severity of illness. In the early 1990s, the National Committee for Quality Assurance began encouraging health plans to report on clinically valid measures of performance for ambulatory care, heavily oriented initially to clinical preventive services [8]. Consequently, over the past three decades, numerous public and private sector organizations have urged greater transparency regarding quality and safety of care, so that patients are better able to make informed choices and the health care profession is inspired to make improvements. Most efforts have started with a small number of measures (voluntarily reported). These reported items are, most frequently, process measures derived from clinical studies that demonstrate a significant link between a specific therapeutic action (e.g., control of hypertension) and outcomes of interest (e.g., stroke, heart attack, chronic renal failure). Strong interest in informing people's choices about which provider to select often meant that selected metrics had to be measurable in a short period of time due to patient turnover within plans and decreasing lengths of stay in inpatient care. These practical realities tended to favor a focus on easily measured care processes, rather than patient outcomes.

Starting in 2005, the Medicare program, in collaboration with many other organizations, shifted the focus of hospital reporting from purely voluntary to one that is referred to as "pay for reporting" (or no pay for no reporting)—i.e., hospitals did not receive a payment update if they chose not to report. Almost all chose to report. Since that time, the number of yearly measures has expanded to include selected measures of avoidable harms, including HAIs, as well as patient experience surveys. At the same time, the payment incentives have also increased (now referred to as value-based purchasing). In this context, reporting is necessary, but not sufficient—i.e., a percentage of a hospital's reimbursement is determined by the actual performance. Numerous public and private sector initiatives also have emerged testing the premise that some portion of payment should be determined by how well providers do.

This link to payment has captured the attention of all hospitals, including those affiliated with AHCs. Notably, few studies have directly examined whether AHCs consistently provide the best care. Instead, the majority of studies have focused on all hospitals. For many of the currently reported measures, performance has been high (i.e., well over 90%) for all institutions. Ratings, such as *US News and World Report*, which includes a measure of "reputation" (among many other factors), have tended to put some AHCs in a far more positive light than other hospitals. The bottom line for all institutions is that the science of measurement remains somewhat immature, and there is no widely accepted system for rating overall safety of care. Instead, patient safety metrics have focused on selected avoidable patient harms considered to be largely preventable (e.g., HAIs, surgical harms, deep venous thrombosis). Notwithstanding measurement challenges, care has shifted considerably over the past several decades from well-remunerated hospital care to ambulatory settings that are often associated with lower payments. Payers and the public expect that care will be integrated and efficient with a strong focus on patient experience (patients' experience ratings of hospital care now comprise approximately 25% of CMS' value-based purchasing payment). The emergence of accountable care organizations, which virtually integrate all aspects of care, represents a growing opportunity for some health care organizations. However, some may not be a good fit with AHCs and their triple mission of clinical care, education, and research.

Throughout the past two decades of increased demands for performance transparency, two theories have been debated regarding how transparency would result in improved performance and better care. One line of reasoning is based on a premise that such transparency would stimulate motivation to improve. Indeed, in an early effort led by a purchasing coalition of small- and medium-sized employers in Wisconsin, hospitals that selected to report performance publicly were found to have substantially increased their quality improvement efforts one year later, compared with their counterparts who viewed their results only with the purchasing group [9]. AHRQ's annual reports on quality have found larger improvements in metrics that are reported publicly.

A second rationale was that informed consumers would make choices based on quality. Initially, it was difficult to demonstrate that consumers were even aware of this information, much less that it informed decisions. Over time, as the Internet has come to serve as a backbone for many consumer activities in other spheres (e.g., banking, travel, news, entertainment), the proportion of patients who are aware of and knowledgeable about publicly available information on quality and safety has increased steadily, reinforced by the efforts of Consumers Union and other organizations. At the same time, many patient groups have formed to advocate for a stronger role in health care, and the recently established Patient-Centered Outcomes Research Institute, created by the ACA, has advocated for patients to be involved in all phases of research on patient outcomes. Hospitals in Massachusetts are now required by law to have patient and family advisory councils, and many organizations are actively involved in improvement efforts. Indeed, patient engagement has been described as the "blockbuster drug of the Century" [10].

However, the patient advocacy movement is still evolving and in search of ways to continue its momentum. As noted by Helen Haskell, President of Mothers Against Medical Error: "In spite of much forward motion over the years, it is clear that many major patient safety problems have remained depressingly constant, and that gains in some areas have been offset by setbacks in others" [11]. Patient advocacy is one that is ripe for study and innovation, yet the impact on, and leadership role for, AHCs has barely been pursued.

AHCs POST-ACA

The active engagement of multiple stakeholders, especially public and private sector purchasers and payers, in health care delivery—through demands for transparency, new payment models, and other activities—poses new challenges for AHCs. Highly regarded for their ability to teach the next generation of clinicians, conduct cutting edge research, and provide excellent clinical care, AHCs now confront new financial pressures and elevated expectations [12]. In addition to a chronic lack of explicit resources for teaching, AHCs today face challenges such as decreased (in real dollars) public funding for research and competition from community hospitals that may be better organized to focus on quality and safety. Some observers have commented that only a few AHCs consistently exemplify integrated, patient care [12]. Although some AHCs have achieved high levels of integration, many others still have a long way to go: "Pockets of clinical excellence may be found in most academic medical centers, but the leaders of these institutions face the challenge of trying to achieve consistently high performance institution-wide" [13]. Keroack et al. identified the organizational factors associated with AHCs that separated them from their peers in quality and safety. Five themes, according to the study team, were especially evident:

1. A shared sense of purpose that includes putting patient care ahead of the other two missions, i.e., "patients first";
2. A leadership style, where the CEO is "passionate about improvement in quality, safety, and service," and has an "authentic hands-on style";
3. An accountability system for quality, safety, and service that is a "blend of central control and decentralized responsibility";
4. A focus on results that shows a zest for measurement across a wide array of programs, not as an end in itself, but as a basis for improvement; and
5. Collaboration among multiple professionals and at every level, i.e., "deference to expertise rather than rank or position" [13].

CHALLENGES FACING AHCs IN PATIENT QUALITY AND SAFETY

One set of challenges pertains to culture. For example, there is the longstanding cultural norm of autonomy. "Esteem for faculty and departmental autonomy often impedes standardization of processes for everything from selection of orthopedic prostheses to agreement on recommended frequency for screening mammography" [12]. Respect for the contributions of leading experts may not easily integrate with contemporary approaches to quality and safety that reinforce the importance of team work. While most clinicians are passionate about providing patient-centered care, achieving superb clinical care requires that all team members share the same vision for patients under their care.

Some of the slow movement toward standardization may be due to some senior faculty giving more deference to their experience, e.g., with specific devices or treatment options, than to evidence that alternative options are equally effective. In addition, many physicians struggle to change their practices when emerging evidence appears to contradict their longstanding past experience. In medicine, as in many fields, increased availability of information via the Internet, social media, and other channels has proved to bring with it cultural challenges where traditional authority and faculty roles have evolved substantially. A shift from what has been coined "eminence-based medicine" to evidence-based medicine challenges all faculty to adjust to a very different environment. In the current context, practice variations attributable to individual physician preferences, rather than evidence, are rightfully seen as potential barriers to seamless, patient-centered care, as well as efficiency.

Further, the longstanding practice of the "teaching physician" (seeing patients one day a week, itself a holdover from the origins of Medicare) continues to influence the way faculty spend their time, despite numerous changes in clinical care and research. In the 1960s and early 1970s, AHCs were awash in funding because Medicare paid "reasonable and customary" charges. AHCs claimed (and reasonably so at the time) that their costs were higher because their physicians were highly paid and spent a great deal of time in the laboratory. (It was also in this time period that clinical departments had large budgets to fund laboratory science and the concept of "protected time" was born.) Fast forward to the present era when "…faculty members and trainees who staff teaching practices frequently spend only one or two half days per week in ambulatory care, complicating timely patient access and continuity…" [12].

Instead of relying upon an outdated concept that only serves to promulgate bad habits in aspiring physicians, a new model of behavior is needed. Today, the term "teaching physician" should apply to the outstanding clinician who teaches at the bedside and in the clinic, and whose expected productivity—e.g., as assessed by Relative Value Unit—is appropriately reduced.

Turning to clinician scientists, the picture is again rapidly evolving. Many clinician scientists whose work has been generously supported by the National Institutes of Health (NIH) now confront unprecedented challenges securing ongoing support. While extramural support has historically favored discovery over application, many of these professionals excel when it comes to ensuring that research findings are translated into patient care. Funding challenges are not new for clinicians trained in health services research, but a growing number now apply their skills to help their institutions achieve improved patient outcomes. These researchers struggle with the issue of whether success, by definition, includes application, and do not consistently find their own organizations to be receptive or prepared to scale-up the innovations they have carefully perfected.

ADDRESSING THE BARRIERS

Few would question that AHCs lack the expertise to address these, or similar, barriers that may be thwarting full-scale attention to patient safety and quality improvement. In fact, a number of medical schools, with grants from the American Medical Association, have embarked on several promising innovative projects specifically aimed at improving education and training related to patient safety and quality improvement. Among these projects are: (1) an effort to better align medical education with health system needs (Penn State College of Medicine); (2) the establishment of a core curriculum that includes a "Teachers of Quality Academy" (Brody School of Medicine at East Carolina University); and (3) the cultivation of the "collaboratively expert physician"—i.e., the physician who is constantly on the lookout for ways "to work within interprofessional teams to continuously improve the safety, quality and value of health care" (University of California, San Francisco) [14].

Implementing these unique projects on a broader scale, however, may require revisiting longstanding practices fundamental to how academic talent is organized/deployed, as well as the methods and structure to reimburse that talent. This section examines three such factors: (1) the multiple roles played by physicians; (2) the reimbursement structure; and (3) faculty as a single unit.

Physicians Play Multiple Roles

One of an AHC's primary strengths is the fact that nearly all of its physicians are also researchers, teachers, or, in many instances, all three. For example, AHC researchers often are at the top of their field. Their dedication to the scientific method and their exacting demand for the best evidence possible are the same attributes which should serve them well as clinical decision-makers. For those who are both clinicians and researchers, the interactions with students and residents benefit both the researcher and the learners and shape the scientific foundation of clinical care. Therefore, care should improve in the presence of researchers and teachers; and in the best teaching institutions, care is often exemplary.

For example, a study that looked at the cost and quality of care for Medicare beneficiaries regarding several common conditions (hip fracture, congestive heart failure, stroke, coronary heart disease) showed a higher survival rate for those who were treated at major teaching hospitals (when compared to patients at nonacademic institutions). For hip-fracture patients at major teaching hospitals, the one-year survival rate was 86%—compared to 85% at for-profit hospitals, 77% at government hospitals, and 81% at nonprofit hospitals [15].

However, a number of challenges stem from these multiple roles. A potential problem for clinician-researchers is the many hours required to be spent in the lab or at a computer terminal, reducing the number of available hours for teaching. The demands of research may be a poor fit with public expectations of care that is timely, safe, and high quality. Further, the skills required to conduct superb research, e.g., generating and testing multiple hypotheses, can be different from those important to seeking and interpreting information from multiple sources and applying it judiciously to patient care. In an era of shrinking margins, testing patients for all possible diagnoses can be both unnecessarily costly and even harmful. Increased difficulty in obtaining funding may result in pressures to spend as much time as possible in research. For those unable to obtain external funding, clinical departments may be forced to reduce their ranks.

Another concern stems from clinicians' perceptions of how teaching is valued. Given, in most cases, the lack of specific reimbursement for both classroom and clinical education, teaching is viewed by some academic physicians as a (an unfunded) burden that holds them back from providing "important" patient care and managing "necessary" administrative paper work. Teachers should be tenured specifically for teaching (not just as part of a portfolio). Tenure is a "bet" on the faculty productivity and relevance in the future. A great teacher will almost always remain a great teacher. Identifying support for outstanding teachers will be a challenge.

Reimbursement Structure

An AHC's physicians are "mostly" salaried. Given this type of compensation structure—one that does not incent extra tests and procedures to be ordered—the care provided, at least in theory, should more closely align with a patient's specific medical needs. However, the word "mostly" is in quotes above, as many in the procedural specialties receive a base "salary" that might be $100,000 (therefore, "salaried") as well as productivity bonuses that may be many times larger than the base salary.

Consistent with public and private sector policies, financial rewards for academic faculty should stimulate outstanding patient care. At this time, notwithstanding new payment models, the majority of health care financing remains fee-for-service, so productivity will remain relevant. However, visible and tangible recognition for high-quality care should have an important signaling effect.

Faculty as a Single Unit

At least on paper, the faculty is a single unit and part of one practice plan. It might therefore be expected that individual faculty members would work as part of teams and that referrals would be seamless. In reality, however, there is little evidence and few explicit incentives for the faculty to behave as a high-performing team. Most institutions place faculty in departments where they are paid as a member of a department and by department rules. The income statements of most departments are maintained separately. While much is made of "teamwork," many current academic physicians were raised in the NIH era of the single Principal Investigator—i.e., a single individual receiving all the credit. In very few cases is teamwork across departments rewarded appropriately. While this is not an easy task, a first step should be to define the specific outcomes of the team that make a difference beyond individual contribution and then reward those outcomes appropriately.

AHCs AND PATIENT EMPOWERMENT

The role of patient engagement at multiple levels of health care delivery, from individual clinical interactions to involvement in health system governance to broader engagement in health policy, has recently been described by Carman et al. [16] A few AHCs have initiated specific initiatives focused on obtaining feedback from patient and families who are frequently hospitalized (e.g., cystic fibrosis, cancer), and more have researchers who have conducted community-based participatory research, a focus of the NIH-supported Clinical Science Translation Award program. Another set of intriguing findings stems from the efforts of several institutions that participated in an initiative called "OpenNotes." Supported in large part by the Robert Wood Johnson Foundation, this initiative encourages physicians to provide patients with electronic access (via a portal) to the physician notes component of their individual electronic health record. Even though a "minority" of the potential 20,000 patients among these institutions took advantage of such access, the results among those who did were impressive. For example, two-thirds of patients on medications who read their physician's notes reported better adherence to prescriptions [17].

What is clear is that, in addition to the weight attributed to patients' perspectives on their hospital care through CMS' value-based purchasing program, a growing proportion of Americans are not only aware of—but increasingly savvy about—quality and safety information. In addition, the phenomenon of patients' arriving for an encounter armed with their own Internet search results is no longer remarkable, and the explosion of health care information available via social and other media has had professional organizations scrambling to craft pragmatic policies.

These developments are both exciting and disruptive—and offer a huge opportunity for AHCs to lead with the patients they serve. Doing so will require the development of new teaching skills—e.g., having experienced patients teach about their experiences with chronic illness management—as well as the courage to be open about performance (both good and less optimal) and what can be expected from an AHC compared with other institutions. In the initial case example, JK and his family asked all the right questions—the clear opportunity for AHCs is to test and evaluate alternative approaches to addressing these and other questions.

AHCs: MOVING FORWARD

What can AHCs do to continue advancing patient safety and quality at their respective institutions? And, what can they do to ensure that patients have timely information regarding patient safety and quality? In other words, what must they do, in light of the barriers discussed above, to help JK and his family to feel more confident that they are receiving the best care possible?

The five themes of successful AHCs offer a solid foundation for moving forward. Below are eight recommendations which flow from these themes, as well as from the additional findings cited in this chapter.

Leadership

1. *Ensure that all academic leadership (Executive Vice President (EVP), hospital Chief Executive Officer (CEO), medical school dean) all "walk the talk."* As Keroack et al. pointed out, leadership must designate and promote quality and safety as top priorities [13]. Clinicians and basic scientists alike must understand that quality and safety are not just important to the patient, but also key to the survival of that AHC. As comparative data become more available, an AHC with low quality, compared to the community hospital "next door," will lose patients and revenue. The people in charge of the academic side must speak with the same voice. It is quite a statement when the basic science administration realizes that quality and safety are important to patients and to the survival of the AHC—as it is that pays their salaries.

Mission Areas

2. *Health Professions Education—Reward faculty and explicitly recognize the skills of the teaching physician as part of their salary.* The incentive should be divided into two parts; the largest should reflect what the faculty can directly control. In many cases, the "what" is not completely obvious. For example, it can be argued that patient satisfaction is due more to the efforts of the nurse who supports the physician than to the physician alone. Nevertheless, appropriate metrics must be identified, incented, and, importantly, agreed to by the teaching physicians as a group to promote and teach health care value. Identifying and demonstrating health care delivery in terms of quality and safety is an important element in health care professions' education. Physician faculty who share best practices and explain the importance of focusing on the quality of the patient experience and safety are the first line to developing a culture of quality of care. Incentives related to overall performance and accountability in the provision of health care and valuing patient safety should be clearly affirmed by the AHC for faculty, students, and trainees.

3. *Clinical care—Seek to be part of systems that have features of a "hub and spoke" with one AHC per system.* Ultimately, some AHCs might consider a path in which the AHC hospital is "quaternary" and cares only for patients with conditions requiring research, referring patients with more common conditions to network partners. This research can be funded in two ways: either competitively (public or private) or by applying a new clinical payment model that recognizes the need for an appropriate number of these institutions and the need for research. This funding would be at a level (set by public and commercial payers) such that the AHC would not need to build patient

volume to do routine care likely more successfully done in tertiary or community hospitals.

As part of the same system, tertiary hospitals would share many characteristics of current AHC hospitals, including Emergency Departments. AHCs justify their funding because of standby capacity for disasters, trauma centers, and burn units, among other services. This funding should be explicit and given to the tertiary centers. The community hospitals should then share routine in-patient care with the tertiary hospitals.

Teaching of students and residents should be practiced at all levels of these systems and funded explicitly by those concerned with physician workforce: federal and state governments, commercial payers, and universities. Ideally, cross-subsidization of education would not be necessary.

4. *Research—Focus on patient-oriented research*, i.e., *clinical or health services research*. Physicians who conduct such research should be given protected time. As detailed above, the "old" model of clinical departments having faculty doing basic research was born when Medicare funded AHC clinical departments; this is no longer the case. "Basic" (i.e., not involving the patient directly) researchers should reside in basic science departments. It is not clear that these departments must be within the medical school structure or budget (as clinical care must pay for clinical care). They should be part of university-wide research enterprises. The teaching of basic science to a medical student can be done in a department that is part of the university structure. They can be explicitly funded by grants and subsidized by the university—the way university science departments are now set up. This is not to say that interaction between physicians and scientists is not vital—it is. But this sort of interaction can better be planned in small groups or conferences when both the researchers and physicians can devote their entire attention to the necessary discussions. There are always exceptions. The individual physician, who truly spends time in the laboratory and then takes those research products to the bedside, while rare, is still real. These individuals should receive a designation similar to "Translational Professor" and be celebrated as the rarities that they are. Their contributions should be expected both in the basic and clinical arenas.

Continuous Learning and Improvement

5. *Recognize the value of the AHC as a* "learning health system" *and demonstrate the value of that contribution.* As observed by Grumbach et al., the concept of the learning health system, as defined by the IOM in 2007 [18], is not only a powerful organizing framework, but one that is ripe for study by AHCs: "Although the IOM has articulated core principles of a learning health system….it has not specifically addressed the uniquely transformative potential of this approach for academic health centers" [12].

As academic institutions, AHCs are uniquely poised to leverage knowledge, findings, and lessons learned to improve outcomes at their respective institutions. The University of California, San Francisco offers a specific example, in the creation of its Center for Healthcare Value, of the unique ability of AHCs to systematically align their missions, foster innovation in patient safety and care, and maintain focus on delivering high-quality health care at a lower cost. Their goal is to achieve and maintain higher standards throughout the health care continuum by having stakeholders in the research, education, and clinical care mission areas adopt strategic changes and policies that align to improve each other.[1]

6. *Think locally, model widely.* As noted earlier, and as evidenced by the medical education projects cited above, there is no shortage of ideas or talent at AHCs when it comes to addressing patient safety and quality improvement. In addition to the American Medical Association-funded medical school education projects above, the Association of American Medical Colleges' Learning Health System Research Awards recognizes "institutions that have implemented or wish to enhance capacity of innovative, system-wide processes that improve the opportunity for research" within three key areas, including quality improvement [19]. For many AHCs, it may be useful to consider these preexisting models and work to adapt them for local implementation.

Yet another source of innovation are new approaches emerging from the federal government. For example, use of AHRQ's TeamStepps (Strategies and Tools to Enhance Performance and Patient Safety) was instrumental to transforming the culture for patient safety at the University of North Carolina School of Medicine [20]. TeamStepps is a three-phased process for developing and sustaining a culture of safety. TeamStepps also was an instrumental component of a successful patient transition program—I-PASS Handoff Bundle—recently implemented at nine academic hospitals. Use of this handoff program was associated with a 23% relative reduction in the rate of all medical errors and a 30% relative reduction in the rate of preventable adverse events [21].

Additionally, new approaches to patient safety are emerging from the Veterans Health Administration. One such approach is the Stop the Line for Patient Safety campaign, which aims to empower employees to speak up if they see something that might compromise patient safety.

---

[1] http://ctsi.ucsf.edu/news/about-ctsi/ucsf-announces-center-improve-health-value.

At the Cincinnati VA Medical Center, for example, one employee "stopped the line" for patient safety when she saw a patient taking the wrong dosage of insulin [22].

Transparency and Patient Engagement

7. *Make information about quality and patient safety readily available.* First, the EVP should publish a periodic dashboard of quality indicators that is publicly available and is in a format that patients, faculty, and staff understand. Further, each AHC itself should produce useful, actionable data on quality and safety. This should include patient stories and reports of their experiences—both positive and negative—which are often more potent motivators of change than statistics. For example, the University of Utah is building metrics to develop value-oriented outcomes, such as making patient satisfaction reviews—conducted by an independent third party (Press Ganey)—publicly available as a step in their strategy to hold themselves to the highest transparency standards, and engage both patients and health care providers in participatory medicine.[2] Potential metrics to measure value-oriented quality and safety outcomes can include emergency room reutilization, readmissions, improved pain management systems, and patient satisfaction surveys.

8. *Harness the energy and dedication of families and patient advocates.* The potential for patient engagement and empowerment to serve as critical catalysts for enhancing all components of AHCs' missions (clinical care, education, research) is both palpable and exciting. However, it is important to keep in mind that most promising initiatives are local, and context-dependent. From engaging patients in feedback on adverse events and sharing clinician notes, to inviting patients to provide feedback to learners, to involving patients and families in all aspects of research, there is no single definition of "success," much less a single strategy for effecting fundamental change. Thus, AHCs have an enormous opportunity to lead, drawing on past work in patient safety, education (e.g., cultural competence, health literacy, clinician-patient communication), and research—including community-based participatory research and community engagement, among other areas. An excellent starting point can be found in the 2014 recommendations made by the Roundtable on Consumer Engagement in Patient Safety of the National Patient Safety Foundation's Lucian Leape Institute [23]. Partnering with or accessing patient-centered networks, advisory groups, and online patient communities that focus on research, such as PatientsLikeMe,[3] can be useful

to build patient quality care and safety models that include patient experience-defined health care quality metrics, thus both empowering patients and acknowledging the value of their input.

CONCLUSION

AHCs have played a vital role in medical education, research, and clinical care. In the post-ACA era, they will provide the important discoveries that advance human health, especially in quality and safety, as well as the education of medical students and many residents. However, the ways in which AHCs function and their reward structures stem from the 1960s. For quality and safety to thrive in the AHC, there must be committed and visible leadership, concentration on patient-oriented research, and transparency in the clinical and basic research careers and clear separation of clinical and basic research careers, with few exceptions.

The landscape of opportunities is immense—and exciting. Transitioning to a context that demands increased efficiency—with detailed focus on all aspects of care and clinical operations—and rewards rapid translation and spread of promising new initiatives, in close partnership with patients, payers, and other stakeholders, will be neither painless nor easy. However, AHCs have a strong tradition of responding to new imperatives (e.g., the HIV epidemic, geriatrics, stunning advances in cardiovascular and cancer care, innovations derived from advances in communications and information technology) and blazing new paths forward. The public should expect no less in the twenty-first century.

REFERENCES

[1] U.S. Agency for Healthcare Research and Quality. National healthcare quality report 2013. Rockville (MD): AHRQ; 2014.

[2] Anderson GF, Reinhardt UE, Hussey PS, Petrosyan V. It's the prices, stupid: why the United States is so different from other countries. Health Aff 2003;22(3):89–105.

[3] Davis K, Stremikis K, Schoen C, Squires D. Mirror, mirror on the wall, 2014 update: how the U.S. health care system compares internationally. The Commonwealth Fund; 2014.

[4] World Health Organization. The world health report 2000-health systems: improving performance. The World Health Organization; 2000.

[5] Lohr KN, Schroeder SA. A strategy for quality assurance in medicare. N Engl J Med 1990;322:707–12.

[6] Institute of Medicine. To err is human: building a safer health system. Washington (DC): The National Academies Press; 1999.

[7] Institute of Medicine. Crossing the quality chasm: a new health system for the 21st century. Washington (DC): The National Academies Press; 2001.

[8] Epstein AM. The outcomes movement – will it get us where we want to go? N Engl J Med 1990;323:266–70.

---

[2] http://healthcare.utah.edu/fad/; http://healthcare.utah.edu/fad/pressganey.php.
[3] http://www.patientslikeme.com/.

[9] Hibbard J, Stockard J, Tusler M. Does publicizing hospital performance quality stimulate quality improvement efforts? Health Aff 2003;23(22):84–94.

[10] Chase D. Patient engagement is the blockbuster drug of the century. Forbes; September 2012.

[11] Haskell H. Patient advocacy in patient safety: have things changed?. AHRQ Web M&M Perspective; 2014. Available at: www.webmm. ahrq.gov/perspective.aspx?perspectiveID=160#tiptop [accessed 09.11.14].

[12] Grumbach K, Lucey CR, Johnston SC. Transforming from centers of learning to learning health systems: the challenge for academic health centers. JAMA 2014;311(11):1109–10.

[13] Keroack MA, Youngberg BJ, Cerese JL, Krsek C, Prellwitz LW, Trevelyan EW. Organizational factors associated with high performance in quality and safety in academic medical centers. Acad Med 2007;82(12):1178–86.

[14] American Medical Association. Accelerating change in medical education. Available at: www.ama-assn.org/sub/accelerating-change/grant-projects.shtml; June 14, 2013 [accessed 08.11.14].

[15] Taylor Jr DH, Whellan DJ, Sloan FA. Effects of admission to a teaching hospital on the cost and quality of care for medicare beneficiaries. N Engl J Med 1999;340(4):293–9.

[16] Carman K, Dardess P, Maurer M, Sofaer S, Adam K, Bechtel C, et al. Patient and family engagement: a framework for understanding the elements and developing interventions and policies. Health Aff 2013;32(2):223–31.

[17] Walker J, Darer J, Elmore J, Delblanco T. The road toward fully transparent medical records. N Engl J Med 2014;370:6–8.

[18] Institute of Medicine. The learning health care system workshop summary. Washington (DC): National Academies Press; 2007.

[19] AAMC. Learning health system research awards. Association of American Medical Colleges. Available at: www.aamc.org/initiatives/rocc/363080/challengeawardwinners.html; 2014 [accessed 10.11.14].

[20] Kirch DG, Boysen PG. Changing the culture in medical education to teach patient safety. Health Aff 2010;29(9):1600–4.

[21] Starmer A, Spector ND, Srivastava R, West DC, Rosenbluth G, Allen AD, et al. Changes in medical errors after implementation of a hand-off program. N Engl J Med 2014;2(371):1803–12.

[22] VA National Center for Patient Safety. Cincinnati VA medical center "I care" awards. Available at: www.cincinnati.va.gov/patients/quality.asp; September 28, 2014 [accessed 10.11.14].

[23] The National Patient Safety Foundation's Lucian Leape Institute. Safety is personal: partnering with patients and families for the safest care. Boston: National Patient Safety Foundation; 2014.

ABOUT THE AUTHORS

Carolyn M. Clancy, MD a general internist and health services researcher, is Interim Under Secretary for Health at the Department of Veterans Affairs, having joined VA in 2013 as Assistant Deputy Under Secretary for Health for Quality, Safety, and Value. Prior to VA, she served as Director of the Agency for Healthcare Research and Quality.

Arthur Garson, Jr., MD, MPH is the Director of the Health Policy Institute of the Texas Medical Center, home to 54 health care institutions in Houston. Prior to assuming his current position, he was the Director of the Center for Health Policy, University Professor, and Professor of Public Health Sciences and Public Policy at the University of Virginia.

Chapter 23

Information Technology and Better Health: Overcoming the Risks

Randolph Hall

INTRODUCTION

"Sebelius Resigns After Troubles With Health Law Website," read the New York Times headline on April 11, 2014. Just a few months after the rollout of the healthcare.gov Web site, this event offered further evidence of how the failure to deploy effective information technology (IT) can bring down health care executives at the highest levels. President Obama himself became the first US President to suffer a huge drop in reputation as a consequence of a Web site, his public approval rating dropping from 53% to 41% between December 2012 and November 2013 as a reflection of the troubles in delivering a promised portal to the health care system mandated by the Affordable Care Act.

IT projects can sap the energy of the best of organizations, drawing significant human and financial resources that might better be deployed toward direct service to patients. They are risky, and create a dependency on software and systems that might feel outside the control of the CEO. When they fail, they can fail big, paralyzing organizations in a way that would never occur using paper systems. New systems can be hard to learn and can bring with them business processes that employees and customers are not accustomed to, creating doubt as to the wisdom of the health center's leadership in launching a project and anger over their failure to appreciate the people who are forced to use a new system. A recent study by the RAND Corporation [8] came to this conclusion regarding electronic health records (EHR):

Physicians noted that EHRs had the potential to improve some aspects of patient care and professional satisfaction. Yet for many physicians, the current state of EHR technology significantly worsened professional satisfaction in multiple ways, due to poor usability, time-consuming data entry, interference with face-to-face patient care, inefficient and less fulfilling work content, insufficient health information exchange, and degradation of clinical documentation.

Given all of these downsides, it is a wonder why any CEO would launch a new IT project. Pen and paper will do just fine. But the upsides cannot be ignored. Failing to keep up with competition with modern user-friendly IT will make an academic health center (AHC) look dated or even obsolete. Patients, students, faculty, and staff have all experienced the empowering effect of online shopping, social networks, information search, finance, etc.; and, from these experiences, they expect similar quality systems in the operation of their health centers.

More than anything else, IT offers transparency across the organization, as well as between the organization and its clients, business partners, and regulators. When done well, IT helps gather data seamlessly and automatically, stores data for use when and where needed, integrates data sources to obtain new insights, and presents information to users in a meaningful form that helps them make better decisions. Through constant monitoring and feedback, IT can raise the quality of products and services, help meet deadlines, manage workloads, achieve financial targets, and improve patient outcomes.

Few of the software options available to AHCs are unique to the enterprise, so the institution will largely acquire tools that are used across a spectrum of organizations, including all types of health enterprises, colleges and universities, and, more broadly, large corporations. Examples include the following:

- Electronic health records
- Patient and staff scheduling
- Financial and billing systems
- Picture archiving and communication for medical imaging
- Personnel and human resources
- Grants and contracts management
- Clinical trials management
- Student records
- Compliance for human subject or animal research
- Specialized analytical and research tools
- Digital repositories and document management
- Business intelligence and reporting

Some of these systems are inherently designed for use across the enterprise, whereas others are specialized for a single function, such as research administration.

The Transformation of Academic Health Centers. http://dx.doi.org/10.1016/B978-0-12-800762-4.00023-2

GOALS AND RISKS

Every IT project should begin with an assessment of goals, risks, and affordability to determine whether the project's benefits outweigh the costs. The initial assessment serves two purposes: (1) to assess whether embarking on a project is worthwhile, and (2) if a project is initiated, to guide the project toward organizational goals, which may include any of the following:

- Promote and attract patients and customers by offering a unique, superior, or perhaps just competitive service, such as telehealth services, online access for medical advice and consultations, or supporting more effective medical decision-making.
- Automate manual labor-intensive processes and reduce costs.
- Offer portability, transparency, and more universal access to data, making information more accessible to patients or improving accountability by making data more visible.
- Meet a regulatory obligation or ensure that violations do not occur.
- Improve the quality of a product or service through standardization, reduction in errors that affect patient safety, increased personalized care to patient preferences, or faster delivery of services to patients.
- Develop new procedures, algorithms, or findings based on the analysis of large data sets collected by the system.
- Identify and act on trends that might be observed across large populations—such as effectiveness of therapies, patterns of health behavior, the onset of pandemics, or misuse of resources.
- Trace access and modification of data or documents, and manage different versions of documents.
- Replace outdated systems that are difficult to maintain and frustrating to use.

All of these goals promise to improve the workplace, the patient experience, and the bottom line. A well-designed and well-maintained system will be something that people want to use, rather than be forced to use.

Weighing against these benefits, IT projects run the risk of making the AHC worse off than when it started. Such risks include:

- Information breach—when data are stolen, released to the wrong person, or misused possibly in violation of the Health Information Portability and Accountability Act (HIPAA, also called the "privacy rule");
- Hacks into systems or sabotage whereby data are destroyed or illicitly modified;
- System failures that prevent access due to hardware or software faults, excessive and unpredictable maintenance, power outages, or disasters;
- User resistance to adoption due to dislike, resistance to change, or job protection;

- Failure to train adequately or support users;
- Lost data due to inadequate backup, retention, and protection;
- Cost overruns, failure to meet time schedules, or failure to meet specifications;
- Inability to maintain the software against changing system-to-system interfaces, obsolescence of programming languages, and availability of trained staff; and
- As a result of any of the above, abandonment of a system after making a substantial investment.

The list of what may go wrong seems daunting, but all can be prevented through adequate planning, anticipation of risk, and investment in areas such as information security, backup, and user support. Most importantly, an IT system requires constant attention well beyond when it is first installed.

Affordability characterizes the costs of supporting an IT system over its entire life cycle, including revenue sources and the savings that may accrue. A system aimed at efficiency should produce savings that exceed costs, from a net present value perspective; whereas systems aimed at quality or revenue growth might be justified by other metrics. No matter what the aim, the funding source (or sources) for any investments must be identified, preferably from the units that accrue the greatest benefits from IT system implementation.

Lastly, from the initial assessment of costs and benefits, it is desirable to define the operational concept for how the system will support desired processes, which can be characterized as a system description [1]:

in terms of the user needs it will fulfill, its relationship to existing systems or procedures, and the ways it will be used.

The operational concept begins to answer the "how" of system performance, including the following:

- How will information be accessed, by whom, where, and when?
- What business processes will be supported by the system and who will conduct these processes?
- How will information be presented to users?
- What technologies might be employed to gather and process data?

CREATING THE SYSTEM

The International Council of System Engineering defines system engineering as [6]

an interdisciplinary approach and means to enable the realization of successful systems. It focuses on defining customer needs and required functionality early in the development cycle, documenting requirements, then proceeding with design synthesis and system validation while considering the complete problem.

System engineering is critical to both the design and deployment of software, along with the integration of software with hardware and devices. Most importantly, a good system does not simply operate reliably and predictably, but fulfills human needs, meaning that users can effectively interact with the system to achieve both personal and organizational goals. System engineering considers how software and technology are used over their life cycle of operation, including development/installation, testing, training, a period of stable maintenance and operation, and eventual retirement and replacement by successor systems.

Unfortunately, even the best-designed information system cannot correct deficiencies in the organization itself, though it may expose the organization's flaws in a way that leads to improvement. Misaligned incentives, reporting structures, cultures, or poor leadership cannot be solved by IT. If an AHC does not integrate medical practices toward common performance goals, a shared medical record system will not suddenly transform the center into a true "system" for the patient; it will still feel like a mishmash of free-agent providers. On the other hand, when Kaiser Permanente deployed the EPIC medical record system, patients experienced the benefits of a standardized approach to care delivery across sites and specialties.

A centerpiece of system engineering is the process of partitioning the system into its components, assigning functionality to these components, defining the interfaces of components to each other, and setting quantitative goals and objectives to be achieved. This process is often called "system architecting" as it resembles the process by which buildings are designed. The design of buildings, like the design of software/hardware systems, begins with an articulation of functionality and goals, then divides the structure into components (imagine sleeping, socializing, and eating components in a home), then sets forth a plan by which the components communicate with each other.

The partitioning process for software/hardware guides both system design and project management. It also guides the management of systems once placed into operation. A component may represent a business function that spans the organization, such as billing or human resource management, or it may represent a unit of the organization, such as a medical specialty (pediatrics, medicine, etc.) or a particular location, such as a clinic, laboratory, or hospital.

A well-designed system aligns technology with function, such that each component serves a well-defined purpose for specified users/customers while achieving measurable aims. The challenge in system engineering is creating components of the right scale so that they can be well managed and respond to their customer base, along with creating the means and incentives to assure that the interfaces between components fulfill the goals of the organization as a whole. When systems fail, it is often because the interfaces have been poorly conceived or executed, or because they result in responsibilities falling "between the cracks" or, the opposite, when organizations battle over turf.

The Department of Defense long ago created standards for engineering complex systems, which are reflected in Military Standard 498 (MIL-STD-498; [2]) and, more recently, the ISO/IEC/IEEE standard 12207-2008 [7]. The value of these standards is that they provide a well-prescribed methodology for translating the needs of the users and organization into template documents that specify how these needs will be satisfied. Documents for system requirements are particularly useful, as they specify exactly what an IT project must achieve to meet the needs of the customer (either internal or external). They can be used both to hold an internal project team accountable and to keep a vendor contractually liable for deliverables. They also provide a mechanism to attain buy-in and agreement among internal stakeholders in the early stages of the project and control the need for changes later on.

IT AS A REFLECTION OF THE SYSTEM

AHCs are premised on the integration of patient care with education and research to improve health, both in the present and for the future. Even in the absence of IT, an AHC—when viewed as a system—must struggle with how the triumvirate of care, education, and research reinforce each other toward the common goal of human health. These primary components are further divided into medical facilities, academic departments, centers and institutes, and administrative and support units. For research to complement patient care, the AHC must establish structures and incentives whereby each supports the other, making innovative research-driven care available to patients and, likewise, informing researchers through exposure to the problems experienced by patients and clinicians. Thus, a well-conceived interface is needed to connect research institutes to complementary medical specialties. IT must reflect the realities of how the AHC, as a system, has been laid out, while guiding it toward the goals of the AHC as a whole.

Viewed on a grander scale, the AHC, while a sizable "system" in its own right, is but a component of still larger systems. Many AHCs are part of a comprehensive university that offers undergraduate education in the arts, sciences, social sciences, humanities, and non-health professions, as well as a myriad of community-focused and athletic programs. Coordination of educational opportunities, admission, shared research services, capital planning, and the like are all aspects of how the university, as a system, is designed. Moreover, the AHC is just one component of the national health system, whereby interfaces are needed with insurers, community hospitals and physicians, specialized care providers, government regulators, financial service companies, pharmacies, drug and device companies, and so on. Each of these interfaces demands technology for the sharing of information and the coordination of action among organizational units.

IT must also reflect the realities of its constituent users, including their education, training, motivation, and patience for learning new pieces of software. Faculty (clinically trained and scientific), nurses, administrative staff, students, patients, family members, and management must all be considered. Whereas a bad user experience can cause patients to vote with their feet and opt to go elsewhere for care, a similar bad experience for staff will more likely lead to poor morale. And, a poorly conceived system roll out affecting faculty may lead to faculty uproar and jeopardize center leadership. No matter what the consequence, systems must be well designed, tested, and user-centered to achieve success. The principle of "first, do no harm" applies as well to IT; new systems should never be deployed if they do not improve the fate of the user community.

ELECTRONIC HEALTH RECORDS

The single largest driver of health care IT today is the EHR (also called the electronic medical record or EMR). The EHR is the repository of clinically relevant information about individual patients, supplanting the traditional paper file. Cerner, Epic, and Allscripts each generate more than $1 billion in annual revenue from their EHR business lines sold to health care providers of all types. The EHR industry has experienced phenomenal growth over the last decade, in part spurred by the general growth in IT, but also by a provision in the American Recovery and Reinvestment Act (ARRA) of 2009, incentivizing the adoption of EHRs by providers of Medicare and Medicaid Services in the United States. Even preceding ARRA, large health maintenance organizations, such as Kaiser Permanente (Kaiser), invested heavily in EHRs. Some national governments, such as the United Kingdom, have invested in nationwide EHRs, and the Veterans Health Administration began development of its EHR in the 1970s. Kaiser completed the installation of its Epic-based system in 2010 for its entire provider network serving nine million patients, at a cost of $4 billion, or a cost of more than $400 per patient. Philip Fasano, Kaiser's Chief Information Officer, cited these advantages [10]:

> We have every piece of information about that patient available to us to draw upon. The primary care physician has all the information about the patient, the specialists have all the information about the patient, and anyone they encounter in any of our hospitals has it as well. He or she can see the patient's health history, diagnosis by other providers, lab results, and prescriptions are all there. X-rays are stored digitally and are there. That information is also available if a patient goes to the ER.

Beyond information access, Kaiser has developed tools to help assure that patient care follows accepted protocols and to avoid duplication of medical tests. But Kaiser's experience as an integrated care-provider network does not readily translate to other health systems, as mentioned in a recent RAND report [4]:

> A team of RAND Corporation researchers projected in 2005 that rapid adoption of health information technology (IT) could save the United States more than $81 billion annually. Seven years later the empirical data on the technology's impact on health care efficiency and safety are mixed, and annual health care expenditures in the United States have grown by $800 billion.

The authors attribute this outcome to a combination of slow adoption, selection of systems that are not interoperable or user friendly, and the failure of health care providers to redesign their systems to capture the full benefits of health IT.

O'Malley et al. [9], examining physician practices, were similarly cautious about the benefits of EMRs. Though such systems made information more accessible during patient encounters and supported electronic messages, physicians found these drawbacks:

- "EMRs are less able to support coordination between clinicians and settings, in part due to their design and a lack of standardization of key data elements required for information exchange;
- Managing information overflow from EMRs is a challenge for clinicians;
- Clinicians believe current EMRs cannot adequately capture the medical decision-making process and future care plans to support coordination;
- Realizing EMRs' potential for facilitating coordination requires evolution of practice operational processes;
- Current fee-for-service reimbursement encourages EMR use for documentation of billable events (office visits, procedures) and not of care coordination (which is not a billable activity)."

Implementing an EHR is challenging and expensive, but the EHR has become the central IT system by which health centers operate and other elements branch off (e.g., scheduling) [5]. Therefore, the question today is not whether to implement an EHR but how to do so effectively so that, as a result, the health system provides better and more efficient care. Nationally, the key question will be how to provide data portability for patients when they visit different health care providers. Equally important to the goals of IT system implementation is how an IT project is managed and executed once decided upon.

PROJECT MANAGEMENT AND EXECUTION

As already mentioned, an IT project begins with an assessment of goals, risks, and affordability along with the creation of an initial operating concept describing how the goals will be achieved. This is the basis for a go/no-go decision, along with an initial guide for how a project should be executed. Subsequently, the IT project goes through these phases:

- User requirements and project charter
- Selection and acquisition
- Development and installation
- Test and refinement
- Maintenance, operation, and support
- Retirement

User Requirements and Project Charter

The **user (or system) requirements document** defines what the IT project should accomplish from the perspective of the user. It does not tell how the specification is achieved. Ideally, the requirements are software/hardware agnostic, thus allowing alternative solutions to be compared against a set of objective standards. A requirements document should include these items:

- Types of users supported by the system and their roles;
- Business processes supported and desired workflows;
- System-to-system interfaces that must be supported;
- Relationship of the new system to existing systems, and which legacy systems might be retired;
- System environment (such as servers, disaster recovery, communication network, and installed hardware and devices);
- General requirements with respect to factors like accessibility, user interfaces, security and information protection, and system availability;
- Functional requirements describing how each business process will be supported;
- Internal interfaces that must be created between system components;
- Operational requirements defining the resources that must be available to support system operation, including user support, desired performance metrics, usage characteristics (e.g., volume and variation of usage), and regulatory standards that must be achieved (for instance, HIPPA compliance); and
- Training and documentation requirements.

Not to be overlooked, the requirements document should specify the types of reports that must be generated for performance monitoring and improvement, quality, and financial management. It is worth considering in advance how reporting and monitoring might be generated and used to achieve managerial objectives, such as improved patient safety, improved utilization of resources, higher patient and employee satisfaction, and reduced delays for providing services.

The **project charter** complements the requirements by specifying the roles and responsibilities with respect to how the IT project is executed and managed, as well as how the completed system is operated and maintained. The charter is the internal contract among all key participants. Critical elements include the following:

Project Manager: the individual designated to have overall responsibility for delivering the project on time, on budget, and to specification, to whomever he or she reports.
Executive Sponsor(s): the organizational leader (or leaders) representing the users for the IT system, to whom the project manager will be held accountable. The sponsor often funds the project, and will be responsible for integrating the system into business processes. When there are multiple sponsors, the project manager may be held accountable to an Executive Steering Committee.
Advisory Committee: a broader group of expert users, representative of the user community as a whole, should be consulted in making design choices. The advisory committee is not a decision-making group, but instead a consultative group to help keep the project on track.

The charter should also describe project finances, budget, controls, and consequences for not meeting targets. These controls should include the process of user acceptance for project deliverables. The charter should also identify the staffing model across the life cycle, considering whether staff is sourced from a central IT department, a business unit, or possibly consultants.

Selection and Acquisition

Once the requirements and charter are established, the selection process can begin. Initially, this may take the form of a cost/benefit assessment of these alternatives:

- Home-grown software, developed internally by the AHC or university;
- Open-sourced software (freely available code that can be modified by the AHC as needed);
- Vendor licensed, institution supported, software that resides on the AHC's servers; and
- Vendor licensed Software as a Service (SaaS), which resides on the vendor's servers.

Depending on the approach taken, a more or less detailed system engineering process will follow. SaaS software has limited flexibility, but requires little in the way of design, development, or support. Organizations that lack IT bench strength can often implement SaaS solutions with ease, a reason why SaaS has grown in usage. At the other extreme, home-grown software can be infinitely customized, but demands skilled programmers and system engineers (and, hence, a much more extensive system engineering process [3]). Larger and more complex systems may be developed with a combination of software solutions, with some components delivered through home-grown software, others vendor-licensed, and others SaaS.

If the choice is made for vendor-supplied software, the next step is selection of a specific product and vendors. Following the AHC's purchasing procedures, project leadership must compare and evaluate the alternatives according to the following factors to select the product that provides the highest value to the organization:

Fulfillment of requirements: the degree to which requirements are met and, through a "fit-gap analysis," assessment of the investments required to fulfill the requirements. The vendor should demonstrate how the software can support specific use cases that reflect the AHC's actual operations. Ideally, requirements can be met through software **configuration**, meaning that users can meet the AHC's needs through setting software parameters. As a less desirable alternative, software **customization** may be required, meaning that new code is written to fulfill the AHC's needs. Alternatively, the AHC may elect to forego certain requirements if they are too costly to satisfy.

Life cycle cost: factors to consider include: (1) initial license cost for both the software and associated databases; (2) annual license maintenance for software and databases (typically about 20% of the initial license); (3) professional service costs for installation, development, and upgrades; (4) staffing costs across all categories, including project management, developers, trainers, and technical writers; (5) hardware, devices, and servers, including their future replacement. These costs must be considered across the entire life cycle, especially considering that different products have different pricing models for up-front versus maintenance costs.

Experience: whether the vendor has successfully installed comparable systems, verified by contacting references at current and former customers and site visits at customer sites.

Vendor team: qualifications of the vendor's team that will support the installation, their availability, and their performance during interviews. In addition, an assessment of the financial health of the vendor is important to gauge whether it will survive through the entire project life cycle.

Terms and conditions: the contractual terms that provide assurance that the vendor will meet its commitments, and that recourse exists should those commitments not be met.

Experience is an especially critical factor. If the vendor or product is not well tested prior to selection, then the AHC must absorb the risk of potentially failed products, or must invest time and energy to help the vendor develop the product to the point where it does meet user requirements. This approach only makes sense if the AHC wants to put itself in front of its competitors with a particularly innovative IT solution.

Development and Installation

Development and installation can be resource-intensive as programmers, systems, and business analysts and engineers work to fulfill the requirements. Tasks during this phase include configuring forms, data entry systems, workflows, data storage, communication, identity management, user interfaces, and reporting. This is when the organization must finalize new business processes that exploit the capabilities of the IT system toward the goals set forth at the project onset. During this phase, software is deployed on a **development server**, meaning it is kept off-line relative to actual business operations.

During development and installation, documentation should be created to support both users and developers, along with a plan to train and deploy the new IT system in the user environment. Documentation should be instructive as to how to use the system to accomplish normal business tasks, as well as provide explanation as to terminology (e.g., a "data dictionary") and the methodology used to complete analyses or generate reports. Training videos may also be created as a supplement to in-person training. Ideally, training and documentation are fully integrated into the system itself so that the guidance can be fully contextualized, when and where it is needed. But prior to launch, in-person training should be offered.

Test and Refinement

Prior to full launch, all effort should be made to ensure that the system has met all requirements and expectations and operates as intended. To do so, the system should be tested from an operational perspective (e.g., stress testing or regression testing under loads that mimic actual operation), and from a user perspective. Operational testing will occur on the development server and focus on finding software flaws or limitations that prevent it from scaling up to desired volumes.

User testing is divided into "alpha testing" and "beta testing." The alpha test engages a limited number of users while still in development, identifying flaws that can be deciphered from test scenarios. Once passing alpha tests and operational tests, the system moves from a development server to a production server, and then beta testing begins by a limited user group for actual business operation. Beta users may be volunteers or may be units that are recruited with the idea that they can become first users and influence the ultimate system. Moving from hypothetical cases to true operation can reveal not just system issues, but problems with documentation and training—all of which can be corrected before moving into full production, or before mandated for all users.

Maintenance, Operation, and Support

Once the system has moved into full production, additional flaws may appear that require correction. After these are rectified and the system has been accepted by the executive sponsors, the system moves into a period of more stable operation. By this time the system ought to be fully documented and users trained, a help desk should be established, and the system should be fully integrated into business processes. From that point, key issues will include data and identity management, information security, and periodic upgrades, including installation of new versions of the software. Upgrades are inevitable as hardware, storage media, communication technology, and user expectations change over time; therefore, an upgrade strategy that reflects all software customizations is essential.

Retirement

If well managed, a system may never need to be retired. But it is more likely that a system will eventually need to be replaced by newer IT that better meets the needs of the organization, or relies on more modern programming or hardware paradigms. The transition from an old system to a new system is challenging, as the systems may operate in parallel for a period of time, each requiring support, as the full functionality of a new system is brought into production. The transition of legacy data into new formats can be particularly challenging and requires careful planning.

Management Approaches

When a project goes over budget, beyond schedule, or fails to meet specification, it is most likely the consequence of inadequate planning, communication, or coordination— and, ultimately, failure of project leadership to anticipate risk. It is always important to place the responsibility for a project in the hands of a single manager who is accountable to the executive sponsor. Depending on the complexity of the project, it may be desirable for the project to also be accountable to, or reside within, the IT organization so that it can draw from its technical expertise or physical resources.

An effective project manager will follow such best practices as:

- Creating a detailed project plan that tracks tasks and resources by date, communicating that plan to the user community, and honestly keeping stakeholders informed of progress or delays;
- Allowing time for testing and refinement prior to widespread deployment;
- Obtaining signoff from stakeholders on specifications prior to initiating work, and obtaining signoff for acceptance upon completion;
- Including all aspects of user support, documentation, and training in the project plan; and
- Following a change-order process to manage new requests as they come up in the course of the project.

CASE EXAMPLE: RESEARCH ADMINISTRATION

In 2011, the University of Southern California (USC) embarked on a project to support research administration across the university, later called TARA, or Total Access for Research Administration. TARA was envisioned as a "system-of-systems," offering a single portal to a variety of tools, along with interfaces that enable data to be shared among software components. TARA includes these elements:

- Grants and contracts management and financial analysis
- Regulatory tools
- Business intelligence and reporting

The project has been guided by a systems requirements document [11], which specifies the needs of the university (not just the AHC) for research administration and follows the format mentioned earlier in this chapter. At inception, the university had fully adopted just one system—the institutional review board application from Click Commerce, which USC called "iStar." Additional systems have since been deployed:

- Click Commerce applications for institutional animal care and use, biosafety, radiation safety, and conflict of interest;
- Kuali Coeus pre-award and post-award for grants and contracts management;
- FiPS, a home grown add-on to Kuali Coeus for financial projections;
- CORES from Vanderbilt University for automated billing and accounting for core laboratories;
- Sophia from Wellspring for Intellectual Property Management; and
- A set of standardized performance reports programmed through Cognos Business Intelligence software.

At USC, we will soon embark on two more projects— installation of EH&S Assistant for laboratory safety and a Clinical Trials Management System (CTMS).

Only the CTMS was not anticipated when the project began. Unlike the other research administration software applications, the CTMS is not designed to support solely the research mission of the AHC, but to support both the clinical mission and the research mission. As the university embarked on the TARA project and simultaneously on a Cerner EMR, the CTMS fell through the cracks. It was only after the university established a clinical trials office under the Office of Research that the CTMS became a priority as a joint effort of the Keck Medical Center, the Keck School of Medicine, USC's Clinical and Translational Science Institute, Children's Hospital of Los Angeles, and USC's Office of Research.

USC is embarking on one more project that spans the clinical and research missions, though the emphasis here is on research itself, not the administration of research. That is the creation of a clinical data research warehouse, a repository from which research can be conducted drawing from data that originate from the Cerner EMR system. The warehouse provides a nexus for an AHC that brings the clinical and research enterprises into full alignment, whereby research insights emerging from clinical data are fed back to the medical center for the benefit of patients.

PREPARING FOR THE FUTURE

Digitized patient records and other health data are the kernel from which future discoveries will occur for the benefit of population health. We are moving to an era of ubiquitous information, big data, simulation, and mobile technology. The doctor/patient encounter cannot be separated from the context of information that the patient has already obtained via Internet resources (and will likely continue to seek out on their mobile phones from inside the examination room). AHCs must determine how clinicians, students, and researchers will work in concert with IT for the benefit of patients.

The AHC ought to be the driver of innovation, combining talents from diverse academic fields across the entire university. The AHC also ought to partner with private industry in the drive for innovation. Whereas the process of system engineering described in this chapter is appropriate for enterprise systems at AHCs, companies often have the agility to introduce new products at a rapid pace, many of which will be adopted by patients long before the AHC itself. Areas of innovation will include the following:

Mobile Health (or mHealth) that couples the smartphone with human-worn or implanted sensors that track movements, risk-exposures, diets, sleep patterns, and other measurements (such as glucose levels). These new data sources offer the ability for real-time feedback and therapies, as well a postanalysis to produce diagnoses and therapies.
Telemedicine offering access to specialists or specialized resources from a distance, helping those who live far from AHCs, or who do not have the mobility to reach an AHC.
Visualization and Simulation based on immersive media that engages participants in new ways, or offers insights or training that would not otherwise be possible.
Big Data Analytics for massive data sets emerging from EHRs, genomic studies, biomarkers, preferences expressed by patients, and other personalized data to improve the accuracy of diagnoses and assess the very best treatment options for individuals.

As a vision for the future, a patient could be offered miniaturized tools that continuously and effortlessly monitor, diagnose, coach, prevent, and treat those following protocols overseen by health care providers. The system would give ready access to clinicians when needed, through an integrated scheduler; and personalized therapies will be created through the analysis of all available data. And the system will present information to clinicians, patients, and family caregivers in a format and manner that they can readily understand and appreciate.

Today, AHCs are struggling with the initial phases of getting EHRs off the ground. We are initially working to attain efficiency, acceptance, and continuity of service. Given the scale of the EHR, the range of users, the need to install both hardware and software, and information security requirements (the list can go on), this alone is a formidable task. But the EHR is just the spring board from which innovation will arise. Turning big data into health is the challenge ahead, a challenge that can be met by combining the talents of clinicians, computer scientists, engineers, social scientists, and other innovators, as only can be accomplished in an AHC.

REFERENCES

[1] Department of Defense. Operational concept description. DI-IPSC-81430. December 5, 1994.

[2] Department of Defense. Software development and documentation – MIL STD 498. December 5, 1994.

[3] Department of Defense. Defense acquisition guidebook. May 15, 2013. Chapter 4 – Systems Engineering.

[4] Friedberg MW, et al. Factors affecting physician professional satisfaction. RAND Corporation Issue Brief; 2013. RB-9740-AMA.

[5] Hall RW, Partyka JG. Scheduling for better healthcare. OR/MS Today 2013;39(3).

[6] INCOSE, International Council of Systems Engineering. What is systems engineering? 2014. http://www.incose.org/practice/whatissystemseng.aspx.

[7] ISO/IEC/IEEE. 12207-2008-ISO/IEC/IEEE standard for systems and software engineering - software life cycle processes. March 21, 2008.

[8] Kellermann AL, Jones SS. What it will take to achieve the as-yet-unfulfilled promises of health information technology. Health Aff 2013;32(1):63–8.

[9] O'Malley AS, Grossman JM, Cohen GR, Kemper NM, Pham HH. Are electronic medical records helpful for care coordination? experiences of physician practices. J Gen Intern Med 2010;25(3):177–85.

[10] Snyder B. How Kaiser bet $4 billion on electronic health records – and won. Infoworld; May 2, 2013.

[11] USC, University of Southern California. USC research administration system (RAS) system requirements – V1.0. 2011. https://research.usc.edu/files/2011/06/RASRequirements_3_7_2011.pdf.

ABOUT THE AUTHOR

Randolph Hall, PhD is the Vice President of Research at the University of Southern California and is responsible for leading research initiatives across all areas of the university, including overseeing research advancement, administration, and ethics activities. He is also the principal architect for the university's TARA (Total Access for Research Administration) information system.

Chapter 24

Market Consolidation and Alignment

Larry R. Kaiser

Hospitals represent the largest source of health care expenditures in the United States, accounting for one-third, or $882 billion. Private health insurers spent approximately $320 billion (35% of premiums) on hospital care in 2012 [1]. The current health care environment has created a situation that is ripe for consolidation in certain markets. Trends prompting hospitals and health systems to consolidate include the downward pressure on revenues, driving the need for scale; the desire to access new geographies and patient populations as well as competencies and the pursuit of greater negotiating leverage with commercial payers; access to capital for financially stressed organizations; capacity management; and cost synergies by consolidating back-office functions, leveraging supply chains, and rationalizing capital projects.

The implementation of the Affordable Care Act (ACA) and a new emphasis on value-based care has also spurred consolidation in a number of markets. A typical region in the United States has three to five consolidated health systems with a smaller fringe of health care centers outside of these systems. Declining reimbursements, significant reduction in inpatient admissions, the potential for limited networks, hospital employment of physicians, and the leverage of payers in a number of markets, among other factors, are all significant contributors to an ever increasing volume of hospital consolidations.

Historically, the primary driver of consolidation has been access to capital, but about five years ago the challenge of physician alignment, employment, and integration became just as important—especially in light of declining reimbursement. The desire on the part of hospitals to capture admissions and prevent "leakage" played a major role in the move toward employing physicians—especially primary care physicians—who potentially would refer to specialists associated with the hospital. A major advantage of consolidated systems is the ability to coordinate care across different practice areas and sites of care in an effort to revamp care delivery and decrease spending. The goal is to build an end-to-end continuum of care. Offsetting the advantages that consolidation may bring is the potential for higher prices secondary to greater leverage when negotiating with payers, and that recognition is attracting increased attention from antitrust authorities—specifically the Federal Trade Commission (FTC) and the Justice Department.

Between 1981 and 2011, inpatient hospital days decreased by 33% despite a growing and aging population. During that same time period, with the decrease in inpatient utilization, more than 15% of hospitals closed. Currently, 60% of hospitals are components of health systems—an increase of almost 10% in the past 10 years [2]. Between 2007 and 2012 there were 432 merger and acquisition deals involving 835 hospitals. The number of hospital deals has risen steadily since 2009, though activity in this area was slightly down in 2013 (Figure 1).

There were 89 mergers and acquisitions in 2013, down from 94 the year before but the total transaction value rose from $1.9 billion in 2012 to $18.6 billion in 2013, primarily due to several large deals in the for-profit sector. The acquisition of Health Management Associates by Franklin, Tennessee-based Community Health Systems by itself was a $7.6 billion deal, and Tenet Healthcare's purchase of Vanguard Health Systems added another $4.3 billion to the total value of transactions.

MERGER CASE EXAMPLES

The acquisition in New York City of the Continuum Health System by the Mount Sinai Health System has created an entity that has become the market leader, surpassing that created by the 1990's merger of New York Hospital and the Columbia Presbyterian Hospital, which created New York-Presbyterian Hospital. It is illustrative to look at the details and the entity that emerged in each transaction [3].

Case One: New York—Presbyterian Hospital

In the case of the New York Hospital/Presbyterian Hospital merger, relationships with two medical schools existed. Specifically excluding the medical schools, the merger resulted in a full asset merger between the two hospitals. The merged entity was incorporated as one institution and operates under unified management. In fact, a unified management structure was put in place even before the effective date of the merger. Despite operating under a single license, the merged institution was challenged by the geography of a 100 block distance between the facilities. The merger was

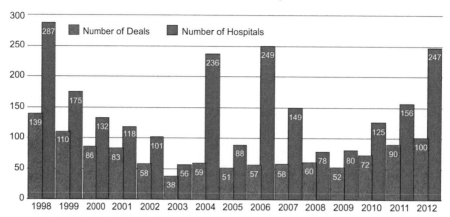

FIGURE 1   Hospital mergers and acquisitions from 1998 to 2012. There has been a steady baseline of merger activity during this time period with 50 or more deals in every year but one, 2003. The number of hospitals involved has varied greatly and more than half of acquired hospitals in recent years had fewer than 150 beds. *From FTI Consulting, Inc., Center for Healthcare Economics and Policy, Hospital Realignment: Mergers Offer Significant Patient and Community Benefits. January 23, 2014.*

predicated on improving quality of and access to care, in addition to achieving and maintaining financial stability. It was well known that both institutions had experienced financial difficulties, but Presbyterian Hospital had only recently emerged from a major financial challenge that pre-existed for several years. The two medical schools independently negotiated affiliation agreements with their former hospitals in order to preserve a funds flow mechanism. The merged hospital was further challenged by having to work with two separate and independent faculties and physician practice organizations. It was agreed that there would be one set of medical bylaws, clinical policies, and standards for the merged institution. The agreement also called for one chief medical officer and a single medical staff office to deal with issues arising with staff members at either institution.

Early on, it was recognized that forcing consolidations among clinical departments would be contentious and likely counterproductive, potentially leading to the departure of key physicians as well as physician leaders. In a number of departments there was pushback from physicians who resisted the notion of having a single program with single leadership. Instead of clinical consolidation, the merged institution formed service lines to foster clinical integration and alignment. Institutional leadership was able to achieve agreement on a set of common protocols as well as quality measures in a given clinical specialty or disease to be used by both programs.

The concept of a service line was defined as the "organization, management, and delivery of a comprehensive continuum of services around a major disease entity, age group, or patient population." The "product" of the service line was to be high-quality, cost-effective, safe, appropriate, and responsive patient care. The goal of the service line concept was to provide the best quality care, establishing common standards and best practices across both hospitals but

without consolidating departments. The service lines were physician-led with equal representation from both facilities. In order to create an incentive for service line organization and performance priority, $60 million of capital investment was made to service line projects over the initial 3 years. This investment served to significantly enhance the clinical strengths of the merged hospital. Service lines were treated as small business units governed by an executive council chaired by physicians who had full responsibility for their "product" and a reporting relationship to the chief medical officer. These service lines accounted for the majority of discharges and most of the revenue of the hospital and significantly increased volume and thus market share, while decreasing the length of stay on certain services despite an increased severity of discharges.

Service lines have allowed the hospital to realize the benefits of clinical integration while avoiding the problems inherent in forced consolidations. Over time, a number of clinical consolidations occurred, including a few departments—especially among some of the surgical subspecialties. And, joint residency programs were established in a few specialties. It is likely that the service lines facilitated the interactions and cultural changes necessary to allow for the consolidations.

The New York-Presbyterian System now includes 11 hospitals in the five boroughs of New York City along with 19 other hospitals in the tri-state area. Not all of these hospitals are owned, and the system defines itself as a "federation of top-quality hospitals, specialty institutes and continuing care centers throughout the tri-state region." In early 2013, New York-Presbyterian stepped in to bail out New York Downtown Hospital, the only hospital in lower Manhattan after St Vincent's Hospital closed in 2010. New York Downtown Hospital had experienced significant financial challenges that threatened its continued viability, but NY-Presbyterian felt it could save the hospital by improving quality, delivery, and efficiency of the existing services. The hospital had been part of the NY-Presbyterian Health System's network since 2006, but with the takeover the losses and assumption of debt is now on NY-Presbyterian's books. Like many other safety-net hospitals, New York Downtown Hospital has a disproportionate share of Medicaid patients who comprise 45% of inpatient discharges. In addition, 20% of patients treated in the Emergency Department are uninsured, and due to higher costs and poor reimbursement rates, the hospital was losing money on maternity and neonatal care.

Case Two: Mount Sinai–Continuum Merger

The Mount Sinai Health System and the Continuum Health System merger involved two large hospital systems that were looking to gain efficiencies through economies of scale that would provide the ability to relocate and consolidate various services and merge administrative and billing operations. Of note, Continuum initially engineered a merger with the New York University Hospital System but Mount Sinai intervened and Continuum opted to proceed with them, thus dropping New York University. The New York University System now finds itself in the unenviable position of competing with two mega-systems in Manhattan.

The Mount Sinai/Continuum merger was somewhat less complex than the previous case, primarily because only one medical school was involved. It would be naïve to think that increased leverage in pricing was not on the minds of the leadership of the two organizations, but there is no question that access to capital, as well as a desire to preserve community benefit, also played a role. In order to preserve and expand those services that classically lose money yet provide significant community benefit, a system has to generate adequate margin. Such services as pediatrics, where the majority of patients are covered by the federal Medicaid program, and psychiatry must be subsidized by margins generated by other programs. Orthopedic surgery, for example, generates a significant margin that is used to cross-subsidize other money-losing, but community beneficial, services. It is far more likely that services providing a significant benefit to the community will survive and thrive in an integrated health care system such as the new Mount Sinai Health System.

The system that has resulted from the merger has a demographic that is approximately 20% Medicaid and 40% Medicare. It includes 3571 licensed beds, 138 operating rooms, 177,000 inpatient admissions, 2.6 million outpatient visits, 35,000 employees, 2784 full and part-time physicians, an additional 3783 voluntary physicians, and 2000 residents and fellows [4]. The dominant position that the Mount Sinai Health System now enjoys in Manhattan also gives it the opportunity and the ability, if they so desire, to add an insurance product in order to form a truly integrated system very much like that of the University of Pittsburgh Medical Center (UPMC). The ability to offer a limited network insurance product provides an opportunity to control costs and effectively compete with the other commercial payers. Mount Sinai currently owns HealthFirst, a Medicaid and Medicare managed-care company that also includes dual eligibles and plans to offer a new Medicare Advantage product. It is clear that for a health system to participate in "disintermediation"—defined as being your own insurer—a certain size and scale has to be obtained in order to be in a position to take on the actuarial risk.

IMPACT OF THE CHANGING CONSOLIDATED HEALTH CARE LANDSCAPE

The implementation of the ACA likely will serve to further accelerate consolidation in local markets because it includes provisions that encourage hospitals and physicians to integrate to facilitate the formation of Accountable Care Organizations (ACOs). ACOs provide the full spectrum of inpatient and outpatient services. By giving physicians and hospitals joint responsibility for patient care, the ACO model fosters alignment and perhaps encourages more consolidation to create a larger, more diversified patient base.

The expansion of coverage, as provided for in the ACA, is to be financed in part by a slowdown in Medicare payment rate updates relative to predicted trends, which may lead to further increases in prices charged by hospitals to private insurers. In addition to the provisions included in the ACA, market consolidation also is being driven by a confluence of other forces, including decreasing reimbursement both by government and commercial payers, changing attitudes of physicians specifically related to employment models, building additional leverage with payers, and the potential to reduce costs. Strategy&, formerly Booz and Company, predicts that at least 1000 of the 5000 existing hospitals could look for merger opportunities in the next five to seven years. Whether mergers and acquisitions can reduce costs and improve quality, or whether they simply result in increased costs, remains the major question.

The future of medical care likely resides in large, integrated health systems that have the ability to manage population health and take on risk in collaborative arrangements with payers. The case for hospital consolidation has shifted away from the revenue side of the equation to improved efficiency and reduction of expenses, though the emphasis has shifted for many systems away from inpatient beds to risk products, clinical integration, and large outpatient centers. No longer is the centrality of medical care the inpatient setting; clearly the shift is not only to the outpatient setting, but more and more is moving into the home. Enlightened health systems and academic health centers are preparing for this reality and recognizing that margins, generated mainly from the inpatient setting, are under severe pressure and thus are looking for ways to deliver efficient care in the most appropriate setting in order to reduce cost.

GEOGRAPHIC MARKET ANALYSIS: PHILADELPHIA

The Philadelphia health care market offers an excellent example of a market that is ripe for consolidation. Philadelphia County, which is mostly an inner city, has challenging demographics, while the other four counties (Bucks, Chester, Delaware, and Montgomery) have more favorable demographic characteristics. The city of Philadelphia is the largest city in the country without a public hospital or health authority that covers care for the indigent. There is no traditional government safety-net hospital, though at least one—Temple University Hospital—has become the de-facto safety-net provider. If we look more closely at Temple University Hospital, which is located in North Philadelphia (an area noted for having the highest incidence of deep poverty, defined as one-half the federal poverty level), approximately 50% of the discharges are covered by Medicaid with another 35% covered by Medicare.

The inpatient market in Philadelphia, like much of the rest of the country, has seen a significant decline; but in Philadelphia the market has shrunk both on an absolute and a relative basis over the past decade—most dramatically in Philadelphia County—while at the same time regional competition has increased. The region has an oversupply of physicians, challenging practice economics due in large part to significantly lower physician professional fees when compared to other regions of the country, and a very contentious professional liability environment. Malpractice coverage in academic health centers is via self-insured plans, due to a lack of carriers interested in selling to academic practices as a result of the history of high malpractice awards to plaintiffs by Philadelphia juries. Malpractice premiums in Philadelphia are among the highest in the country.

Impact on Payers

Even though the hospital sector in Philadelphia has shrunk over the past decade with fewer hospitals and fewer beds, the market remains one of the most overbedded in the country and among the least concentrated. Looking at the Herfindahl–Hirschman Index (HHI), defined as the sum of the squares of an organization's market share, what we find is a significant dichotomy between the commercial payers and health systems within the market [5]. Two payers in Philadelphia, the Blue Cross entity Independence Blue Cross (IBC) and the for-profit insurer Aetna, cover more than 80% of the commercial market. The Department of Justice and the FTC define markets with an HHI of >2500 as concentrated, 1500 to 2500 as moderately concentrated, and <1500 as unconcentrated. The HHI for the commercial payers in the Philadelphia market is 3600, which we can confidently describe as "superconcentrated." On the provider side, the HHI for Health Systems is around 800. Thus, it is fairly obvious where the leverage lies in the current fee-for-service environment. Comparable situations exist in a few other cities including Chicago, Baltimore, and Oklahoma City.

This contrasts with essentially the opposite situation that is present in cities such as Austin, Dayton, Ohio, and Worcester, MA, where the HHI for health systems is in the superconcentrated range while the commercial payers have an HHI only in the moderately concentrated range. The majority of hospital markets in the United States now fall within the moderate to highly concentrated range with an average HHI of 2520 (Figure 2).

The extent of hospital concentrations has increased over time with the average HHI increasing by 40% since the mid-1980s [2] (Figure 3). Across the 306 hospital referral

FIGURE 2 Degree of hospital ownership concentration in metropolitan statistical areas (MSAs). In 2009 hospital ownership was highly concentrated, as defined by the Herfindahl–Hirschman Index (HHI), in over 80% of the MSAs identified by the American Hospital Association. *2009 AHA Annual Survey.*

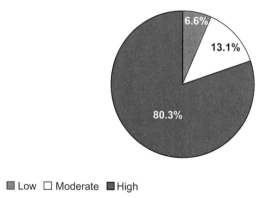

Degree of hospital ownership concentration in MSAs

6.6%

13.1%

80.3%

■ Low □ Moderate ■ High

· Categorization is based on revised thresholds issued by the DOJ and FTC in August 2010
- Low concentration: HHI < 1500
- Moderate concentration: HHI between 1,500 and 2,500
- High concentration: HHI > 2500

regions, the entity with the largest market share accounted for a mean of 42% of all inpatient days (median = 38%).

In almost 70% of these hospital referral regions the largest entity was or included at least one academic health center. Currently in the United States, there are no highly competitive markets as defined by an HHI less than 100. Conversely, nearly one-half (n = 150) of hospital markets are highly concentrated, another one-third (n = 98) are moderately concentrated, and the remaining one-sixth (n = 58) are unconcentrated (Figure 4).

Technological innovations have made it possible to offer more complex life-saving procedures, and this increasing complexity of care is becoming concentrated in fewer inpatient

institutions, though many patients, if given a preference, still prefer to stay closer to home. For these more complex procedures and technologies, academic health centers are seen by many as the place to be treated; and there is a wealth of data supporting that improved quality follows increased volume for a number of procedures. This recognition, and the perception overall of higher-quality care offered in academic health centers, tends to give these centers greater leverage with payers in a given market, leading to higher prices.

In Philadelphia, the largest market share had, until recently, been held by the Jefferson Health System, a system that combined the Thomas Jefferson University Hospital System with the Main Line Health System, a suburban system comprised of

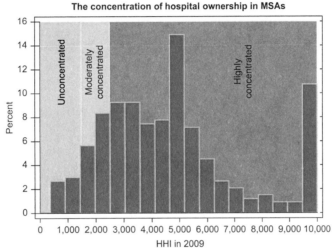

A more granular look at year 2009 MSA-level HHIs shows

■ Slightly more than 10% of MSAs have only one hospital owner

■ The majority of the highly concentrated MSAs have HHIs between 2,500 and 5,000

◆ 2,500 ≡ four equal-size firms

◆ 5,000 ≡ two equal-size firms (duopoly)

FIGURE 3 Average Herfindahl–Hirschman Index (HHI) in metropolitan statistical areas (MSAs) from 1997 to 2009. Between 1997 and 2009 the average HHI increased from 4222 to 4697, an increase of 11%. Note an HHI above 2500 is in the highly concentrated range. *Capps C. Price implications of hospital consolidation, The Healthcare Imperative: Lowering Costs and Improving Outcomes, Chapters 5, Institute of Medicine of the National Academies of Science 2010: pp. 177–187.*

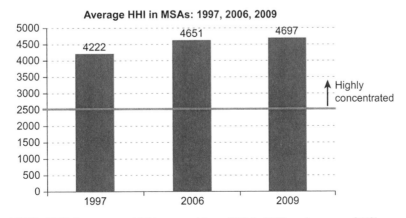

· From 1997 to 2009, the average HHI increased from 4222 to 4697, an increase of 11%.

FIGURE 4 Distribution of hospital ownership concentration as measured by the Herfindahl–Hirschman Index (HHI) in 2009. Slightly more than 10% of metropolitan statistical areas have only one hospital owner (HHI = 10,000). *2009 AHA Annual Survey.*

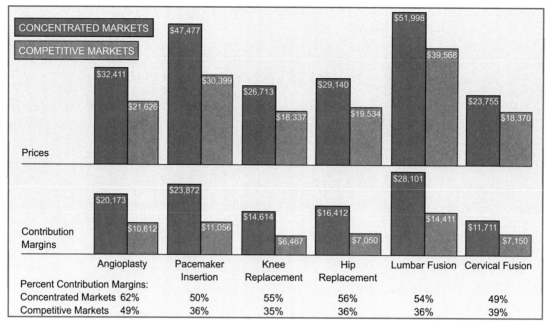

FIGURE 5 Prices and contribution margins in concentrated and competitive markets. Orthopedic and interventional cardiology procedures cost significantly more in consolidated markets and margins in those markets are markedly greater. *From Robinson, Ref. [7].*

four hospitals and multiple outpatient facilities with a highly desirable payer mix. In 2010, that system controlled approximately 17% of the market, and the University of Pennsylvania Health System accounted for another 10%. Catholic Health East, now merged with the Trinity Health System, also controlled about 10%. The rest of the market was composed of systems or hospitals each having 6% or 7% market share.

The recent separation of the Main Line Health System from the Jefferson Health System caused that system to cease to exist and thus created further fragmentation within the provider community. Prior to the dissolution of the Jefferson Health System, Independence Blue Cross had greater than three times the market share of that system. That multiple now likely exceeds five. Most of the hospital systems in Philadelphia lack significant capital, which may offer the opportunity for further incursion of for-profit systems in the market. Already, both Tenet and Community Health Systems (CHS) have a presence in the market, with Tenet being the only one with ownership of a hospital that is part of an academic health center—specifically Drexel. CHS has a significant presence in the Philadelphia region with several community hospitals.

Impact of Consolidation on Pricing

The issue of hospital market concentration and pricing has always been a focus for federal agencies concerned with antitrust, specifically because of concern regarding rising costs to the consumer. The issue is gaining greater attention with implementation of the ACA, as noted earlier. Most observers agree that consolidations among hospitals, in addition to hospitals actively purchasing physician practices, have led to price increases for hospital care.

Depending on the particular market, price increases ranging from 10% to 40% have been observed as a result of mergers [6]. In Massachusetts, for example, prices vary significantly across hospital systems with little correlation between quality and price but a strong correlation between institutional reputation and price. Ownership status, whether for-profit or not-for-profit, is not a deterrent for price increases. Prices charged by not-for-profit institutions can be just as high.

James Robinson looked at hospital market concentration as a function of pricing and profitability in two specific clinical areas—orthopedic surgery and interventional cardiology [7]. Data were obtained on patients admitted to 61 hospitals in 2008. The specific procedures for which data were collected included coronary angioplasty with drug eluting stent, insertion of a cardiac rhythm management (CRM) device (pacemaker or AICD), total knee replacement, total hip replacement, and lumbar or cervical spine fusion. The market for each hospital was identified by the Hospital Referral Region, one of 306 markets as defined by the Dartmouth Atlas based on patient flow data for Medicare patients. The 61 hospitals used in the study were distributed across 27 of these markets encompassing 8 states. Consolidation of the local market was determined by the previously referenced HHI.

Robinson found the average price per procedure was significantly higher in concentrated markets than in competitive markets for all six procedures (Figure 5). The differences in pricing as a function of market structure ranged from 29.3% for cervical fusion to 56.2% for CRM device insertion. This translated into significant market-related differences in contribution margins based on payments from commercial payers.

The average difference in contribution margins between concentrated and competitive markets was $9561 (90%) for

TABLE 1 Procedure-Specific Prices and Contribution Margins for Commercially Insured Patients in Consolidated and Competitive Hospital Markets

Type of Market	Angioplasty	CRM Device Insertion	Knee Replacement	Hip Replacement	Lumbar Fusion	Cervical Fusion
			Mean			
Consolidated Markets						
Price (insurance payment)	$32,411	$47,477	$26,713	$29,140	$51,998	$23,755
Contribution margin	$20,173	$23,872	$14,614	$16,412	$28,101	$11,711
Percent contribution margin	62%	50%	55%	56%	54%	49%
Competitive Markets						
Price (insurance payment)	$21,626	$30,399	$18,337	$19,534	$39,568	$18,370
Contribution margin	$10,612	$11,056	$6467	$7050	$14,411	$7150
Percent contribution margin	49%	36%	35%	36%	36%	39%

CRM indicates cardiac rhythm management.

angioplasty, $12, 816 (116%) for CRM device insertion, $8147 (126%) for knee replacement, $9363 (133%) for hip replacement, $13,690 (95%) for lumbar fusion, and $4561 (64%) for cervical fusion ($p < 0.01$ for all procedures) (Table 1). Hospitals in concentrated markets were able to charge higher prices to commercial payers than similar hospitals in competitive markets for all 6 procedures. However, even in some competitive markets, reputation for quality or possession of major market share allows some hospitals to achieve "must-have" status and thus be able to charge more than other institutions in the market.

With increasing complexity of medical care driven by technological innovation, there is a tendency for care to be concentrated in fewer inpatient institutions, especially flagship academic health centers, offering what is perceived to be higher-quality care and thus greater market power with commercial payers. Consumers value and choose insurance products that allow access to these institutions. Partners HealthCare in Boston, a system anchored by two of the nation's leading academic health centers, is a truly integrated system that includes community and specialty hospitals, a managed-care organization, a large physician network, community health centers, home care, and other related services. Partners is the largest private employer in Massachusetts with approximately 60,000 employees and a research budget that exceeds $1.4 billion. The size of Partners, in addition to its academic health center reputation, has endowed it with considerable market clout and the ability to negotiate very favorable reimbursement rates from commercial payers; they clearly are a "must have." This is in line with what other hospital referral regions are experiencing when one dominant system emerges, especially when it comes to the care of patients who require more complex procedures.

Trends Driving Consolidation

There are a number of trends that are moving hospitals and health systems to consolidate, but perhaps the major driver is declining reimbursements both from government and commercial payers and the certainty of evolution from a fee-for-service model to alternative methods of reimbursement—including bundled payments and assumption of risk contracts (Figure 6).

Population management can no longer be considered a side business, but, for many systems, is a powerful force prompting a major focus on the formation of fully integrated systems. This focus must be on building an end-to-end continuum of care in an effort to redefine how care is delivered, with the ultimate aim of providing high-quality care at less cost. The downward pressure on revenues drives the need for scale to take advantage of the economies that accompany scale. Many observers note that the successful hospital systems will be those that have revenues in the $3–$5 billion range. To get to this level, hospital and health system mergers will need to occur.

In addition, many systems are looking to expand into new geographies to access additional patient populations, and it is becoming increasingly common for academic health centers to seek out community hospitals to either purchase or collaborate with in joint ventures. These ventures provide access to capital for financially challenged organizations and provide a more appropriate and cost-effective venue for those patients whose illness does not require inpatient care at the flagship academic health center.

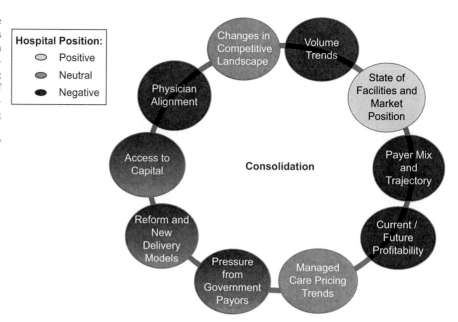

FIGURE 6 Drivers of consolidation in the health care market. The color of the various drivers is associated with the outcome in many community hospitals. Note other significant factors in addition to the three big drivers: access to capital, the challenge of physician alignment, employment and integration, and health care reform and declining reimbursement. *Ponder and Co., Q1 2014, Update on Mergers and Acquisitions Activity in the Hospital Industry.*

There are cost synergies to be gained by consolidation of back-office functions, leveraging supply-chain management based on scale and rationalizing capital expenditures. Independent hospitals, including those currently with solid finances, know they will need size and scale in order to confront the changes coming in the way care is reimbursed and delivered, and many are prospectively looking for a partner before their finances deteriorate to the point where closure is imminent.

Merger Alternatives

Alternatives to a full asset merger do exist and are being used in a number of markets. Academic health centers have been particularly active in developing joint ventures, in most cases with other nonprofits but in some cases with for-profit entities. Not-for-profits desire to join with academic health centers to leverage the clinical and quality strength as well as the brand name of the academic partner without having to provide the capital and operating expenses. Vanguard Health Systems, now part of Tenet Health care, formed a joint venture with Tufts Medical Center in order to pursue appropriate mergers or acquisitions in order to build an ACO infrastructure. Community Health Systems and the Cleveland Clinic formed a joint venture to pursue various acquisitions, though their first attempt—the acquisition of the Akron General Hospital—ultimately failed due to a lack of agreement on the part of the two acquiring partners. The Cleveland Clinic on its own has recently announced a joint venture with Akron General where the Clinic will become a minority owner of Akron General and make a substantial capital investment. A key component of the deal is the Cleveland Clinic's Quality Alliance, an integrated network of employed and independent physicians that allows them to connect and share information. The deal gives

the Clinic a presence in Summit County where their main competitor, University Hospitals, already has a presence with physician practices and outpatient clinics.

The Duke University Health System formed a joint venture with the LifePoint Hospital System to form Duke LifePoint, a venture that recently announced the acquisition of the largest hospital system in West Central Pennsylvania—the Conemaugh Health System. The acquisition includes the System's three hospitals, outpatient centers, and Conemaugh Physician Group Practices. Duke Life-Point has committed to invest over half a billion dollars in the system over the next 10 years.

Short of a joint venture where significant investment of capital may be involved, joint operating agreements—where there is a desire to partner and combine efforts but little or no desire to fully contribute operating assets or combine balance sheets—may provide an alternative. Bay-Care Health System in Tampa, Florida and Promina Health System in Atlanta, Georgia are both successful examples of joint operating agreements.

Relationships and alignments tend to migrate over time along the continuum of deal structures. That continuum ranges from the least change in governance—a service line affiliation—to full asset acquisition where there is the greatest change in governance.

INTERNAL CONSOLIDATIONS AND ALIGNMENT

Driven by multiple factors and, most recently, by provisions in the ACA, hospital systems increasingly are becoming more vertically integrated in addition to the horizontal integration characterized by mergers or acquisition of other hospitals.

Vertical integration or consolidation is characterized by the acquisition of other health care provider entities including, but certainly not limited to, physician practices. From 2004 to 2011, hospital ownership of physician practices increased from 24% to 49%. Today, approximately 80% of physicians have at least some type of financial relationship with a hospital. Hospitals also are acquiring, or are affiliated with, post-acute entities and services with 60% offering home health service, almost 40% have skilled nursing facilities, 62% own hospice services, and 15% offer assisted living facilities.

As previously noted, the ACA has accelerated the trend of hospitals purchasing physician practices in order to achieve the closer alignment necessary to maximally benefit from value-based care. However, a secondary motivation for hospitals to purchase physician practices, as opposed to simply being better able to coordinate care and control costs, is the ability to negotiate higher prices with insurers and build referrals—thus increasing admissions.

Inpatient admissions are declining across the country and hospitals are scrambling to build "loyalty" by owning physician practices, although by law the hospital cannot force or even encourage physicians necessarily to use the hospital that "owns" them. Paying for referrals or even giving the perception of paying for referrals is illegal, and there have been multiple instances when penalties from government regulators have been assessed for such violations. When hospitals buy physician practices they are able to charge higher prices for the same service that was previously performed in the physician office, resulting in increased spending by payers.

A look at 2.1 million claims from workers of self-insured employers between 2001 and 2007 showed prices were most likely to increase when hospitals bought physician practices, as opposed to forming looser contractual relationships with physicians [8]. As they purchase physician practices, hospitals are discovering that these practices almost always have to be subsidized because the professional fees collected rarely cover the cost of the practice when salary, benefits, and overhead costs are taken into account. This usually results from the fact that the physician salary guaranteed by the hospital when purchasing the practice almost always exceeds what the physician was taking home prior to the acquisition. In addition, unless appropriate incentives are put into place, productivity could decrease significantly once the practice is owned and the physician is being paid a guaranteed level of compensation.

The FTC and other government agencies are closely monitoring vertical integration activity within hospital systems but have only intervened when one entity controls so many physicians that an anticompetitive environment is likely. Most recently, the FTC blocked Idaho's largest hospital system, Boise-based St Luke's Health System, from buying the state's largest physician practice, the 43-member Saltzer Medical Group, which would have resulted in control of over 80% of the doctors in Nampa, Idaho [9]. The FTC filed no cases regarding hospital mergers between 1999 and 2007 and filed only three since, until just recently. The agency has indicated that it will step up monitoring of the health care sector, recognizing the increasing pace of consolidation—both horizontal and vertical—and their concerns regarding hospital–physician consolidations. The question to be answered is whether a hospital, combined with a large physician practice or a large hospital system, can be so dominant in a market that it could control pricing.

The effects of consolidation differ depending on the payer, whether government or commercial. Public payers set prices and do not negotiate. Private payers must negotiate rates and may find they have less leverage when trying to negotiate with a consolidated system, especially in a highly concentrated market. An integrated health system, by virtue of covering the entire continuum of care, may be able to reduce costs by 10–20%, but likely this reduction will not translate into lower charges to private payers. As a result, antitrust regulators are paying increased attention to the effects of internal consolidation on private payers. For a consolidated entity to be in the safe zone it must have less than 30% of the market [10]. The major issue for regulators is whether consumers will be better or worse off as a result of consolidation. With the ACA's encouragement of consolidation, by promoting ACOs as a strategy to increase efficiency, the Centers for Medicare and Medicaid Services (CMS) must follow through with the oft-discussed—but not yet implemented—alternative payment mechanisms, such as bundled payments; and private insurers must follow.

HEALTH INSURANCE INDUSTRY CONSOLIDATION

The health insurance industry also has been engaging in significant consolidation in a number of markets around the country [11]. The American Medical Association, in its biannual survey of the health insurance marketplace in 2011, found that 60% of the nation's metro areas had two insurers with a combined share of 70% or more of the market—up from 53% in 2009. This high concentration on the payer side is largely the result of consolidation, which has resulted in the exercise of market power with a potential for harm to consumers and care providers.

PAYER–PROVIDER MERGERS

Another trend developing in the health care industry is that of hospitals and health systems purchasing insurance companies or vice versa. The purchase of the West Penn Allegheny Health System by Highmark, the dominant payer in Western Pennsylvania, was driven mainly by the dominance of the UPMC on the provider side, with its ownership of over 20 hospitals and its own insurance product.

The leverage gained by its dominance in the marketplace has allowed UPMC to benefit both by competing with Highmark on the insurance side of the business and also being able to demand better rates from Highmark. In order to more effectively compete, Highmark considered it necessary to enter the provider side of the business and purchased the financially challenged West Penn Allegheny System, which was in dire need of a capital infusion. A dispute as to whether those insured by Highmark will be able to utilize UPMC facilities is a current concern.

Other health care organizations are looking to become both payer and provider in order to be in a better position to focus on population health management and better design quality-care programs within a unified organizational structure. The Geisinger Health System is a case in point of an organization that has become highly successful as both payer and provider. In addition to the potential for better care, there is also the opportunity to reduce costs by focusing on preventive care with the goal of keeping patients healthy. Offering appropriate incentives might be one means to an end in this regard. More and better preventive care would be expected to lead to dramatic cost savings. Several hospitals or health systems have entered the insurance market. Detroit Medical Center, owned by a for-profit system, purchased ProCare Health Plan, a Medicaid HMO based in Detroit. The largest nonprofit health care system, Ascension Health, recently announced its intention to purchase an insurance company.

Hospital and health system entry into the insurance business has attracted the attention of antitrust regulators since these combinations may limit the entry of other companies into a market and thus create a situation where the unified companies charge higher premiums. At least one study has shown that when hospitals purchase insurers, higher premiums result [12]. Unfortunately, the investigators found that the higher premiums did not necessarily correlate with higher quality of care.

It should be noted that consolidation on either the payer or provider side creates a situation where providers can demand higher prices and insurers, facing less competition, pass the costs along in the form of increased premiums with the net effect of increased costs for individuals and families, as well as employers and government agencies. Excessive market power on either side seemingly is not in the best interest of the consumer. Despite the aim of the ACA to encourage formation of ACOs, based on a premise of cost savings, one has to ponder whether the savings accrued from government payers will be offset by higher pricing in the private sector.

CONCLUSION

Hospital consolidation is being driven by a number of factors, including policy reform promulgated by the

ACA, the changing economic landscape, and declining reimbursements.

There are a number of benefits in hospital consolidation and realignment including, but not limited to, increased access, higher value, and greater efficiency with cost savings as the potential end result. A number of studies, however, have shown that consolidation in many markets has been associated with increased prices that ultimately are passed on to the consumer. In hospital referral regions that are highly concentrated, the leverage gained by a dominant health system usually results in pricing favorable to the health system—specifically higher reimbursement rates. Often these higher rates are not just limited to the flagship institution but are enjoyed by the peripheral members of the system as well. Even in less concentrated markets, the reputation of a provider based on the physicians associated with that entity often results in significant leverage with a payer that recognizes a system as a "must have," resulting in higher rates. Consumers make it known that they wish to have access to such a system or hospital, and thus the payer has to be willing to submit to the demands of such a provider.

The combination of horizontal and vertical integration should give health systems the opportunity to deliver care more efficiently and at less cost, but these savings have not necessarily been passed on to individuals, employers, or the government. Alignment with physicians in order to fully realize efficiencies and control costs has become a major consideration for many health systems and underscores the importance of physician leadership within these systems. In most markets, consolidation among hospitals has remained under the radar of antitrust regulators—including the FTC and the Justice Department—underscoring the fact that the majority of mergers have not yet significantly stifled competition.

REFERENCES

[1] Capps C. Price implications of hospital consolidation. The healthcare imperative: lowering costs and improving outcomes. Institute of Medicine of the National Academies; 2010. Chap. 5: pp. 177–187.

[2] Cutler DM, Morton FS. Hospitals, market share and consolidation. JAMA 2013;310:1964–70.

[3] Corwin SJ, Cooper MR, Leiman JM, Stein DE, Pardes H, Berman MA, et al. Model for merger: New York-Presbyterian's use of service lines to bring two academic medical centers together. Acad Med 2003;78:1114–20.

[4] Otis GA. Mount Sinai, continuum complete merger to create huge non-profit health system. NY Daily News; 2013. Retrieved from: http://www.nydailynews.com/new-york/mount-sinai-continuum-complete-health-system-merger-article-1.1472091.

[5] Wikipedia. Herfindahl Index. http://en.wikipedia.org/wiki/Herfindahl-index.

[6] Gaynor M. Health care industry consolidation. 2011. [statement before the Committee on Ways and Means Health Subcommittee, US House of Representatives]. http://waysandmeans.house.gov/upload-edfiles/gaynor_testimony_9-9-11_final.pdf.

[7] Robinson JC. Hospital market concentration, pricing, and profitability in orthopedic surgery and interventional cardiology. Am J Manag Care 2011;17:e241–8.

[8] Baker LC, Bundorf MK, Kessler DP. Vertical integration: hospital ownership of physician practices is associated with higher prices and spending. Health Aff 2014;33:756–63.

[9] Galewitz P. As medical providers consolidate, questions about effects on costs, quality of care. The Washington Post; 2014. Retrieved from: http://www.washingtonpost.com.

[10] Kendall B. FTC gets more muscle in policing hospital mergers. Wall Str J; 2013. http://online.wsj.com/articles/SB10001424127887323495104578314041654081664 [accessed 23.10.14].

[11] Dafny L, Duggan M, Ramanarayanan S. Paying a premium on your premium? consolidation in the US health insurance market. Am Econ Rev 2012;102:1161–85.

[12] Frakt AB, Pizer SD, Feldman R. Plan-provider integration, premiums and quality in the medicare advantage market. Health Serv Res 2013;48:1996–2013.

ABOUT THE AUTHOR

Larry R. Kaiser, MD, FACS is the President and CEO of the Temple University Health System, Dean of the Temple University School of Medicine, and Senior Executive Vice President for the Health Sciences. Prior to his current position he served as the President of the University of Texas Health Science Center at Houston.

Section V

Conclusion

Chapter 25

Academic Health Center Transformation: Future Shock or Future Success?

Steven A. Wartman

As is evident from the content and flow of the previous chapters of this book, academic health centers have entered into what may be euphemistically described as a period of "mission disruption." The factors leading to this disruption are numerous and, in the big picture, include a combination of societal, scientific, and economic forces (Table 1).

Each of these factors, in and of itself, is formidable; when put together they constitute a substantive challenge to each and every academic health center. For example, the explosion of consumer empowerment created by the Internet and related technologies challenges the hegemony of care givers with regard to medical knowledge. Science is leaping forward with the "omics" revolution, which, along with entrepreneurial advances in health and Internet-related technologies, is creating a new scale of "personalized medicine." The long-standing, highly successful model of biomedical research in the United States, in which clinical revenues subsidize research, is increasingly fragile given the downward pressures on reimbursement and lack of real growth in many funding agencies, including the US National Institutes of Health (NIH). And the ongoing consolidation within the health marketplace, especially in the United States, raises serious concerns about the ability of individual academic health centers to compete with far larger national or international health systems. These examples, when coupled with the other factors in Table 1, constitute both a significant challenge and an opportunity for academic health centers. This outlook is reflected in an Association of Academic Health Centers (AAHC) member survey conducted in early 2014, in which 41% of respondents indicated that they are undergoing major expansions of their hospital or physician network, 37% are embarking upon large-scale cost reduction initiatives, 36% are opening a new health professions school or new branch campus, and 31% are changing their governance structures or significant reporting relationships [1].

The authors of this volume have set out clearly the impact of these and other challenges and opportunities

with examples of how academic health centers can address them. This concluding chapter aggregates their findings and suggests an overall transformative path forward. Applying a selective overview and response, I will first focus on a specific challenge in each of the traditional academic health center mission areas of health professions education, biomedical/clinical research, and patient care/population health. The areas I have chosen to discuss represent, in my view, difficult and paradigm-shifting issues: *interprofessional education, the single-laboratory-funded investigator,* and *the "new physics" of patient care.* Next, I will discuss the overwhelming importance of finding and selecting the best leaders for the academic health center enterprise. And finally, I will conclude with some overarching *suggested guidelines* to manage the much-needed, and unavoidable, institutional transformation.

SPECIFIC CHALLENGES FACING ACADEMIC HEALTH CENTER MISSIONS

Interprofessional Health Professions Education

Health professions students are fully acclimatized to the era of the Internet and, given the ready accessibility of information, expect the latest information presented in the most user-friendly fashion. They are in the relatively unique position of being able to find information wherever in the world it is available; and this in itself creates new scales of educational expectations for their individual institutions. Teaching methodologies, as described by Combs et al. (Chapter 7), such as the "flipped classroom," adds a new and important element to problem-based learning and other standard methodologies. The role of the traditional "curriculum committee" is being challenged to apply higher standards of quality and accountability. While information overload is not a new problem with regard to the curriculum, what and how to teach have become increasingly difficult and important

The Transformation of Academic Health Centers. http://dx.doi.org/10.1016/B978-0-12-800762-4.00025-6

TABLE 1 What Is Leading Mission Disruption?

• Changes in societal needs and values	• Population demographics
• Disease patterns	• Market consolidation
• Economics	• Consumer empowerment
• Globalization	• Policy changes
• Politics	• Entrepreneurism Science and technology

questions, especially with the realization that today's students have decades of professional practice ahead of them. These issues are doubly difficult when one considers them in the context of interprofessional education.

The Case for Interprofessional Care[1]

In medicine writ large, a strong business case can be made for interprofessional health care. Worldwide, the increasing dominance of noncommunicable diseases—to say nothing of the pressing need for better access to health care in general—as well as our growing understanding of the social determinants of health, all argue for an approach to patient care that maximizes the value that can be gained from input and collaboration across the full diversity of health care professionals. In the United States, baby boomers will be spiking demand for health services for at least the next two decades. Conventional wisdom suggests that effective interprofessional care will lead to decreased demand for acute care services, but the reality is more likely to be a shift in needs for acute care. For example, while the demand for acute care of diabetes and hypertension may diminish, the need for care in other critical areas, such as cancer and Alzheimer disease, will increase as people live longer. Robust and well-integrated collaboration among diverse health professionals will be needed to meet these needs for health care.

Another compelling reason in favor of interprofessional health care is that, fundamentally, patients deserve better access to core provider competencies. Increasingly, they will demand that such care be accessible as readily as any other service they currently receive. The ability of the health care system to provide that kind of access will be predicated on more widespread use and acceptance of interprofessional health care. Providing such access will lead to increased effectiveness of care and improving health care outcomes and quality while lowering costs—all of which is in the best interests of the health care system.

---

[1] See Cerra and Brandt, Chapter 9.

Barriers to Interprofessional Health Professions Education: A Baseball Metaphor

Having had the opportunity to participate in discussions regarding interprofessional education, I have often reflected on the barriers to interprofessional education and practice. It may have been because one such conversation occurred during the baseball World Series that I started to think about these impediments in terms of baseball. It occurred to me that too often those of us who seek better ways to integrate interprofessional education and practice spend inordinate energy pursuing home runs—in other words, trying to find and implement large-scale fixes that might solve many of the problems related to this challenge all at once. It also occurred to me that our attention might be more productively focused on incremental fixes—or, to extend my baseball metaphor, to round the bases one by one, addressing challenges and advancing incrementally in ways that might eventually accrete into more wholesale reform (Figure 1).

First base, in theory the easiest base to get to, represents in this baseball metaphor the "guild mentality" of the health professions. Our propensity to silo health care disciplines not only helps divide health practitioners but also creates competition and duplication where today we urgently need collaboration and efficiency. The guild mentality inhibits an integrated, interdisciplinary approach to a full spectrum of health care and population health. As we endeavor to promote and enhance better integration of interprofessional education and practice, we first need to develop strategies that overcome the attitudes and beliefs that inhibit true interprofessional learning and practice and often divide health care professionals arbitrarily and inefficiently.

As a second step—or second base, if you will—we need to carefully review current university and hospital structures and procedures. Traditionally, professions, disciplines, hospitals, and health systems are separated administratively into departments and other units, each with its own schedules, operating principles, and policies—such

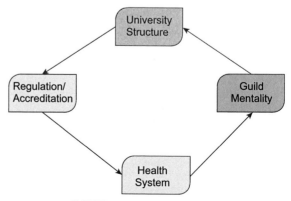

FIGURE 1 Lessons from baseball.

as those for promotion and tenure—that deeply impact behavior. Often, they compete with each other for limited resources. Overall, a lack of alignment between management and infrastructure drives a wedge between types of professionals. How can we restructure institutions to ensure better integration and alignment? Addressing that challenge would move us that much farther along a path to broader reform—and to "home plate."

The third major impediment—or third base—concerns the regulation and accreditation requirements that the various health professions must meet. As laid out in Chapter 11, licensure requirements, scope of practice laws, accreditation requirements, and other compliance regulations complicate coordination and collaboration across professions. Such strictures limit, for example, who is qualified to serve as an educator. They overburden some clinicians and undervalue others. Generally, the lack of harmonization among regulators and accreditors impedes innovation and the efficient delivery of health care. A deep look at this body of regulations—with reform in mind—would greatly facilitate the process of moving us closer to true integration of interprofessional education and practice.

Ultimately, in baseball we want to score. To reach home plate, however, is perhaps the most formidable task. In the area of interprofessional education and practice this involves aligning the incentives of the health care delivery system to support and promote the kind of system we would like to envision.

It is becoming readily apparent that interprofessional education and practice will serve an increasingly important role in health care in the years ahead. It is therefore incumbent upon academic health centers and health systems to begin an organized process of aligning curricula and policies to support and nurture true collaboration among health practitioners at all levels. The "four bases" scenario described above, which considers the guild mentality of the health professions, university policies and procedures, accreditation and regulatory bodies, and the incentives of the health care system, offers an approach to this important and challenging issue.

The Future of the Single-Laboratory-Funded Investigator

The image of a brilliant, single-minded scientific researcher who produces astounding insights is indelible. From Archimedes to Galileo to Newton to Einstein, the dazzling accomplishments of lone scientists have reinforced society's preconceptions of how science is done. In contrast to that paradigm, though, breakthroughs in scientific research today increasingly derive not from lone researchers but from teams of scientists collaborating across disciplines.

As a careful reading of Section III of this book indicates, the research enterprise faces significant challenges

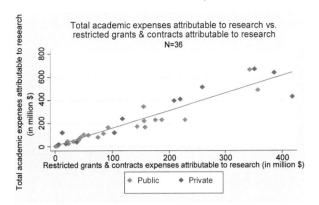

FIGURE 2 Total academic expenses attributable to research versus restricted grants & contracts attributable to research. *AAHC Metrics and Benchmarking Data, 2014.*

as a result of a "perfect storm" of sociopolitical, scientific, and economic forces. Sociopolitical forces include an ever-increasing set of demands on researchers to produce work that more quickly leads to disease breakthroughs. The rise of mega data sets, along with the implications of cloud and crowd sourcing, suggest that the "control" of research is beginning to shift from tightly controlled, peer-reviewed mechanisms, to a more open framework. With the possibility of data aggregation open to individuals through new medical applications and body sensors, for example, individuals may choose to consult millions of their peers when faced with a medical issue rather than consult the medical establishment. For the purposes of this section, the rise of team science coupled with the economic realities of supporting research is challenging the classic model of the single-laboratory-funded principle investigator, which has historically been the backbone of much biomedical research at academic health centers. This is a game changer for the traditional science paradigm.

Economics of Research and the Changing Research Model

Research has rarely been a profit center for institutions, whether academic or commercial in their orientation. Data collected from AAHC's Research & Analytics Program indicate that, on average, external grants and contracts are the largest funding source for US medical schools, and that 35% of total research expenses are funded using internal funds [2]. Specifically, this means that for every $1 increase in research expenses funded by external grants and contracts, US medical schools pay an additional 52 cents (Figure 2).

It is clear that much of the additional support needed for research at U.S. academic health centers has traditionally come from patient care revenues. The rise in NIH funding, for example, has been largely paralleled by the rise of nontenure track clinical faculty. However, if, as anticipated, clinical margins shrink coupled with the decreased purchasing power of and increased competition for traditionally available research funding resources, new sources to support research are aggressively being sought in the public and private sectors as well as internationally.

In addition, more attention is being paid to research efficiency and research emphases. Accordingly, and as described in Section III of this book, institutions are increasingly moving to shared resource models that offer the promise of lower overhead and increased economies of scale. As a result, it is no longer feasible for many institutions to consider any grant a good grant. Rather, institutions are in the early stages of adopting a more businesslike approach to research and development with careful budgeting that mirrors those areas that have been chosen for emphasis. It is likely that these economic forces will drive further differentiation among academic health centers in the extent and reach of their research portfolios.

Budgeting for research in academic health centers, however, seems not to have caught up with this fundamental shift. Too regularly, academic health centers support budgets and allocate space based on the long-standing tradition of the lone-funded investigator heading up his or her laboratory. This model has served science well, but it is significantly challenged in three ways: (1) by the erosion in traditional funding sources (e.g., the NIH), both in terms of the decrease in the number of successful applications—now for the NIH at record low levels—and the reality that funding overall has not kept pace with inflation and has thus lost purchasing power [3]; (2) by the growth of team science mentioned above, which demands different levels of intra- and interinstitutional cooperation and funds flows; and (3) by a variety of sociopolitical forces that seek to focus research more sharply and demand "success" in a considerably shortened time frame. These factors challenge the fundamental utility and sustainability of the single-laboratory-funded investigator model.

The "New Physics" of Patient Care

As is evident in Section IV of this book, the health care landscape is a challenging mix of change in process. In general, this is reflective of the observation that care is moving from a fixed setting to wherever the patient may be, given new technologies for monitoring and following patients. As the clinical data available on each patient increases exponentially, a new interface between "medicine and machine" is needed to manage it and will present a challenge to the traditional role of the provider, especially the physician. Further adding to these changes is the expectation of an enlarging stream of entrepreneurially invented medical devices that will be directly sold to individuals, providing them with new information and

TABLE 2 Impact of Disruption on Patient Care

- Care is moving from a fixed setting to wherever the patient may be
- Real-time continuous monitoring of patients
- Big clinical data sets: how to manage and operationalize them
- The Pandora's box of technology and new entrepreneurial income streams
- Patient empowerment
- Shifting from provider control to "crowd control"

resources that may or may not be in conflict with established medical care norms. The result is increasing patient empowerment and a gradual shift from provider control to crowd control (Table 2).

In a sense, a "new physics" of patient care is being created that is illustrated somewhat whimsically in Table 3.

The three key factors in this formulation, noted by c^1, c^2, and c^3, are *care anywhere, care in teams, and care by large data sets* [4]. Technology is moving with and within the patient's body wherever he or she may be. The traditional large, fixed medical structures are certainly necessary, but will need to be configured differently. And health care consumers will want convenience and one-stop shopping. The sacrosanct one-to-one doctor–patient relationship is being replaced by relationships with multiple health professionals and, to an increasing extent, by machines. Patients and providers will have to learn how to gain the most value from team and machine-driven care, and the scope of practice will need careful redesign. As the collection of metadata sets become routine for individuals, a new interpretive infrastructure will be required to manage them; the locus of decision making is shifting to a new marriage of medicine and machine.

TABLE 3 The New Physics of Patient Care: $E = mc^3$

The Emerging model of healthcare°, where:

➤ m = the population, both individually and collectively

➤ c^3 =

    c^1 = care anywhere

    c^2 = care in teams

    c^3 = care by large data sets

°Inspired by Eric Dishman's Ted Talk at
http://www.ted.com/talks/eric_dishman_health_care_should_be_a_team_sport.htm.

Machines and Patient Care

The practice of medicine is increasingly taking place at the nexus of patients and machines. From diagnostic modalities, such as magnetic resonance imaging to rapid data analysis and robotic surgery, computer-assisted advances are transforming the delivery of health care. Couple that evolution with patients' expanding access to medical information via the Internet, and the traditional role of all health providers is quite challenged. The doctor may no longer be seen as the sole principal expert and possessor of unique skills. We are entering an era in which machines are fundamentally changing the nature of the provider–patient connection—and, ultimately, what it means to deliver health care. If physicians, for example, are going to be essential and effective in the future, the professions need to refine approaches to the intensifying marriage of medicine and machine.

Foremost is the issue of how physicians trained today—who will still be practicing medicine in 2050—will develop the expertise needed for the future. Currently, we educate and train health professionals quite well for practice *as it was*, but less well for *how it is*. Curricula *today* must address the development of a new kind of proficiency which I call *professional intelligence*, defined as the *confluence of professional values and expertise*.

The curriculum for professional intelligence has yet to be written, but we need to get started. That curriculum needs to reflect the reality that no human can effectively process the exploding volume of medical knowledge and data. Moreover, it needs to acknowledge the implication that machines will know more and be able to perform more tasks than physicians. Scientific advances and entrepreneurial inventions are already creating devices that outperform human capacity in both the cognitive and physical senses. Computer algorithms, for example, offer rapid analyses and suggest both diagnostic and therapeutic avenues to pursue, far outperforming the recognized limits in the number of facts an expert can review at one time to reach a reasonable decision.

The imperative of *professional intelligence* demands that we accelerate the preparation of health professions students for practice *as it will be*. When coupled with the inherent difficulty of predicting the future, it is not surprising that health professions schools concentrate more on training for the development of skills and competence than on *expertise*. Skill is the ability to perform a concrete act. In medicine, this could be the insertion of an intravenous line, laparoscopic surgery, or a colonoscopy. Competence is the level at which you are able to perform that skill. Expertise, however, refers to the ability to see the big picture, to understand all the unique elements involved, and to draw appropriate conclusions. These distinctions are increasingly important because the first two (skills and competence) will be largely taken over by machines, while the latter (expertise) is uniquely human.

Evolving Payment Models

Academic health centers that serve as comprehensive care providers, and often as community health care safety nets, are vulnerable in this environment to the market forces driving consolidation and attaining provider power. They will need to form new alliances and strategic partnerships, allowing them to flourish while preserving their fundamental missions in this challenging environment.

Against this complex backdrop, we need to recognize that, although there are many and varied health systems around the world, there is no perfect payment model. As put rather understatedly in a 2001 paper: "There are many mechanisms for paying physicians; some are good and some are bad. The three worst are fee-for-service, capitation, and salary." [5] The contention here is that each payment methodology has its flaws: fee-for-service leading to overuse of health services, capitation leading to underuse, and salary creating an incentive to do less work less efficiently. As health systems evolve, there will be ongoing hybrids of payment methodologies that reflect health system priorities and political contingencies. These methodologies will be calibrated according to the degree of risk for population health that is assumed by the care provider. Managing this risk will eventually be a defining characteristic of the "new physics" of patient care.

THE SIGNIFICANCE OF FINDING OPTIMAL LEADERSHIP

The selection of high-level appointments—including presidents, chancellors, vice presidents, deans, institute/center directors, and department chairs—is arguably one of the most important actions to be undertaken at academic institutions. In working with and visiting more than 100 academic health centers, however, I have found that finding and keeping the caliber of leadership necessary for these demanding positions is challenging. Too often, and too regularly, academic health centers find themselves wondering how to reach the next level and searching, yet again, for new talent to fill key leadership positions.

In part, the problem is often attributable to two types of lapses: lack of understanding of the necessary characteristics of successful leaders for these complex institutions and breakdowns in the recruitment process itself. Through a collaborative effort with academic health center leaders and leading search firm executives, AAHC analyzed the search process in depth and made a series of recommendations to improve the likelihood of a successful outcome [6]. These suggestions apply broadly for searches, not just in academic health centers, but throughout academe.

TABLE 4 Academic Skills versus Leadership Skills

Academic	Leadership
• Intellectual capacity	• Emotional intelligence
• Narrow knowledge base	• Broad range of interests
• Strong work ethic	• Highly institution motivated
• Gets individual results	• Gets institutional results
• Rises up the academic ladder	• Manages 360°

Successful Leaders

Being a successful academic is not necessarily the key to being a successful leader. While a strong academic track record is often an important prerequisite for top leadership posts, other factors—such as humility and emotional intelligence—may be as important or even more important to successful leadership [7]. Similarly, candidates with a high level of narcissism and/or arrogance may be perceived—incorrectly—as well suited for the job [8]. A charismatic personality may obfuscate a candidate's actual managerial and administrative capabilities. In the quest for top-level talent, therefore, those participating in the search process should be advised about the importance of distinguishing confidence from competence, among other leadership traits. Table 4 compares academic skills with the characteristics of successful leaders.

Finding Successful Leaders[2]

The search process for institutional leaders has been only intermittently successful. In part, this is due to lack of detailed organization of the search itself and the lack of a "pathway" to becoming an academic health center leader. Ideally, a search process consists of three distinct phases and proceeds in an orderly and efficient manner.

Phase I, the *presearch phase*, establishes the foundation for a successful search. *Phase II*, the *active search phase*, involves screening, interviewing, and ultimately selecting the final candidate. *Phase III*, the *transition/on-boarding period*, introduces the successful candidate to the institution and is designed to help the new hire adapt successfully to his or her new role.

Phase I, the presearch phase, lays the groundwork, carefully defines the roles of participants in the search, and sets the tone and expectation for the search. In some respects, this phase is perhaps even more important than the ones that follow. In this initial phase, thoughtful reflection on key questions is fundamental. Why is this search necessary? What does the institution need in terms of leadership in this

---

[2]More details may be found in the book ***Searching for Leadership*** listed as reference [6].

area at this point in time? What skills sets are essential to this job? What style of leadership and personality is likely to succeed in the job? If this is a new position, how does it fit into the current governance and fund flow structures? If the position is an existing position, should the position be reframed, and if so, how? Does the existing governance and culture support the successful placement of a chosen candidate? Is the institution organized to conduct efficient and successful searches? If not, how can this be improved?

Phase II, the *active search phase*, includes the identification and selection of candidates, ultimately leading to the final choice. While the structure and requirements for searches may vary from institution to institution, several principles should be followed in order for the search to function effectively and efficiently. The plan for the search strategy should be clearly described and the evaluative role of the search committee, if one is constituted, should be clearly defined. After initial interviews are completed, a high-level vision statement should be drafted by the short-listed candidates, the purpose of which is to confirm directionality and that the new hire's expectations about resources align with the institution's needs and abilities. When the final decision is made, a letter of offer should include, among the usual items, a well-crafted, explicit explanation of expectations for the position and an expected time line for them to be achieved.

Phase III, *onboarding*, begins from the moment the candidate agrees to take the position and extends about 1 year into the job. This often overlooked period is vital for facilitating a smooth and effective transition. The goal during this period is to help the new hire develop a more in-depth knowledge of the institution and to establish an effective list of job priorities. A formal "on-boarding" plan enhances the ability of the individual to get up to speed quickly. There should be an objective review of the new hire's strengths and weaknesses (no candidate is perfect) to identify where early developmental work would be useful.

Changing the Dim View of Leadership

Among the faculty there is often the vague impression that the administration (or "suits") represent something adversarial to the academic ethos. Part of the problem is that there is insufficient understanding of what the leadership role actually entails. Faculty may have the misapprehension that the leader is "sitting on a pile of money and not giving me any of it." Because of the lack of deep appreciation for the leadership role, faculty are often not inclined to seek these positions, but rather find themselves in the position of "accidental leader" when they happen to be chosen. As a result, there is not a clear preparatory pathway to obtaining leadership positions. Academic administration needs to be "demystified" through open and transparent leadership styles that clearly demonstrate the realities and challenges of leadership, along with establishing programs (e.g., leadership academies) to promote the development of effective leaders.

CONCLUSION: OVERARCHING GUIDELINES FOR THE PATH FORWARD

As is evident from this and the preceding chapters, academic health centers are undergoing significant evolution as they transform themselves to be successful in the highly complex but ultimately worthwhile environment of health care. In this regard, I suggest seven broad guidelines that leadership should adopt as they lead their institution into this new era.

1. *Decide on the best mission balance for the institution*

As difficult as it can be, institutions need to take a hard look at their strengths and weaknesses and focus on those areas where it can truly make the most difference and greatest contribution. Moving up on various ratings (e.g., NIH, World University rankings), while good for public relations, is not necessarily good for society. Rather, focused contributions that make real differences should be emphasized and supported via a well-designed budgetary process.

2. *Prepare for the era of no more open-ended funding*

As new funding sources become harder to find and there is downward pressure on existing resources, it is essential to develop methodologies and tools to assess efficiency throughout the institution and especially in the areas chosen for emphasis. This is not easy because it involves consideration of institutional definitions of research, education, and administrative full-time equivalents. Establish how much the institution is willing to invest in current and new areas and gauge how to account for optimization.

3. *Develop an integrated, interprofessional vision*

Improve alignment and connectivity at all levels throughout the institution to best capture the combined power of the various components. Shift the operations and leadership from a highly-siloed enterprise to an aligned organizational structure. Develop networked and interconnected consortia with other institutions on a national and international basis.

4. *Broaden the understanding of what your institution does*

Actively incorporate disciplines previously viewed as external to the academic health center where such disciplines have unique and important contributions to make to health system change. Examples include engineering, business management, social sciences, humanities, and so forth.

5. *Understand the operational implications of assuming more financial risk for population health*

Shifting the view of the organization from a near-exclusive focus on the management of individual patients to management of community and population health is challenging.

It is important to gain a deep understanding of the operational implications of assuming more financial risk for population health, knowing that risk goes beyond the cost and quality of individual procedures. Develop an approach to addressing the social determinants of health for your patient population in order to better achieve positive health outcomes.

6. *Develop strong international networks and alliances*

Globalization is an increasingly powerful force. Every level of the academic health center enterprise experiences its impact. Student bodies are increasingly international, research often spans several countries, and patient care and population health in the era of rapid communication and transportation knows no boundaries. Academic health centers must view themselves increasingly as world institutions and work to develop their international portfolio of activities.

7. *Find the right leaders*

The value and importance of leadership must be made known to all who work at the institution through communication and transparent management. Cultivate a new cadre of leaders to move the institution ahead. When searching for new leaders, evaluate actual leadership skills—not just academic skills.

Every new century (some would argue every new decade) offers challenges and opportunities. The twenty-first century promises to be one of extraordinary scientific achievement coupled with patient empowerment through knowledge, demographic and population growth, climate change, and significant economic and sociopolitical factors. I believe that academic health centers can succeed and thrive in the coming environment by remaining true to their underlying value propositions:

- *Applying knowledge to improve health and well-being*
- *Building the knowledge economy and applying it in patient care*

In so doing, the academic health center will be well positioned to respond if it functions as an organization that aligns academics (teaching and research) with the care of patients; focuses on the next generation of education, research, and patient care; and has the transformational leaders to change culture and behavior.

REFERENCES

[1] Association of Academic Health Centers (AAHC). 2014 survey of CEOs. Washington, DC: AAHC; 2014.

[2] AAHC. AAHC research data. Washington, DC: AAHC. http://www.aahcdc.org/Resources/BenchmarksandMetrics.aspx [accessed 20.08.14].

[3] Center for American Progress. Erosion of funding for the national institutes of health threatens US leadership in biomedical research. http://www.americanprogress.org/issues/economy/report/2014/03/25/86369/erosion-of-funding-for-the-national-institutes-of-health-threatens-u-s-leadership-in-biomedical-research/; 2014. [Last accessed 23.10.14].

[4] Dishman E. Health care should be a team sport. Ted Talk 2013. Accessed at http://www.ted.com/talks/eric_dishman_health_care_should_be_a_team_sport.htm.

[5] Robinson JC. Theory and practice in the design of physician payment incentives. Milbank Q 2001;79(2).

[6] Wartman SA. Searching for leadership: best practices for academic institutions. Washington, DC: Association of Academic Health Centers; 2014.

[7] Goleman D. What makes a leader? Harv Bus Rev 2004;82(1). Accessed at http://hbr.org/2004/01/what-makes-a-leader/ar/1.

[8] Chamorro-Premuzic T. Why do so many incompetent men become leaders? Harv Bus Rev Blog. Aug 22, 2013. Accessed at http://blogs.hbr.org/2013/08/why-do-so-many-incompetent-men/ on September 17, 2013.

ABOUT THE AUTHOR

Steven A. Wartman, MD, PhD is the President and CEO of the Association of Academic Health Centers in Washington, DC. Prior to assuming this position, he was Executive Vice President for Academic and Health Affairs and Dean of the School of Medicine at the University of Texas Health Science Center in San Antonio.

Index

Note: Page numbers followed by b, f, and t indicate boxes, figures, and tables respectively.

Printed in the United States
By Bookmasters